AF749279

Smart Nanomaterials Technology

Series Editors

Azamal Husen, Wolaita Sodo University, Wolaita, Ethiopia

Mohammad Jawaid, Laboratory of Biocomposite Technology, Universiti Putra Malaysia, INTROP, Serdang, Selangor, Malaysia

Nanotechnology is a rapidly growing scientific field and has attracted a great interest over the last few years because of its abundant applications in different fields like biology, physics and chemistry. This science deals with the production of minute particles called nanomaterials having dimensions between 1 and 100 nm which may serve as building blocks for various physical and biological systems. On the other hand, there is the class of smart materials where the material that can stimuli by external factors and results a new kind of functional properties. The combination of these two classes forms a new class of smart nanomaterials, which produces unique functional material properties and a great opportunity to larger span of application. Smart nanomaterials have been employed by researchers to use it effectively in agricultural production, soil improvement, disease management, energy and environment, medical science, pharmaceuticals, engineering, food, animal husbandry and forestry sectors.

This book series in Smart Nanomaterials Technology aims to comprehensively cover topics in the fabrication, synthesis and application of these materials for applications in the following fields:

- Energy Systems - Renewable energy, energy storage (supercapacitors and electrochemical cells), hydrogen storage, photocatalytic water splitting for hydrogen production,
- Biomedical - controlled release of drugs, treatment of various diseases, biosensors,
- Agricultural - agricultural production, soil improvement, disease management, animal feed, egg, milk and meat production/processing,
- Forestry - wood preservation, protection, disease management,
- Environment – wastewater treatment, separation of hazardous contaminants from wastewater, indoor air filters.

Hind Abdellaoui · Sanjay M. R. · Suchart Siengchin
Editors

Mechanics of Nanomaterials and Polymer Nanocomposites

Springer

Editors
Hind Abdellaoui
Mohamed VI Polytech University (UM6P)
Benguerir, Morocco

Suchart Siengchin
King Mongkut's University of Technology
North Bangkok (KMTUNB)
Bangkok, Thailand

Sanjay M. R.
King Mongkut's University of Technology
North Bangkok (KMTUNB)
Bangkok, Thailand

Smart Nanomaterials Technology
ISBN 978-981-99-2351-9 ISBN 978-981-99-2352-6 (eBook)
https://doi.org/10.1007/978-981-99-2352-6

© The Editor(s) (if applicable) and The Author(s), under exclusive license to Springer Nature Singapore Pte Ltd. 2023
This work is subject to copyright. All rights are solely and exclusively licensed by the Publisher, whether the whole or part of the material is concerned, specifically the rights of translation, reprinting, reuse of illustrations, recitation, broadcasting, reproduction on microfilms or in any other physical way, and transmission or information storage and retrieval, electronic adaptation, computer software, or by similar or dissimilar methodology now known or hereafter developed.
The use of general descriptive names, registered names, trademarks, service marks, etc. in this publication does not imply, even in the absence of a specific statement, that such names are exempt from the relevant protective laws and regulations and therefore free for general use.
The publisher, the authors, and the editors are safe to assume that the advice and information in this book are believed to be true and accurate at the date of publication. Neither the publisher nor the authors or the editors give a warranty, expressed or implied, with respect to the material contained herein or for any errors or omissions that may have been made. The publisher remains neutral with regard to jurisdictional claims in published maps and institutional affiliations.

This Springer imprint is published by the registered company Springer Nature Singapore Pte Ltd.
The registered company address is: 152 Beach Road, #21-01/04 Gateway East, Singapore 189721, Singapore

Contents

About the Editors

Dr. Hind Abdellaoui is a research scientist in the field of nanocellulose and nanocomposites. She received her Ph.D. in Mechanical Engineering from the National Superior School of Electricity and Mechanics, Hassan II University, Casablanca, Morocco (2015). Since 2016, Dr. Abdellaoui is a professor of Composite Materials, Material Sciences, Mechanics and Structure Analysis in Engineering School in Rabat, Morocco. In 2020, she joined the University Mohamed VI Polytech as a post-doctoral fellow in the field of Supramolecular Nanomaterials, Composite and Nanocomposite Materials and Nanocellulose Extraction. In addition, Dr. Abdellaoui joined the Institute of Nanomaterials and Nanotechnology, Moroccan Foundation for Advanced Science Innovation and Research (MAScIR), Rabat, as a over twenty publications of articles and book chapters, and a dozen contributed national and international conference papers. Her research fields of interest include reinforced/filled polymeric composites (thermoset, thermoplastic and natural matrices), lignocellulosic reinforced/filled polymeric composites, nanocomposites as well as manufacturing processes (liquid composite molding), characterization and prediction of composite behavior, treatment and surface modification of natural reinforcement, preparation of blend and foam composites, etc. Dr. Abdellaoui was a keynote speaker at the 1st International Symposium on Composite and Nanocomposite Materials: ISymposite I, Rabat, April 8–9, 2019. She was a member of the organizing committee of the International Conference on Advanced Technologies for Humanitarian Sciences: ICATHS, Rabat, Morocco, July 25–26, 2019.

Dr. Sanjay M. R. is currently working as a Senior Research Scientist/Associate Professor and also 'Advisor within the office of the President for University Promotion and Development towards International goals' at King Mongkut's University of Technology North Bangkok, Bangkok, Thailand. He received the B.Engg (Mechanical Engineering) in the year 2010, M.Tech (Computational Analysis in Mechanical Sciences) in the year 2013, Ph.D (Faculty of Mechanical Engineering Science) from Visvesvaraya Technological University, Belagavi, India in the year 2018 and Post Doctorate from King Mongkut's University of Technology North Bangkok, Thailand, in the year 2019. He is a Life Member of Indian Society for Technical Education

(ISTE) and an Associate Member of Institute of Engineers (India). He is an Associate Editor of Heliyon Elsevier (Materials Science) and Frontier Materials Journals. Also acting as a Board Member of various international journals in the fields of materials science and composites. He is a reviewer for more than 120 International Journals (for Nature, Elsevier, Springer, Sage, Taylor & Francis, Wiley, American Society for Testing and Materials, American Society of Agricultural and Biological Engineers, IOP, Hindawi, NC State University USA, ASM International, Emerald Group, Bentham Science Publishers, Universiti Putra, Malaysia), also a reviewer for book proposals, and international conferences. In addition, he has published more than 250+ articles in high-quality international peer-reviewed journals indexed by SCI/Scopus, 10 editorial corners, 65+ book chapters, one book, 35 books as an Editor (Published by lead publishers such as Elsevier, Springer, Taylor & Francis, Wiley), and also presented research papers at national/international conferences. In 2021, his 17 articles got top-cited in various top journals (Journal of Cleaner Production, Carbohydrate Polymers, International Journal of Biological Macromolecules, Journal of Natural Fibers, Journal of Industrial Textiles). He is a lead editor of Several special issues. Based on google scholar, the number of citations amounts to 12700+ and his present H-index is 58 with i10-Index of 185. In addition, 1 Thailand Patent and 4 Indian patents are granted. He has delivered keynote and invited talks at various international conferences and workshops. His current research areas include Natural fiber composites, Polymer Composites, and Advanced Material Technology. He has received a 'Top Peer Reviewer 2019' award, Global Peer Review Awards, Powered by Publons, Web of Science Group. The KMUTNB selected him for the 'Outstanding Young Researcher' Award 2020 and 'Outstanding Researcher' Award 2021. He is recognized by Stanford University's list of the world's Top 2% of the Most-Cited Scientists in Single Year Citation Impact 2019 and also for the year 2020. In 2021, he is recognized by Stanford University's list of the world's Top 2% of the Most-Cited Scientists in Single Year Citation Impact and also in Career-long Citation impact. He is listed in 'Top 100 Scientists' in Thailand, by AD Scientific Index (Top 4th Rank among Mechanical Engineering in Thailand).

Prof. Dr.-Ing. habil. Suchart Siengchin is President of King Mongkut's University of Technology North Bangkok. He has received his Dipl.-Ing. in Mechanical Engineering from University of Applied Sciences Giessen/Friedberg, Hessen, Germany in 1999, M.Sc. in Polymer Technology from University of Applied Sciences Aalen, Baden-Wuerttemberg, Germany in 2002, M.Sc. in Material Science at the Erlangen-Nürnberg University, Bayern, Germany in 2004, Doctor of Philosophy in Engineering (Dr.-Ing.) from Institute for Composite Materials, University of Kaiserslautern, Rheinland-Pfalz, Germany in 2008 and Postdoctoral Research from Kaiserslautern University and School of Materials Engineering, Purdue University, USA. In 2016 he received the habilitation at the Chemnitz University in Sachen, Germany. He worked as a Lecturer for Production and Material Engineering Department at The Sirindhorn International Thai- German Graduate School of Engineering (TGGS), KMUTNB. He has been full Professor at KMUTNB and became the President of KMUTNB. He won the Outstanding Researcher Award in 2010, 2012 and 2013 at KMUTNB.

His research interests in Polymer Processing and Composite Material. He is Editor-in-Chief: KMUTNB International Journal of Applied Science and Technology and the author of morethan 350 peer-reviewed Journal Articles, 12 editorial corners, 80 book chapters, one book, and 40 books as an Editor. He has participated with presentations in more than 60 International and National Conferences with respect to Materials Science and Engineering topics. He has been recognized and ranked among the world's top 2% scientists listed by prestigious Stanford University. He has received the National Excellence Researcher Award (Engineering and Research Industry) of the Fiscal Year 2021 by the National Research Council of Thailand.

Introduction of Nanomaterials and Polymer Nanocomposites

Asmaa Dghoughi, Marya Raji, Souad Nekhlaoui, Hamid Essabir, Rachid Bouhfid, and Abou el kacem Qaiss

Abstract Polymeric nanocomposites (PNCs) have been widely investigated recently owing to their high performance and physical properties that can be readily controlled by several factors. The performance of these systems is not only controlled by the characteristics of the polymers or nanofillers but also by the techniques used to incorporate nanomaterials into polymer matrices. In this chapter, a brief classification of PNCs that emphasizes the importance of the shape, size, and distribution of nanofiller, as well as the polymer matrix type, is given to better understand the pathways for improving their properties. Then, a highlight of the adopted techniques for the evaluation of PNCs mechanical behavior, starting from tensile testing to fatigue testing, is performed. Finally, the focus is on the theories and models defining the mechanical behavior of PNCs.

Keywords Nanomaterials · Polymer nanocomposites · Mechanical testing · Nanofillers · Mechanical

1 Introduction

Polymer nanocomposites have sparked significant scientific and technological interest over the past two decades. PNCs are always referred to as a combination of polymers and organic or inorganic nanoparticles, in which the nanoparticle component has at least one of their dimensions below 100 nm. If even a little amount of filler particles is added, the characteristics of a polymeric material can be drastically altered. Some of the parameters that affect the properties of a composite are the size, volume fraction, shape, and mechanical characteristics of the filler particles. The high surface to volume ratio of the nanoscale filler particles leads to enhanced polymer-filler interactions. Previous researchers have used many different types of filler particles including graphene [1], carbon nanotubes (CNT) [2], silica [3], and

A. Dghoughi · M. Raji · S. Nekhlaoui · H. Essabir · R. Bouhfid · A. Qaiss (✉)
Composites and Nanocomposite Center, Moroccan Foundation for Advanced Science, Innovation and Research, Rabat, Morocco
e-mail: a.qaiss@mascir.ma

© The Author(s), under exclusive license to Springer Nature Singapore Pte Ltd. 2023
H. Abdellaoui et al. (eds.), *Mechanics of Nanomaterials and Polymer Nanocomposites*, Smart Nanomaterials Technology,
https://doi.org/10.1007/978-981-99-2352-6_1

metal nanoparticles [4]. In terms of application of these materials, they are present in a wide range of industries such as in textiles, clothing, and cosmetics, as well as in the pharmaceutical, electronic, and paint industries due to their enhanced properties including physical and mechanical properties. This chapter aims to underline the fundamentals of polymer nanocomposites starting from their classification and the procedures used to obtain them, to highlight their mechanical properties and the techniques used to measure them.

2 Introduction to Nanomaterials

As their name indicates, nanomaterials are known as materials that have at least one dimension within the nanoscale domain, specifically in the 100 nm range [5]. Nanomaterials include materials occurring in nature, as well as materials that have been developed in laboratories using certain engineering techniques to meet specific application purposes. They can be found in single, fused, aggregated, or agglomerated forms, and they can have irregular, spherical, or tubular geometries [6].

This novel class of materials has drawn a lot of attention from science and technology because of unique characteristics that are rarely observed in bulks of the same compounds. These unique characteristics include special optical, magnetic, and electrical properties [7]. The various outstanding features of nanomaterials over their bulk counterparts are assigned to their high specific surface area and the effects of quantum confinement [8]. The size, shape, and particle structure all have a significant impact on these characteristics. For instance, nanoparticles' size allows them to absorb other materials remarkably well [9]. Furthermore, by adjusting the nanomaterial's size, these properties can be tailored to meet the needs of particular applications [8]. Such unique properties provide these nanomaterials the capacity to be employed in a wide range of applications, mainly in textiles, clothing, cosmetics, pharmaceutical, electronic and paint industry [10].

Nanomaterials can be classified according to their origin, size, shape, chemical composition, source, dimension, and other factors. Nevertheless, dimension is very important, thus their classification is mainly based on their dimensions. Nanomaterials can be categorized into four important subfamilies according to their dimensions as illustrated in Fig. 1.

2.1 *Zero-Dimensional*

The nanomaterials in this class have all three dimensions in the nanoscale range. Some examples are noble metal nanoparticles such as silver (Ag), silica (SiO_2), and gold (Au); magnetic nanoparticles such as pure metals (Fe, Co, Ni), and oxides (Fe_3O_4, Fe_2O_4); gold nanoparticles; organic quantum dots like CD; and inorganic quantum dots such as inorganic semiconductor dots like CdSe, CdS, ZnS; carbon

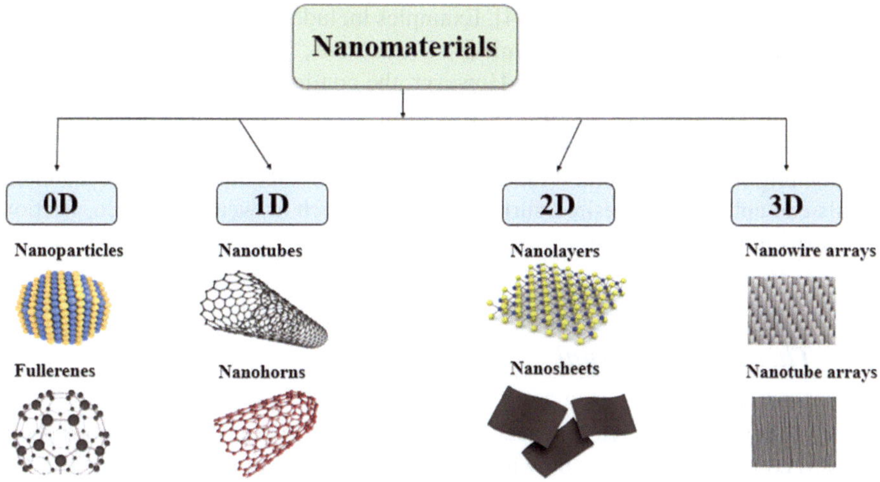

Fig. 1 Classification of nanomaterials based on their dimensions

quantum dots, graphene quantum dots, and fullerenes [11]. Due to their extremely small size, quantum confinement effect, superior physical and chemical characteristics, and strong biocompatibility, 0D nanomaterials have demonstrated significant promise in the fields of ion detection, biomolecular recognition, illness diagnostics, and pathogen detection [12].

2.2 *One-Dimensional*

The second category of nanomaterials, called 1D nanostructures, is dedicated to materials which have just one dimension >100 nm. Similar to 0D materials, one-dimensional nanoscale materials can be metallic, ceramic, or polymeric, as well as amorphous or crystalline, and single or polycrystalline. The chemical purity or impureness of 1D nanomaterials can vary, though (such as in doped semiconductors) [13]. These 1D nanomaterials, such as nanotubes, nanowires, and nanofibers, are highly appealing as the primary components of electron sources that emit electrons in a weak electric field [10].

2.3 *Two-Dimensional*

They are a subclass of two-dimensional objects (2D) whose two dimensions are orders of magnitude larger than their third, nanometer-scale, dimension [15]. Two-dimensional nanomaterials have thin layers that are at least one atomic layer thick and have forms similar to plates [14]. Examples include platelets, layered materials, and ultrathin films on surfaces. Unlike 1D materials, 2D nanomaterials are made of various chemical compositions. However, the confinement and delocalization of electrons in 2D materials are similar to that in 1D nanomaterials [13]. One of the areas in which this class of nanoparticles is used is bulk structural materials. While maintaining the bulk structural material's overall characteristics, these 2D nanomaterials can enhance the desired surface attributes, such as wear resistance, friction, and corrosion resistance [16].

2.4 Three-Dimensional

These materials have three dimensions greater than 100 nm. Box-shaped graphene nanostructures, bundles of nanowires, and nanotube arrays are all included in this class. 3D dimensional nanomaterials can exhibit distinctive properties, such as the crucial shift from mechanical strength to biostability in a square twist origami. Thus, they can be used as components of micro-electromechanical systems, biomedical devices, robotics, and solar cells [14].

In many cases, nanomaterials can be blended with other bulk materials to create nanocomposites with outside dimensions higher than 100 nm. A brief introduction to polymer nanocomposites based on nanomaterials is presented in the next paragraph.

3 From Nanomaterials to Polymer Nanocomposites

In the nanotechnology domain, one of the most common areas for ongoing research and development is polymeric nanocomposites. A polymer nanocomposite is a mixture of one or more polymers with one or more nanomaterials (other additives may also be present), where at least one phase in the final product is still in the nanometer range (within 100 nm). Depending on the nanomaterials used, the presence of nanomaterials in a polymer matrix can enhance the desired characteristics (such as mechanical, thermal, gas barrier, flame retardant, biodegradability, etc.) of polymers as well as produce a new set of features [11]. The reinforcing nanomaterials may have any shape, from straightforward fibers and particles to intricately layered materials and clusters, and they may be dispersed in either a natural or artificial polymeric matrix. The primary reason for using nanoscale fillers is to capitalize on the huge surface area for a fixed volume that these nanomaterials possess [17]. When the nanoparticles are well dispersed, the resultant nanocomposites tend to display better emergent properties compared to either traditional microcomposites, or the inadvertent microcomposites resulting from poorly dispersed nanoparticles.

The inclusion of nanomaterials in polymer matrices alters the surface chemistry and tailors the physicochemical intricacies, which are directly related to the performance of the resultant polymeric materials [11].

3.1 *Classification of Polymer Nanocomposites*

Polymer Nanocomposites can be classified by several criteria.

3.1.1 Based on the Nanofiller Type

The polymer nanocomposites can be categorized as carbon, metallic, natural, and clay-based groups depending on the type of nanofiller used. In addition to enhancing its traditional properties, the reinforcement of a polymer with carbon nanofillers (such as carbon black, fullerene, carbon nanotubes, graphene, and carbon nanofibers) expands the scope of possible applications for these nanocomposites by giving the polymer an electrical conductivity [18]. Metallic particles (like gold, silver, copper, and aluminum) lead to the enhancement of electric and thermal conductivity. This also generates an increase in material density that limits the requirements related to the low weight [19]. Natural nanofillers, such as natural nanofibers, natural nanowires, natural nanocrystals, and nanofibers are all included in this class. Nanoparticles are widely inserted into polymers in order to enhance their physical and mechanical properties. However, the weak interfacial attractions between the matrix polymer and these fibers and the high characteristics moisture absorption properties of many natural fibers remain the drawbacks to overwhelm in order to not hamper their applications [20]. Layered mineral silicate nanoparticles make up clay-based nanofillers. Their incorporation into the polymeric matrix can result in various forms of morphology dependent on the ability of polymer chains to pass through clay galleries (the concept of intercalation) or to disrupt the stacked structure of clays by delaminating sheets (the notion of exfoliation) [20]. Montmorillonite, halloysite, sepiolite, and laponite are widely used as reinforcement agents (Fig. 2).

3.1.2 Based on the Nanofiller Shape

The classifications according to the shape of the nanofiller are particulate composites (composed of particles that have a sphericity factor ≤ 1), fibrous composites (composed of fibers), and tubular composites (composed of tubes) [19].

3.1.3 Based on the Nanofiller Distribution

The manner in which the nanofiller is distributed inside the continuous phase leads to layered and dispersed nanocomposites. In the case of nanotubes or nanofibers, the distribution of such fillers can be isotropic or anisotropic [19].

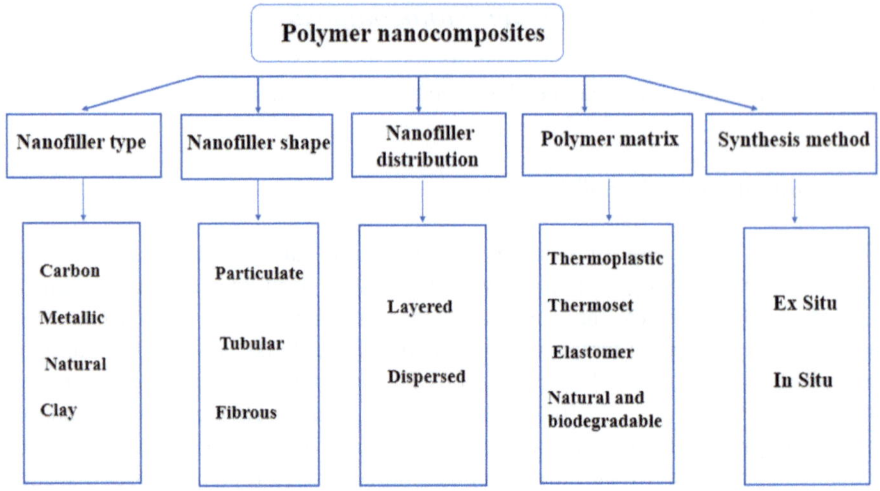

Fig. 2 Classification of polymer nanocomposites

3.1.4 Based on the Polymer Matrix

Based on the origin of both phases, one can distinguish natural and synthetic polymer nanocomposites. Starting from the matrix properties, polymer composites can be divided into thermoplastic, thermo-resistant, and elastomeric. They can be also divided on the basis of matrix, that is, a non-biodegradable matrix and biodegradable matrix. Bio-based composites made from natural/biofiber and biodegradable polymers are known as green composites [21].

3.1.5 Based on the Synthesis Method

They can be categorized into four synthesis methods as follows:

Ex situ addition of nanoparticles to the polymeric solution or melted polymer matrix. The nanoparticles inserted are synthesized separately.

In situ synthesis of nanoparticles occurs within a polymer. It involves the addition of precursors to the interstices of the polymer which may be converted to nanoparticles later on. For example, metal ions may be reduced to metallic form within the polymeric matrix.

Polymerization of monomers may be carried out in the presence of pre-synthesized nanoparticles. Simultaneous synthesis of the polymer matrix through monomers and nanoparticles using precursors [22].

3.2 *Methods for the Incorporation of Nanomaterials in a Polymer Matrix*

For the purpose of providing or enhancing the mechanical, thermal, and barrier properties of polymers, nanofillers are added to polymer matrices to create nanocomposites. The processing strategies and environmental factors have an impact on the final properties of the nanocomposite. Synthesis of polymer nanocomposite materials can be done using a variety of methods. Four techniques, including in situ polymerization, melt intercalation, electrospinning, and sol gel will be covered in detail in this section.

3.2.1 In Situ Polymerization

The basis of this technique is to penetrate the monomer solution through the swollen platelets. The monomer then undergoes polymerization underneath the presence of heat, an initiator, or UV light. Consequently, exfoliated nanocomposite or intercalated nanocomposite production occurs [23]. By using this technique, multidimensional structures with clearly defined shapes and unique features can be produced. This method allows for homogeneous dispersion in a polymer matrix and regulates the size, shape, and morphology of the nanomaterials [24]. Nevertheless, it has the drawback of producing high-temperature polymer chain breakdown [22] (Fig. 3).

3.2.2 Melt Intercalation Method

The conventional standard method for creating thermoplastic polymer nanocomposites is melt intercalation. As shown in Fig. 2, it entails adding the filler, annealing the polymer matrix at high temperatures, and then churning the composite to ensure uniform dispersion. Due to the absence of solvent use, this method has the benefit of being environmentally benign as well as cost-effective [20]. However, several requirements must be met for this technique, most notably the compatibility of the matrix and nanofillers. As a result, attaining effective dispersion and exfoliation depends greatly on the adjustment of the processing parameters. For instance,

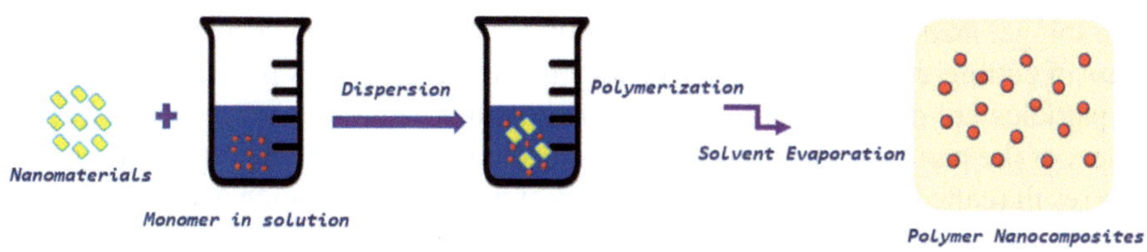

Fig. 3 In situ polymerization process

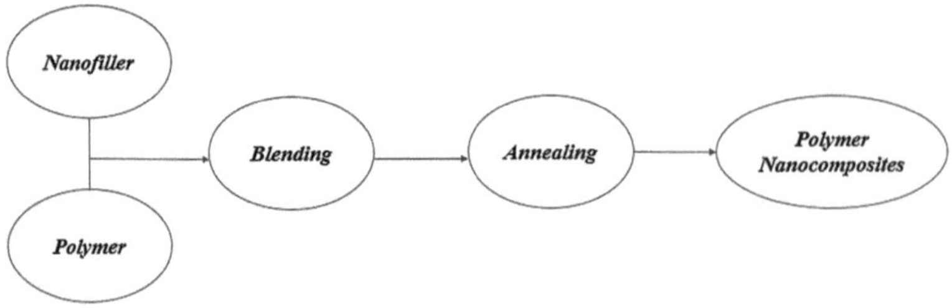

Fig. 4 Melt intercalation process

adopting more thermally stable modifications or working at lower temperatures can reduce the degradation and damage of the filler's modified surface [21] (Fig. 4).

3.2.3 Electrospinning

Making electrostatic fibers via electrospinning is a flexible process. In order to create fibers from a polymer solution or melt, electrospinning uses an electric force. Fibers that are electrospun range in diameter from nanometers to microns and have a substantial surface area. As seen in Fig. 5, an electrospinning system typically consists of a spinneret, collecting plate, and high voltage source to jet a polymer solution or melt towards the oppositely polarized collector [26]. During this process, the polymer liquid droplet ejected at the needle tip deforms into a cone shape during the electrostatic operation, which is known as a Tylor cone. The polymer solution of this Tylor cone drop extends as a result of these repelling forces acting in the opposite direction of the surface tension, which is a critical step in the surface tension's beginning [27]. The syringe's hydraulic pressure, flow rate, tip diameter, electric field strength, tip-to-collector distance, as well as other spinning parameters all affect the nanofibers' uniformity, shape, and sizes [11]. Applications for desalination membranes, biomaterials, and energy devices all show great promise for this process [26].

3.2.4 Sol Gel

The sol–gel method (Fig. 6) enables scientists and engineers to develop a brand-new class of cutting-edge materials with distinctive features. In this procedure, the fillers or particles are disseminated in a sol, or monomer solution, which permeates the layers of the particles. An intricate network of polymer and nanoparticles is created as a result (called gel). Consequently, polymer aids in the initiation and development of nanomaterials [22]. The narrow particle size distribution, great product purity, and ability to produce a homogeneous nanostructure at low temperatures are the primary advantages of the sol–gel method. This method is commonly used to synthesize metal nano-oxides [28]. Although this technique potentially has the ability to encourage

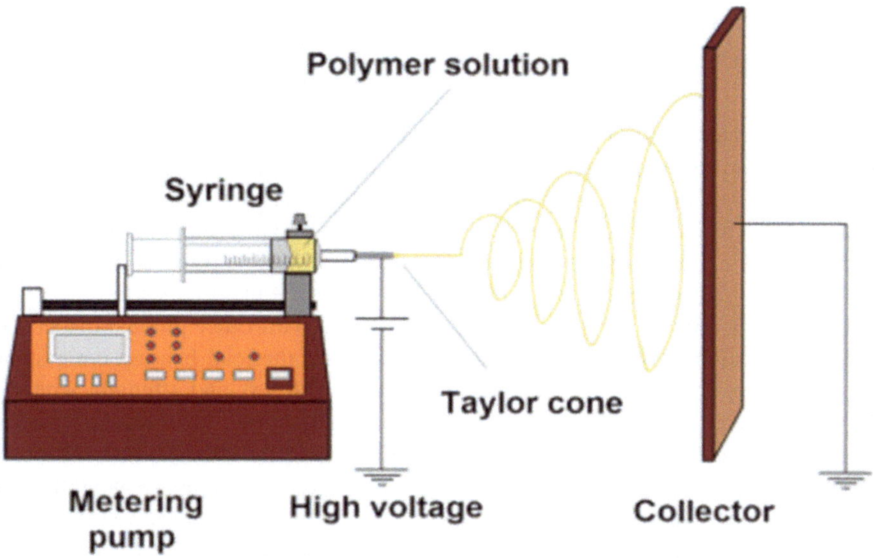

Fig. 5 Schematic diagram of a horizontal type electrospinning setup [25]

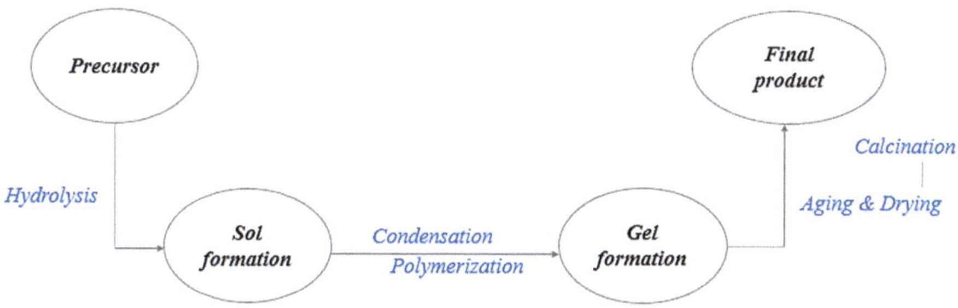

Fig. 6 Schematic diagram of different stages of sol–gel process

the dispersion of the nanomaterial in a single step without the usage of additional energy, it has significant drawbacks. For instance, the processing of clay minerals in the case of clay/polymer nanocomposite necessitates a high temperature, which may cause the polymer matrix to break down. A severe processing setting may also cause the nanomaterial to agglomerate. Consequently, this technique is less widely used than the earlier techniques [11].

4 Mechanics of Nanomaterials and Polymer Nanocomposites

A material's reaction to a physical stressor is the subject of the research of mechanics of polymer nanocomposites. This is typically thought to relate to the investigation of material failure. However, non-failure experiments and analysis can also be covered by this [29]. The mechanical characteristics of polymer nanocomposites are strongly influenced by a variety of factors, including the filler volume fraction, filler length, packing arrangement, and filler orientation, as well as the material's structural characteristics, manufacturing techniques, compatibility with other phases, dispersion method, adhesion, etc. [30]. In structural composites, the load-bearing component is provided by the reinforcement, which has high stiffness and strength, whereas the matrix typically has a moderate modulus and small strength. However, it makes a substantial contribution since it serves as a load transfer medium into the fibers and provides some kind of protection. The contribution of each component must be considered when constructing expressions to forecast mechanical behavior [31]. In order to get a consistent measurement of the mechanical properties of nanomaterials, it is necessary to use appropriate equipment, which we will discuss in the next paragraph.

4.1 Techniques and Equipment to Characterize the Mechanical Properties

4.1.1 Tensile Testing

Tensing testing is knowing as the most well-known method for identifying the mechanical characteristics of various materials [32], and has been used to characterize polymer nanocomposites as well [33–36]. The procedure involves grasping the sample's ends with a fixture and pushing it apart until it fails or reaches a specific strain [37]. The sample's stress resulting from the external load is determined using the sample's cross-sectional area and the tensile force applied to it. The stress–strain curve can then be derived by measuring the applied load as a function of the displacement. Many mechanical parameters, including Young's modulus, Poisson's ratio, yielding and ultimate strengths, can be determined using such a curve. In addition, it is feasible to locate tensile stages for creep tests in the literature [37, 38].

4.1.2 Hardness Testing

Hardness testing, described as a test procedure to measure the resistance of a material to continuous indentation, was designed for metals but may be performed on polymers, even though they display a viscoelastic response that renders the analysis difficult [39, 40]. Though, it can only be suggested for the determination of permanent deformation of a polymer's response to indentation following adequate calibration and parameter selection. Typically, an indenter is pressed into the surface of the material being tested while being subjected to a specified load for a predetermined amount of time, and the size or depth of the indentation is measured. The main goal of the hardness test is to ascertain whether a material is suitable for a certain application or the particular treatment to which it has been put [40]. There are several different methods of hardness testing that can be categorized using two fundamental defining characteristics, namely the loading method and the indenter shape. For metals, hardness values obtained using various test techniques are compared to specific hardness scales to determine their equivalency. However, it is more difficult to determine the equivalency between hardness scales for polymers because of their viscoelastic response and temperature response. It is not recommended to attempt to use different scales' polymer hardness numbers interchangeably [41].

4.1.3 Bending Testing

For many years, tests in which specimens are bent have been commonly utilized for the mechanical testing of fiber-plastics. The most popular mechanical test for composites and nanocomposites, along with the uniaxial tension test, involves loading simply supported specimens at a location centered between the supports during the three-point bending test [42]. Since bending tests require less forces and produce greater displacements, they are sometimes preferred to tension testing. Continuum mechanics models are used to deduce the mechanical properties in this situation. In specifically, these models relate the imposed load, the displacement field, and the sample's mechanical and geometrical characteristics. The mechanical characteristics can be back-calculated by measuring the first two quantities and using the sample geometry as a reference [38].

4.1.4 Fatigue Testing

The purpose of fatigue tests is to evaluate stress versus the number of failure cycles (S–N curves), fatigue damage, and the beginning and growth of fatigue cracks [38]. Composite materials' anisotropy makes their fatigue behavior more complex than that of isotropic materials, mostly because of the various types of damage that can happen and how they interact. Among other factors, the fiber type, matrix type, reinforcing structure, stacking order, environmental factors, loading factors, and boundary factors affect the fatigue performance of composites [43]. A wide range of fatigue tests can be performed on nanofiber-reinforced composites due to the large number of parameters: I the amplitude control (stress or strain); (ii) the testing frequency; (iii) the loading direction (axial, bending, biaxial); and (iv) the load ratio (tension/tension, tension/compression, compression/compression). Nevertheless, only one of these

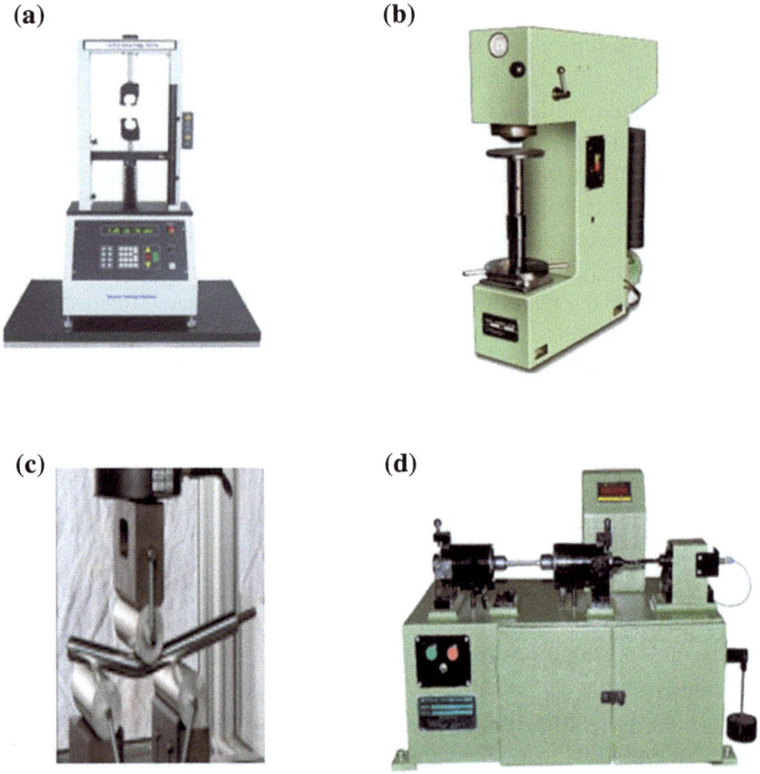

Fig. 7 Illustration of some mechanical tests equipment: **a** Tension machine, **b** Hardness testing device, **c** Bending testing equipment and **d** Fatigue testing machine

tests—the tension-tension fatigue test with a constant-amplitude load—has been standardized in both ASTM and ISO standards [44] (Fig. 7).

4.2 Mechanical Investigation

In this section the discussion will be focused on understanding the mechanics of nanomaterials and polymer nanocomposites, examining stresses, strains, and creep phenomenon.

4.2.1 Stress

Under the mechanical definition, the term stress is defined as the distribution of force acting on a given cross-sectional area. It is expressed as force intensity, i.e., as force per unit area (Eq. 1) [45].

$$\text{Stress} = \frac{\text{Force}}{\text{Area}} = \frac{\text{F}}{\text{A}} \tag{1}$$

In CGS systems, the stress is often stated in pounds per square inch (psi), whereas in SI units, it is expressed in newtons per meter squared (MPa). The amount of stress at a specific site depends not only on its geographic position but also on the passing plane. As a result, it is a second-order tensor. There are two major categories of stress: include normal which characterizes dimensional changes, and shear strain, which describes distortion (changes in angles) [45].

4.3 Strain

A body deforms when a set of forces acting on it change certain points' positions in relation to one another. Displacement is the phrase for the movement of the provided point to its new position, and it causes strain on the body. However, when a body is rotated or translated, there is no strain experienced. This is because the body's relative location has not changed [48]. The elongation indicated by the symbol occurs when an axial force is applied to a body, when a body expands in response to an increase in temperature, or even when force and temperature increase interact. Like stress, strain is classified into 2 types including normal and shear strain. Normal strain is designated by a Greek letter epsilon (ε). It is defined as the elongation per unit length and is represented by Eq. 2 [45]:

$$\text{Normal Strain} = \frac{\text{Elongation}}{\text{Length}} = \frac{\delta}{L} \tag{2}$$

Shear strain, on the other hand, is described as the angular length change in the x and y axes from their original perpendicular and parallel positions. Variations in the two angles add up to the overall shear stress. The shearing strain, γ_{xy}, is defined as (Eq. 3):

$$\gamma_{xy} = \theta_1 + \theta_2 = \left(\frac{\partial v}{\partial x} + \frac{\partial u}{\partial y}\right) \tag{3}$$

4.3.1 Creep

According to traditional definitions, creep is a process that occurs at a high temperature and under constant load, which assures that strains are typically within the range specified by Hooke's law. When loads are applied to products operating under creep conditions, stresses and strains are created that have variable values over time, primarily when the load remains constant and stresses lower than the yield strength are produced [46]. The fact that the strain diminishes and what is known as strain recovery occurs after the material in use has experienced strain relief is a defining aspect of creep [47].

Both plastic and reversible strains arise in materials during creep. They undergo a mechanism that causes continual form and size changes while in service. The components decay and can no longer be used after a certain amount of time working under creep circumstances. When the rate of strain shifts from its minimal and proportionate values to disproportional ones as it increases, the degradation processes begin [45].

5 Conclusion

The basic principles of nanomaterials and PNCs were thoroughly illustrated in this chapter. In addition to being distinctive in terms of their structure and characteristics, these novel types of materials are also used in a variety of applications, including textiles, clothing, and cosmetics, as well as in the pharmaceutical, electronic, and paint industries. This chapter clearly revealed that the properties of nanomaterials as well as polymer nanocomposites are strongly influenced by the size, shape, and aspect ratio of the nanomaterial. Additionally, it has been shown that the choice of an appropriate production method, which in turn relies on the type of nanoparticles being employed as nanofillers, is necessary for the development of polymer nanocomposites. In-situ polymerization process, melt intercalation method, electrospinning technique, and sol gel are among the most fabrication techniques used to incorporate nanofillers in PNCs. To characterize these fabricated nanocomposites, mechanics is among the tools the most used. Through the few illustrated techniques used for the mechanical testing of PNCs, we found that the mechanical performance of the final product is extremely sensitive to each component of the PNCs.

References

1. Cruz-Aguilar A, Navarro-Rodríguez D, Pérez-Camacho O, Fernández-Tavizón S, Gallardo-Vega CA, García-Zamora M, et al (2018) High-density polyethylene/graphene oxide nanocomposites prepared via in situ polymerization: morphology, thermal, and electrical properties. Mater Today Commun 16:232–241. https://doi.org/10.1016/j.mtcomm.2018.06.003
2. Lecocq H, Garois N, Lhost O, Girard PF, Cassagnau P, Serghei A (2020) Polypropylene/carbon nanotubes composite materials with enhanced electromagnetic interference shielding performance: properties and modeling. Compos B Eng 189. https://doi.org/10.1016/j.compositesb.2020.107866
3. Pribyl J, Benicewicz B, Bell M, Wagener K, Ning X, Schadler L et al (2019) Polyethylene grafted silica nanoparticles prepared via surface-initiated ROMP. ACS Macro Lett 8:228–232. https://doi.org/10.1021/acsmacrolett.8b00956
4. Črešnar KP, Aulova A, Bikiaris DN, Lambropoulou D, Kuzmič K, Zemljič LF (2021) Incorporation of metal-based nanoadditives into the pla matrix: Effect of surface properties on antibacterial activity and mechanical performance of pla nanoadditive films. Molecules 26. https://doi.org/10.3390/molecules26144161

5. Jordan J, Jacob KI, Tannenbaum R, Sharaf MA, Jasiuk I (2005) Experimental trends in polymer nanocomposites—a review. Mater Sci Eng, A 393:1–11. https://doi.org/10.1016/j.msea.2004.09.044
6. Kumar NKS (2016) Essentials in nanoscience and nanotechnology. Wiley, New York
7. Hong NH (2019) Introduction to nanomaterials: basic properties, synthesis, and characterization. Nano-sized multifunctional materials: synthesis, properties and applications. Elsevier, pp 1–19. https://doi.org/10.1016/B978-0-12-813934-9.00001-3
8. Ojijo V, Sinha Ray S (2018) Introduction to nanomaterials and polymer nanocomposite processing. Springer Series in materials science, vol 277. Springer, pp 1–14. https://doi.org/10.1007/978-3-319-97779-9_1
9. Ranjan S, Dasgupta N, Rajendran B, Avadhani GS, Ramalingam C, Kumar A (2016) Microwave-irradiation-assisted hybrid chemical approach for titanium dioxide nanoparticle synthesis: microbial and cytotoxicological evaluation. Environ Sci Pollut Res 23:12287–12302. https://doi.org/10.1007/s11356-016-6440-8
10. Abdellaoui H, Raji M, Bouhfid R, Qaiss A el kacem (2019) Investigation of the deformation behavior of epoxy-based composite materials. In: Failure analysis in biocomposites, fibre-reinforced composites and hybrid composites, pp 29–49. https://doi.org/10.1016/B978-0-08-102293-1.00002-4
11. Karak N (2019) Fundamentals of nanomaterials and polymer nanocomposites. Nanomaterials and polymer nanocomposites: raw materials to applications. Elsevier, pp 1–45. https://doi.org/10.1016/B978-0-12-814615-6.00001-1
12. Wang Z, Hu T, Liang R, Wei M (2020) Application of zero-dimensional nanomaterials in biosensing. Front Chem 8. https://doi.org/10.3389/fchem.2020.00320
13. Bashir S, Liu JL (2015) Nanomaterials and their application. Advanced nanomaterials and their applications in renewable energy. Elsevier Inc., pp 1–50. https://doi.org/10.1016/B978-0-12-801528-5.00001-4
14. Barhoum A, García-Betancourt ML, Jeevanandam J, Hussien EA, Mekkawy SA, Mostafa M, et al (2022) Review on natural, incidental, bioinspired, and engineered nanomaterials: history, definitions, classifications, synthesis, properties, market, toxicities, risks, and regulations. Nanomaterials, p 12. https://doi.org/10.3390/nano12020177
15. Schmid G (2004) Nanoparticles: from theory to application. Wiley-VCH, p 444
16. Koch CCOISSVS (2007) Structural nanocrystalline materials: fundamentals and applications. Cambridge University Press, New York
17. Hiremath A, Murthy AA, Thipperudrappa S, Bharath KN (2021) Nanoparticles filled polymer nanocomposites: a technological review. Cogent Eng 8. https://doi.org/10.1080/23311916.2021.1991229
18. Thomassin J (2013) JC. Les nanocomposites a' base de polymères : vers de nouveaux matériaux légers ultraperformants. Liège 82: Bull. la société R. des Sci
19. Barzic AI (2016) Nanocomposite polymeric-based coatings: From mathematical modeling to experimental insights for adapting microstructure to high-tech requirements. Industrial applications for intelligent polymers and coatings. Springer International Publishing, pp 355–71. https://doi.org/10.1007/978-3-319-26893-4_17
20. Essabir H, Raji M, Bouhfid R, Qaiss A el kacem (2019) Rheological properties of functionalized graphene and polymeric matrices-based nanocomposites. In: Functionalized graphene nanocomposites and their derivatives: synthesis, processing and applications, pp 109–120. https://doi.org/10.1016/B978-0-12-814548-7.00006-4
21. Parameswaranpillai J HNKTYY (2016) Nanocomposite materials: synthesis, properties and applications. ISBN 9781482258189: Taylor and Francis
22. Sharma AK, Priya, Kaith BS (2019) Polymer nanocomposite matrices: classification, synthesis methods, and applications. Handbook of polymer and ceramic nanotechnology. Springer International Publishing, pp 1–26. https://doi.org/10.1007/978-3-030-10614-0_51-1
23. Alexandre MDP (2000) Polymer-layered silicate nanocomposites: preparation, properties and uses of a new class of materials. Mater Sci Eng 28(1–2)

24. Basavegowda N, Baek KH (2021) Advances in functional biopolymer-based nanocomposites for active food packaging applications. Polymers (Basel) 13. https://doi.org/10.3390/polym13234198
25. Riazi K, Kübel J, Abbasi M, Bachtin K, Indris S, Ehrenberg H, et al (2016) Polystyrene comb architectures as model systems for the optimized solution electrospinning of branched polymers. Polymer (Guildf) 104:240–250. https://doi.org/10.1016/j.polymer.2016.05.032
26. Kausar A (2021) Polymeric nanocomposite via electrospinning: assessment of morphology, physical properties and applications. J Plast Film Sheeting 37:70–92. https://doi.org/10.1177/8756087920937344
27. Kailasa S, Reddy MSB, Maurya MR, Rani BG, Rao KV, Sadasivuni KK (2021) Electrospun nanofibers: materials, synthesis parameters, and their role in sensing applications. Macromol Mater Eng 306. https://doi.org/10.1002/mame.202100410
28. Bokov D, Turki Jalil A, Chupradit S, Suksatan W, Javed Ansari M, Shewael IH, et al (2021) Nanomaterial by sol-gel method: synthesis and application. Advances in materials science and engineering 2021. https://doi.org/10.1155/2021/5102014
29. Callister WD (2007) DGR. Materials science and engineering: an introduction, 7th edn. Wiley, New York
30. Woldemariam MH, Belingardi G, Koricho EG, Reda DT (2019) Effects of nanomaterials and particles on mechanical properties and fracture toughness of composite materials: a short review. AIMS Mater Sci 6. https://doi.org/10.3934/MATERSCI.2019.6.1191
31. Mechanics of material behavior. Studies in applied mechanics 1984 https://doi.org/10.1533/9781845698560.49
32. Scholze M, Safavi S, Li KC, Ondruschka B, Werner M, Zwirner J, et al (2020) Standardized tensile testing of soft tissue using a 3D printed clamping system. HardwareX, p 8. https://doi.org/10.1016/j.ohx.2020.e00159
33. Bouhfid N, Raji M, Boujmal R, Essabir H, Bensalah M-O, Bouhfid R, Qaiss A el kacem (2019) Numerical modeling of hybrid composite materials. In: Modelling of damage processes in biocomposites, fibre-reinforced composites and hybrid composites, pp 57–101. https://doi.org/10.1016/B978-0-08-102289-4.00005-9
34. Kumar Singh S, Kumar A, Jain A (2018) Improving tensile and flexural properties of SiO2-epoxy polymer nanocomposite. Mater Today Proc, vol 5, Elsevier Ltd., pp 6339–6344. https://doi.org/10.1016/j.matpr.2017.12.243
35. Hou D, Yu J, Wang P (2019) Molecular dynamics modeling of the structure, dynamics, energetics and mechanical properties of cement-polymer nanocomposite. Compos B Eng 162:433–444. https://doi.org/10.1016/j.compositesb.2018.12.142
36. Taghipoor H, Fereidoon A, Ghasemi-Ghalebahman A, Mirzaei J (2022) Experimental assessment of mechanical behavior of basalt/graphene/PP-g-MA-reinforced polymer nanocomposites by response surface methodology. Polym Bull. https://doi.org/10.1007/s00289-022-04420-x
37. Navindaran K, Kang JS, Moon K (2023) Techniques for characterizing mechanical properties of soft tissues. J Mech Behav Biomed Mater 138. https://doi.org/10.1016/j.jmbbm.2022.105575
38. Alfano M, Pagnotta L, Pantano MF (2011) A review of patented works on the mechanical characterization of materials at micro-and nano-scale. Recent Pat Nanotechnol 5(1):1
39. Abdellaoui H, Raji M, Essabir H, Bouhfid R, Qaiss A el kacem (2019) Mechanical behavior of carbon/natural fiber-based hybrid composites. In: Mechanical and physical testing of biocomposites, fibre-reinforced composites and hybrid composites, pp 103–122. https://doi.org/10.1016/B978-0-08-102292-4.00006-0
40. Prasad A, Fotou G, Li S (2013) The effect of polymer hardness, pore size, and porosity on the performance of thermoplastic polyurethane-based chemical mechanical polishing pads. J Mater Res 28(17):2380–2393. https://doi.org/10.1557/jmr.2013.173
41. Wu H, Dave F, Mokhtari M, Ali MM, Sherlock R, McIlhagger A, et al (2022) On the application of Vickers micro hardness testing to isotactic polypropylene. Polymers (Basel), p 14. https://doi.org/10.3390/polym14091804

42. ISO 178 | Three-Point Bending Test | Measurlabs (n.d.) Retrieved May 11, 2023, from https://measurlabs.com/products/three-point-bending-iso-178/
43. Dittenber DB (n.d.) Fatigue of polymer composites: life prediction and environmental Fatigue of polymer composites: life prediction and environmental effects effects
44. Sevenois RDB, van Paepegem W (2019) Fatigue testing for polymer matrix composites. Creep and Fatigue in Polymer Matrix Composites, Elsevier, pp 403–437. https://doi.org/10.1016/B978-0-08-102601-4.00013-8
45. Ashter SA (2014) Mechanics of materials. Thermoforming of single and multilayer laminates. Elsevier, pp 123–145. https://doi.org/10.1016/B978-1-4557-3172-5.00006-2
46. Wójcik Z (2006) Stanislaw Staszic's concept of science. Dissertations for the History of Education
47. Zieliński A (2015) DJPHGG. Properties, structure and creep resistance of austenitic steel Super 304H. Materials Testing
48. Zieliński A, Sroka M, Tański T, Golański G (2020) Introductory chapter: creep—an overview of new research results. Creep characteristics of engineering materials, IntechOpen. https://doi.org/10.5772/intechopen.85477

Carbon Nanotubes Particles: Processing, Mechanical Properties and Application

Houda Maâti, Othmane Amadine, Said Sair, Soumia Abouelhrouz, Boubker Ouadil, Hassan Mahi, Younes Essamlali, and Mohamed Zahouily

Abstract Due to its unusual and amazing features, carbon nanotubes (CNT), an allotrope of carbon atom, exhibits exceptional mechanical characteristics, including extraordinary tensile strength and Young's modulus. This has encouraged the scientific community to use the CNT as a suitable material for polymer reinforcement. However, attention must be paid for the choice of the preparation method in order to create a CNTs/polymer composite with superior mechanical properties. In this context, this chapter provides an overview of carbon nanotube (CNT) as a filler for the enhancement of physico-chemical properties of polymer composites. First, the chapter begins with a summary of the structure and synthesis methods of CNTs. Then it describes the mechanical, electrical, and thermal properties of CNTs. The next section reviews different processes for the development of CNT based polymer composites. After that, a brief description of the electrical and thermal properties is given with an emphasize on the mechanical properties. At the end of this chapter, different applications of CNT based polymer composites are presented, in which enhanced mechanical properties are needed.

Keywords Synthesis methods of CNTs · Mechanical properties · Electrical properties · Thermal properties of CNTs · CNTs based polymer composites · CNTs applications

H. Maâti · O. Amadine (✉) · S. Sair · S. Abouelhrouz · H. Mahi · Y. Essamlali · M. Zahouily
Mohammed VI Polytechnic University, Lot 660-Hay Moulay Rachid, 43150 Ben Guerir, Morocco
e-mail: o.amadine@mascir.ma

Rabat Design Center, MASCIR Foundation, Rue Mohamed El Jazouli, Madinat El Irfane, 10100 Rabat, Morocco

B. Ouadil · M. Zahouily
Laboratory of Materials, Catalysis and Valorization of Natural Resources, Hassan II University, 20000 Casablanca, Morocco

© The Author(s), under exclusive license to Springer Nature Singapore Pte Ltd. 2023
H. Abdellaoui et al. (eds.), *Mechanics of Nanomaterials and Polymer Nanocomposites*, Smart Nanomaterials Technology,
https://doi.org/10.1007/978-981-99-2352-6_2

1 Introduction

The need for innovative materials has been growing for three decades and this field is constantly evolving [1, 2]. The main properties sought in these materials are the important mechanical reinforcement. The major areas where the properties of mechanical reinforcement are favored are sports and leisure items [3], industry [4] or even defense aeronautics and automobiles [5]. Much research is carried out at the fundamental level to understand the phenomena governing the properties materials and at the application level to improve existing materials.

This evolution in materials comes both from the recent development of a family of composites from nanoparticles, called nanocomposites as well as important achievements in polymer chemistry. Many particles have been tested but carbon nanotubes are among the most studied, thanks to their very promising characteristics [6, 7]. Research on carbon nanotubes and significant advances in chemistry naturally led to combining the two systems to form innovative materials based on carbon nanotubes.

However, the incorporation of carbon nanotubes, especially on an industrial scale, presents many constraints and limitations that need to be identified [8, 9]. Thus, a better knowledge and better control of the properties of composites is essential to optimize the performance of these materials and use them for specific applications.

In this context, the main objective of this chapter is to review the published literature on carbon nanotubes, and their applications.

2 History of Carbon Nanotubes

Carbon nanotubes CNTs were first demonstrated by transmission electron microscopy (TEM) in 1952 and were published in the Russian Journal of Physical Chemistry as long carbonaceous filaments having an internal tubular structure [10]. Then, the study of carbon nanotubes was continued in the 1960s. They were first thought to be a by-product of a chemical industrial process to produce fullerenes by Bacon [11]. At that time, they were called filaments or carbon fibers. Then, thanks to advances in the development microscopy techniques, in particular the transmission electron microscope (TEM), Endo [12] was able to estimate, in the 1970s, their diameter to be less than 100 nm. However, this observation did not attract much attention because the scientists did not have any experimental tools allowing the study and manipulation of such small objects. This scientific context completely changed in the 1990s, which explains the enthusiasm of the scientists after the publication of the TEM observations at high resolution by Ijima in 1991 [13]. This article, whose main subject is based on the structures close to that of fullerene, a very active field of research at this time, presents TEM images of tubular filamentous carbonaceous structures (Fig. 1).

Since his first report on multi-walled carbon nanotubes (MWNTs) [13], followed by their single-walled counterparts (SWNTs) [14, 15], CNTs have emerged as one

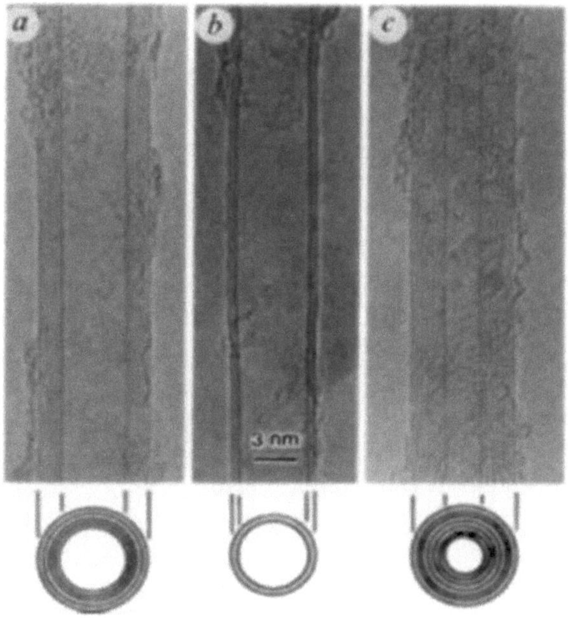

Fig. 1 Electron micrographs of different CNTs: **a** CNT consisting of five graphitic sheets with a diameter of 6.7 nm **b** a double wall CNT with a diameter of 5.5 nm, **c** seven sheet CNT with an innermost shell having a diameter of 2.2 nm [13]

of the most intensively investigated nanostructured materials [16], with thousands of papers published every year, promising new applications that have attracted both academic and industrial interest.

3 Structure and Models of CNT

CNT can be arranged as Single-wall carbon nanotubes (SWCNT) or double wall carbon nanotubes (DWCNT) made up of a single or a double graphene sheet. However, a multiwall carbon nanotube (MWCNT) is composed of an array of coaxial nanotubes [17]. Different orientations of the chiral vector regulate the wrapping of graphene sheets, resulting in diverse CNT geometries. The SWCNT is called armchair when the chiral indices are equivalent ($n = m$), and the chiral angle is 30°. If one of the chiral indexes is zero (n, 0) or (0, m), the SWCNT is referred to as zig-zag, and the chiral angle is 0° (achiral nanotubes). In the other situations ($n \neq m$), the nanotube is chiral, with a chiral angle of $0° < \theta < 30°$ (Fig. 2).

SWNTs have a diameter of 0.7 to 2 nm and a length of a micron. As a result, these nanotubes have a high length to diameter ratio also called aspect ratio, ranging between 10^3 and 10^5. This anisometric structure is part of what makes carbon nanotubes appealing for use as mechanical reinforcement. After their synthesis, SWNTs are not individualized but rather in bundle form. As an example, an assembly

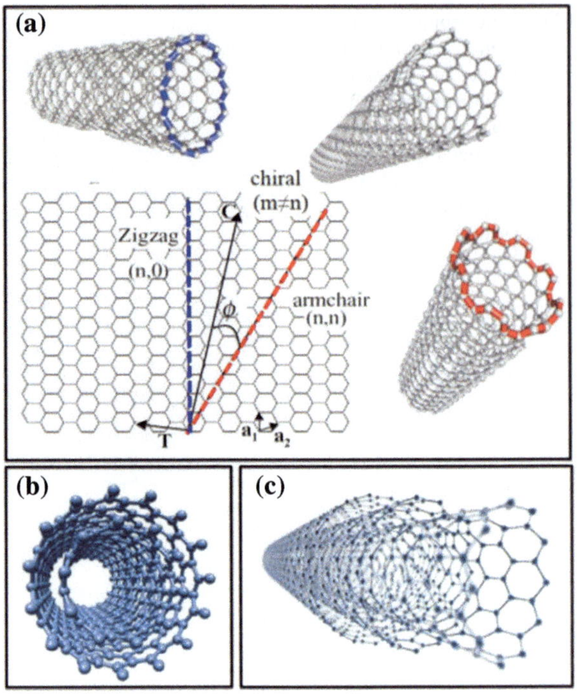

Fig. 2 Structure and models of carbon nanotubes in function of their number of walls. **a** Single-walled carbon nanotubes (SWNTs) structures in function of their chirality (zigzag, armchair, and chiral). **b** Model of double-walled carbon nanotubes (DWNTs). **c** Structure of multi-walled carbon nanotubes (MWNTs) made up of several concentric shells [17]

of ten to a hundred SWNTs aligned in a hexagonal network have a density ranging from 1.33 to 1.40 $g.cm^{-3}$ [18].

On the other hand, MWNTs have a diameter varying from a few tens to a few hundreds of nanometers depending on the synthesis conditions. The internal diameter is on the order of a few nanometers and the distance between the walls, corresponds approximately to the distance between two graphene planes in graphite, which is 3.4 Å [19]. Furthermore, they have a density of 1.75 g/cm^3. These nanotubes have all have a metallic character and their use is sought in the manufacture of conductive composites.

4 Synthesis Methods

There are different ways for the synthesis of CNTs in which high, medium, or low temperature can be used. The first CNTs were synthesized starting from a solid carbon source by the sublimation of graphite using temperatures above 3200 °C [20]. Nowadays the commonly used method in liquid or gaseous phases is by low or medium temperatures ranging from 350 to 1000 °C. The growth of the nanotubes can be achieved by the decomposition of the gaseous carbon precursor using metal nanoparticles as catalysts.

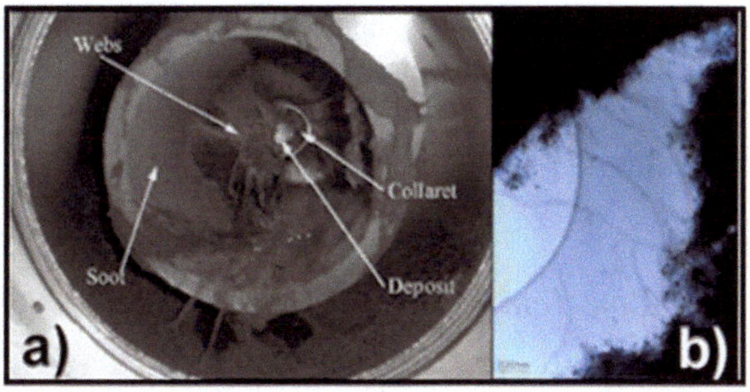

Fig. 3 **a** View of the reactor after the synthesis where one can distinguish the various products when using a mixture of carbon and catalyst **b** TEM images of the collaret containing a large amount of bundles of SWNTs [27]

4.1 High Temperature Methods

All the methods using high temperature are based on the sublimation of graphite and can be differentiated by each other by the process used for this purpose. These methods are electric arc discharge [21], laser ablation [22, 23] and vaporization induced by a solar beam [24, 25]. By using a pure graphite anode or cathode, in both electric arc discharge and laser ablation, multi-walled carbon nanotube MWCNT can be produced. The addition of metals in small quantities can produce SWCNT with high quality crystalline quality [26]. The main disadvantage of these methods is the formation of undesired carbon by-products such as fullerenes, soot, and amorphous carbon [27] (Fig. 3). For this reason, these approaches are replaced by "medium temperature methods".

4.2 Medium Temperature Methods

Since the early 70's, the method used for the formation of carbon filaments is catalytic chemical vapor deposition (C-CVD) that was realized on metal surfaces at temperatures around 600–1200 °C [28]. However, the popularity of the method didn't start until the 90's when Jose-Yacaman et al. [20] reported that the production of carbon filaments could be realized in the same crystalline structure than reported by Lijima in 1991 [13].

Catalyst chemical vapor deposition (C-CVD) consists of the catalytic decomposition of carbonaceous precursors (hydrocarbons for example) on nanometric catalytic particles, based on transition metals (Co, Fe, Ni) [29]. Depending on the used catalysts, this process can lead to SWNTs or MWNTs. However, it is possible to produce only one of the two types of nanotubes. By adapting the catalysts, it is also possible to control the number of walls and obtain double or triple nanotubes walls [30]. Although the produced nanotubes contain structural defects which cause a more or less pronounced longitudinal waviness, this method has the advantage of producing very long carbon nanotubes. Another notable advantage of the C-CVD process lies on the possibility of a continuous supply of a carbon source, which allows large-scale production of nanotubes, and this can make the process scalable. Medium temperature methods have many advantages over the high temperature ones, especially when it comes to the growth control which can help reach a high level of purity and selectivity [31].

4.3 Low Temperature Methods

Some studies have focused on the synthesis of carbon nanotube using mild conditions under low temperatures between 160 and 450 °C. Xiao et al. [32] have reported the synthesis of MWCNT on an insulating substrate at low temperature using a complementary metal–oxide–semiconductor (CMOS) compatible process. By using a new designed Ni–Al–Ni multilayer catalyst and ZrO_2 as a substrate, the authors were able to reduce the activation energy of the reaction compared to other Fe or Co based catalysts. Moreover, this Ni–Al catalyst was preferred to Fe since this later is not CMOS compatible due to contamination issues. The low activation energy makes it easier to disassociate the carbon–hydrogen source and supply sufficient carbon atoms at low temperatures. This process allowed the authors to reach a high density with a value of 1 × 1010 tubes/cm^2 and an average diameter of 26 nm.

Another low temperature method was the hydrothermal synthesis which was reported by Murali et al. [33] using ethanol, polyethylene glycol (PEG) as precursors and sodium hydroxide as activating agents with a reaction time of 20 h. The authors were able to reach an average particle size of 39 nm and a dislocation density was calculated as 6.5746 × 1014 m^{-1}.

Another type of CNT which has not gained much attention is the amorphous CNTs (a-CNTs) (Fig. 4). It is anticipated that a-CNTs with an amorphous wall structure, resulting from defects in the carbon network, may result in more intriguing features and the creation of new potential nanodevices. In this context, Liu et al. [34] have developed a novel and simple chemical route to synthesize a-CNTs via the reaction between ferrocene and ammonium chloride in air at only 200 °C.

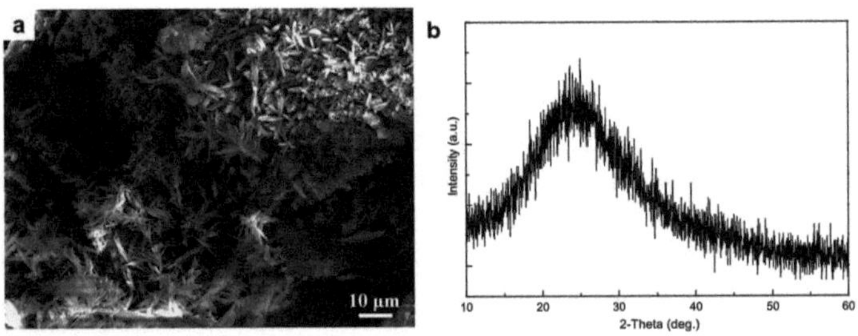

Fig. 4 **a** SEM images of (**a**) low magnification of the as-prepared a-CNTs on the surface of NH_4Cl particle, **b** XRD pattern of the a-CNTs

5 Properties of Carbon Nanotubes

5.1 Electrical Properties of Carbon Nanotubes

CNTs are known for their good electrical conductivity, thanks to their structure that make them good electrical conductors. Based on their chirality, SWNTs can be metallic, semi-metallic or semi-conductors. Their conductivity can be a function of their chirality, the crystallinity of their tubular structure as well as their diameter [35]. The CNT fibers have a wide range of conductivities ranging from 10 to 67,000 $S.cm^{-1}$ [36]. These values are still below aluminum and copper [37] which currently represent a major obstacle impending the application of CNT for electrical wiring. This is mainly due to the limited control over the morphology of the CNTs and CNT fibers.

The advantage of nanotubes lies on their low density coupled with their high anisometry [38], which decreases the amount of filler to be incorporated to make the material conductive and achieves a very low percolation threshold. Thus, one of the most promising facts for carbon nanotubes is the contribution of electrical conduction in polymers by using a very small quantity of nanotubes while retaining the properties of the polymer matrix. Carbon nanotubes can support a high current density of $400MA/cm^2$ [39] and have a resistivity of $10^{-4}\,\Omega/cm$ in the case of MWNTs compared to that of copper which is $10^{-6}\ \Omega/cm$ [7].

5.2 Mechanical Properties of Carbon Nanotubes

5.2.1 Young's Modulus of CNTs

The Young's modulus of a material is the most important parameter that defines its use as a structural element for various applications. CNT is considered as the strongest and stiffest material yet discovered in terms of elastic modulus and tensile strength [40, 41]. This is mainly due to the sp^2 bond strength formed between the individual carbon atoms. This covalent bond, often known as the "σ-bond," is the key parameter that significantly contributes to the remarkable mechanical properties of CNTs. Each atom is bound to three adjacent atoms owing to the periodicity. This hybridization, in which one s-orbital and two p-orbitals combine to generate three hybrid sp^2-orbitals at 120° to each other within a plane [42]. Many methods and instruments were used to calculate the young's modulus of CNTs.

Tracey and his co-workers [43] were the first to publish the Young's modulus of MWCNT by thermally inducing the vibration of cantilevered MWCNT and ranging from which exceeds any conventional metals used in electrical engineering. The Young's modulus of SWCNTs was calculated by Krishnan et al. in 1998 [44] using the same method, and they came up with an average value of 1.3 TPa.

In their experiment, Lourie and Wagner [45] used the bar model to measure the compressive response using micro-Raman spectroscopy. For SWCNT and MWCNT, they reported Young's moduli of 2.8–3.6 TPa and 1.7–2.4 Tpa, respectively. Yu et al. [46] have conducted direct tensile loading testing on SWCNT and MWCNT. For SWCNT, the obtained Young's modulus varies between 320 and 1470 GPa, and for MWCNT, it ranges from 270 to 950 GPa.

5.2.2 CNT's Strength

The strength of a material is not as well understood as the Young's modulus since it depends on the material's properties. It is closely related to potential structural defects and imperfections in the solid, and materials have in very few cases strengths that are close to their theoretical maximum [47].

Different model systems exist to study the intrinsic strength of the sp2 bonded CNT and how it relaxes mechanical energy at high strain [42]. At low temperatures, CNTs, like other covalent materials, are brittle regardless of diameter or helicity. Carbon nanotubes are flexible at room temperature because of their great strength and the hexagonal network's great ability to distortion for stress reduction. This has been established both theoretically and experimentally. CNTs appear to behave as ideal carbon fibers that can be stiff yet flexible, associating extremely high modulus with extremely high strength [48]. As a result, carbon nanotubes have enormous potential for applications requiring high-modulus and high-strength materials. In a study conducted by Zhao et al. [49], the authors showed that CNT fibers studied at 1000 °C and 2400 °C preserved 82% and 54% of the strength, respectively, when compared to room temperature strength. Whereas CNT fibers tested at −196 °C increased in strength by 68% (Fig. 5).

5.3 Thermal Properties of Carbon Nanotubes

CNTs, as rolled graphitic structures, are significant not only for their electronic and mechanical properties, but also for their thermal properties [4]. The thermal conductivity of individual MWCNTs at room temperature was measured by Kim et al. [50] and found it to be 3000 W/K (higher than graphite) using a microfabricated suspended device (Fig. 6). This value was two orders of magnitude more than prior findings by using macroscopic mat samples. However, the direct and quantitative measurement of individual nanotube thermal transport properties remains difficult. As a result, CNTs' thermal conductivity calculated by theoretical simulations and calculations from indirect experiments gave values in the order of 10^3 W/mK [51, 52]. After these speculations, the scientific community was motivated to verify these predictions empirically [52, 53]. However, the thermal conductivity of fiber or films of CNTs was less thermally conductive or even insulating [54]. Additionally, CNT ensembles are frequently made up of CNTs of varied chiralities, each with a unique

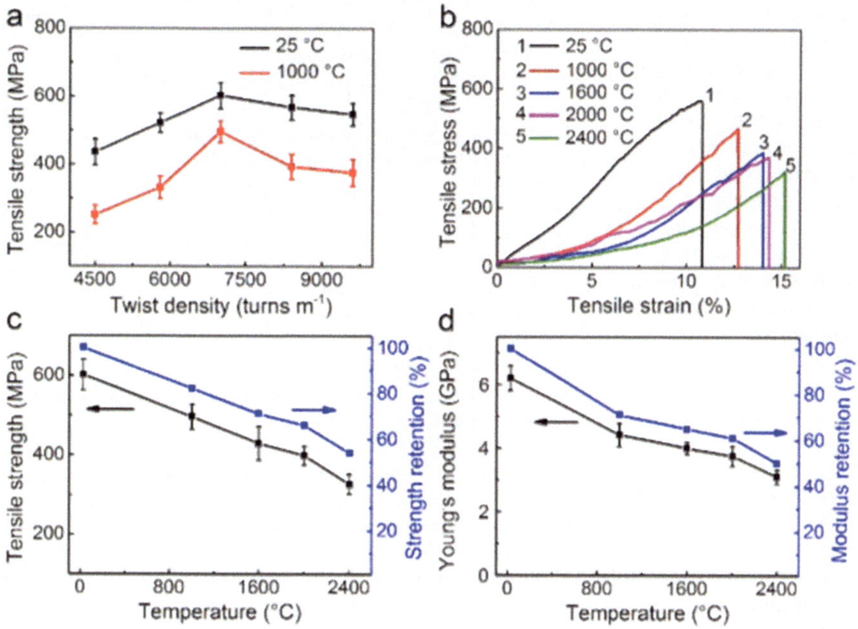

Fig. 5 **a** Tensile strength of CNT fibers tested at 25 °C and 1000 °C as a function of twist density that was inserted into CNT fibers. **b** Typical tensile stress–strain curves for CNT fibers tested at different temperatures. **c** Tensile strength and strength retention of CNT fibers as a function of test temperature. **d** Young's modulus and modulus retention of CNT fibers as a function of test temperature [49]

band structure. As a result, it may be anticipated that this relationship will have a considerable impact on the electronic contribution to thermal conductivity [55].

The thermal conductivity of carbon nanotubes is influenced by different parameters such as: atomic arrangement, the diameter and length of the tubes, the quantity of structural defects, the morphology, and the presence of impurities.

6 Functionalization of CNTs

It has been well documented that the efficient dispersion of CNTs into a polymer matrix is a fundamental challenge in the preparation of CNT-reinforced composites [55, 56]. For this reason, the functionalization method plays a crucial role in the interaction between the polymer matrix and CNTs. Several methods exist to achieve this functionalization, the most common ones are the chemical for covalent attachment of different organic groups and the physical by a noncovalent functionalization. These functionalizations allow the improved dispersion and interaction with various

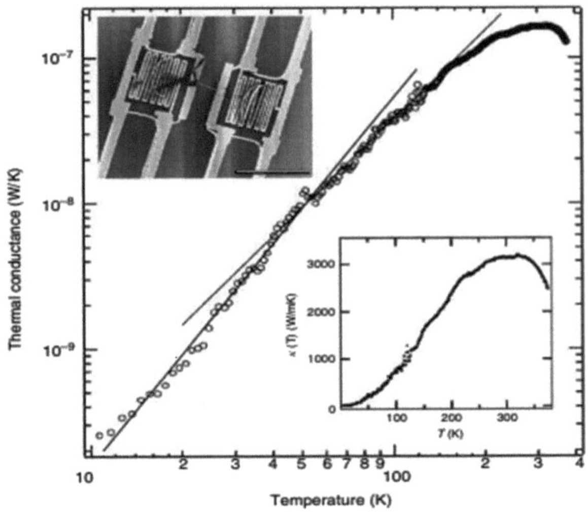

Fig. 6 Thermal conductance of an individual MWNT of 14 nm diameter. The power law slope decreases from ~2.5 to ~2.0 above ~5 K. Saturation occurs at 340 K, the thermal conductance decreasing at higher T. Lower inset shows the thermal conductivity, for which some assumptions about effective area are required [50]

polymer matrices, resulting in an improvement of the acquired characteristics of the nanocomposites.

6.1 Covalent Functionalization of CNT

Due to weak interfacial contact and difficulties in homogeneous dispersion, the full potentiality of CNTs when put in polymer matrix has been severely constrained [57–59]. To address all these issues, certain surface modifications have been developed in order to modify CNT surface properties [60, 61]. These modifications can be divided based on two different approaches, chemical (covalent) and physical (noncovalent) functionalization.

Depending on the functional group, these modifications can involve oxidation [62, 63], halogenation [64–67], amination [68], cycloaddition [61, 69, 70]...etc. Among, these methods, oxidation is the most known technique that can improve the dispersion of CNTs in liquids, organic solvents, and polymers. It is also used for the final treatment to improve the functionality of SWCNT and MWCNTs, or as the first treatment of a sequence of chemical modifications [71] (Fig. 7).

Oxidation can be performed by inorganic acids such as nitric acid [72], sulfuric acid [73] or a mixture of both [74] at elevated temperatures. Oxidation agents such as hydrogen peroxide, ammonium persulfate [75], and potassium permanganate [76] can also be used. And in some cases, the gaseous approach can also be effective and is based on the treatment of carbon nanotubes with air [77], ozone [78] or even acid vapors as oxidants [79, 80] at elevated temperatures. Several functional groups, including carboxyl, carbonyl, and hydroxyl groups, can be formed at the open end or defect sites of the CNT structure [81] (Fig. 8).

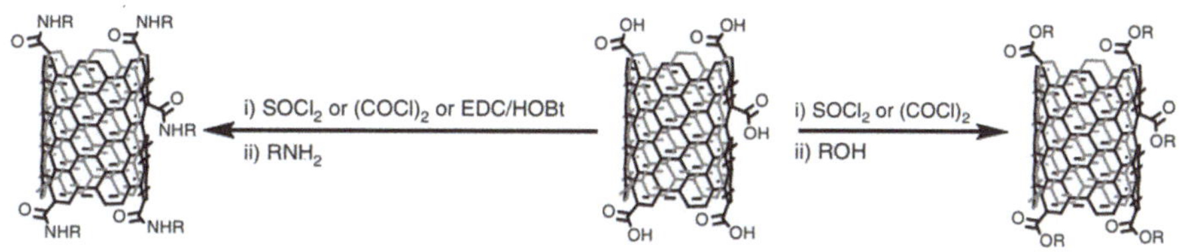

Fig. 7 Oxidized carbon undergoing esterification and amidation reactions [71]

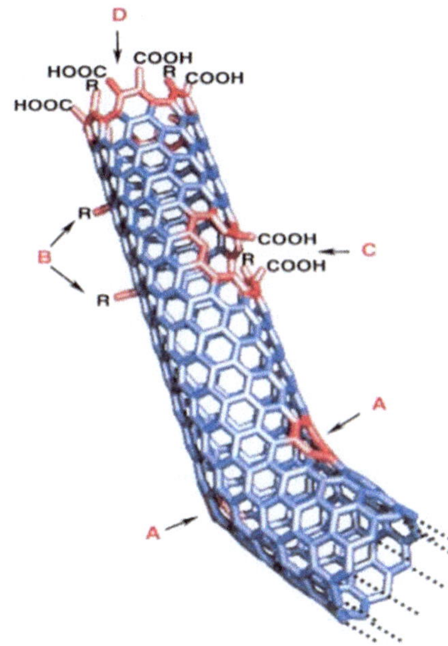

Fig. 8 Typical defect in SWCNT. **a** five or seven membered rings in the carbon frameworks, leads to a bend in the bube, **b** sp3 hybridized defects, **c** carbon framework damaged by oxidative conditions, which leaves a hole line with –COOH groups, **d** open end of SWCNT terminated with COOH groups [81]

Because of the existence of multiple polar and non-polar functional groups, such functionalized CNTs are extremely stable in water and many organic solvents. Furthermore, functionalized carbon nanotubes exhibit a considerable increase in interfacial bonding between CNTs and polymer matrices, resulting in a stronger nanotube-polymer interaction that might contribute to improved mechanical characteristics in nanocomposites [82, 83]. However, the covalent functionalization can represent some drawbacks because it breaks the usual graphene-type structure [84] affecting the intrinsic qualities of CNTs such as conductivity and mechanical strength.

6.2 Non-covalent Functionalization

The non-covalent functionalization of CNT is preferred to the covalent one due to the aggressive nature of it that might introduce structural defects resulting in inferior properties [85]. Non-covalent functionalization can consist on interactions with molecules through π-π, CH-π or Van Der Waals interactions. Different type of molecules can be used for this purpose such as aromatic (Fig. 9) or bio-molecules [86], polymers [87] and surfactants [88]. Aromatic molecules have a strong affinity for graphitic surfaces through π-stacking. The adsorption of several polycyclic aromatic hydrocarbons on SWCNTs has been widely studied, including pyrene [89], anthracene [90] and ferrocene [91] Polymers produced by repeated units of alkyl chains and aromatic moieties that involve both π-stacking on the CNT surface and van der Waals interactions between the hydrophobic nanotube surface and alkyl tails are the commonly used dispersants [92]. Despite their weakness, these interactions between long carbon nanotubes and macromolecules can be quite powerful due to their large number.

Noncovalent functionalization of CNTs with polymers is an efficient method for dispersing the tubes in aqueous and non-aqueous liquids while keeping their intrinsic characteristics.

7 Polymer/Carbon Nanotubes Nanocomposites

Thanks to their remarkable properties such as high strength and stiffness (54,55), CNTs are great candidates for structural applications. The use of carbon nanotubes as filler materials has aided in the development of CNT-polymer nanocomposites as next-generation advanced structural materials [94]. Since 1991 [13], a new class of carbon nanotube (CNT)-based polymer nanocomposites has been gaining popularity due to its notable mechanical and electrical capabilities, as well as their strong electron transport properties. This demonstrates that the development of CNT-reinforced

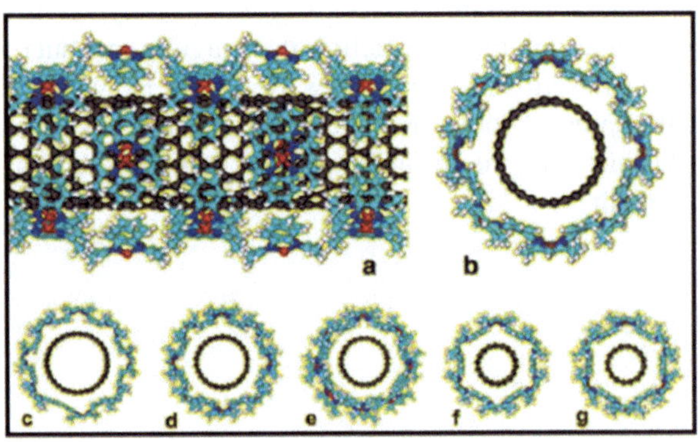

Fig. 9 Aromatic annulene adsorbed on SWCNT walls. Exemplified for NiTMTAA and zigzag SWNTs: **a**, **b** (14, 0); **c** (16, 0); **d** (13, 0); **e** (12, 0); **f** (9, 0); and **g** (8, 0). Side view (**a**) and cross sections (**b**–**g**). Atom coloring: carbon, black (nanotube) and light-blue (complex); hydrogen, white; nitrogen [93]

polymer composites has the potential to contribute to the expansion of numerous areas of usage, ranging from energy-related gadgets to structural components.

7.1 Preparation of Polymer/Carbon Nanotubes Nanocomposites

Several methods exist for the preparation of carbon-based polymer nanocomposites depending on the thermal or chemical properties of the matrix polymer. The appropriate synthesis method for a particular case can be selected. Preparation technique can influence the properties of the obtained CNTs, such as the dispersion and the length of CNTs, the matrix and the chirality.

7.1.1 Solution Processing

Solution processing is considered as a useful process for developing composites with homogeneously dispersed CNTs and is frequently used to create CNT's films. Usually, CNTs are typically dispersed in a solvent before being combined with polymer solution by shear mixing [95], magnetic stirring, or sonication in solution processing [91]. Recovery of the nanocomposite is realized by precipitating or casting a film. In this context, Saleh et al. [96] revealed a perfect dispersion and selective localization of CNT in acrylonitrile–butadiene–styrene (ABS) in the styrene–acrylonitrile (SAN) phase of the ABS matrix. The electrical conductivity of the nanocomposite as a function of filler content revealed that the nanocomposite exhibits typical percolation behavior with a percolation threshold of only 0.06 vol.%. In order to avoid the agglomeration of CNTs inside the composite film caused by the casting/evaporation process, Du and colleagues [97] proposed coagulation as an alternative method for producing composites containing individually dispersed CNTs.

7.1.2 Melt Processing

Melt intercalation is considered as environmentally beneficial and a far superior substitute for solution mixing. However, the melt-mixing process is governed by several parameters like mixing time, screw speed, and mixing temperature, that can directly affect the homogeneity in the dispersion of CNTs inside the polymer phase. Kasaliwal et al. [98] investigated the effect of altering screw speed and mixing temperature throughout a five-minute mixing period on the dispersion state of CNTs in a polycarbonate (PC) matrix. The amount of dispersion of CNTs in PC matrix was found to be the same at high mixing speeds regardless of temperature or melt viscosity. However, at low melt mixing speeds, excellent dispersion was achieved at either low temperature or high melt viscosity of the PC matrix.

Alig et al. [93], Basiuk Golovataya-Dzhymbeev et al. [99] investigated the relationship between processing conditions and CNT nanocomposites morphologies. According to the authors, the preparation procedure can be divided into four steps: (1) Polymer wetting of initial agglomerates, (2) infiltration of polymer chains into the agglomerates to weaken them, (3) agglomeration dispersion by rupture and erosion, and (4) distribution of individualized nanotubes into the matrix. Mezghani et al. [100] on the other hand, described the synthesis of linear low-density polyethylene (LLDPE)/MWCNT nanocomposite fibers prepared using melt extrusion and spun through a spinneret die. The effects of CNT loadings on the characteristics of LLDPE/MWCNT nanocomposite were explored, and it was discovered that a small addition of CNT improves the properties overall.

7.1.3 In Situ Polymerization

To avoid the dispersion concerns between CNTs and polymers, it has been proposed to develop CNT composites by in-situ polymerization. This approach is also suited for polymers that are insoluble and thermally unstable and cannot be synthesized through solution or melts processing [101]. CNTs are often combined with monomers before being polymerized under certain conditions to produce CNTs/Polymer composites [102]. To induce polymerization in thermosets such as epoxies or unsaturated polyesters, a curing agent or peroxide is applied.

Keteklahijani et al. [103] reported the synthesis of conducting nanocomposites of poly (anilineboronic acid) (PABA) based sensors for detection of neurotransmitter dopamine by in situ polymerization of 3-aminophenylboronic acid monomers with the presence or absence of CNT. Among all the synthesized materials, the hybrid structure of DNA-functionalized carbon nanotube with nitrogen-doped graphene (DNA_CNT_NEG) generated highly ordered and polyconjugated PABA nanocomposites with better electrical conductivity (14,300 $S.m^{-1}$ at 3.0 wt% filler content) compared to pure PABA (4×10^{-9} $S.m^{-1}$) or polymer composites synthesized with only CNT (1.75×10^{-4} $S.m^{-1}$ at 3.0 wt% filler content).

Wang et al. [104] used the in-situ polymerization to design a multiwalled carbon nanotube with phenylenediamine (MWNT-NH2)–polyimide (PI) matrix interface. A high-performance nanocomposite was obtained exhibiting a 50.5% increase in the tensile strength and an 83.1% increase in the Young's modulus by adding a 3.0 wt% of MWNT-NH2. Similarly, Wu et al. [105] investigated the mechanical and thermal properties of hydroxyl functionalized MWCNTs/acrylic acid grafted polytrimethylene terephthalate (PTT) nanocomposites and discovered a notable improvement in the thermal and mechanical properties of the PTT matrix due to the formation of ester bonds between the –COOH groups of acrylic acid grafted PTT and the –OH groups of MWCNTs.

Overall, in-situ polymerization may be used to create practically any polymer nanocomposites comprising CNTs that are either non-covalently or covalently bonded to the polymer matrix [106–108].

7.2 *Mechanical Properties of Polymer/Carbon Nanotubes Nanocomposites*

Thanks to their unique structure and features, the usage of carbon nanotubes as nanofillers in polymeric matrices can result in drastic improvement of physicochemical properties. The incorporation of carbon nanotubes (CNTs) into polymer matrices resulted in a considerable change in the mechanical, electrical, and thermal characteristics of the polymer matrices [109, 110]. Processing procedures [111], aspect ratio [112] and CNT content [113] are all factors that impact property modification. Several research have been conducted to assess the mechanical, electrical, and thermal characteristics of CNT/polymer nanocomposite under various conditions and filler loading.

Since carbon nanotubes are stronger than steel and lighter than aluminum, mechanical reinforcement of nanotubes for polymer/CNT-based composites is one of the most researched disciplines. Several studies have been conducted in order to elucidate the outstanding mechanical performances of these composites.

7.2.1 Young Modulus

Zhao et al. [114] have shown in their study that by combining polydimethylsiloxan (PDMS) composites with boron-doped CNT sponge, mechanical parameters such as compressive modulus (70%), viscous modulus (243%), and damping capacity (50%) were significantly improved. According to Liu et al. [115], the mechanical properties of CNTs/polymer nanocomposite materials are highly dependent on CNT content. Mechanical characteristics of nanocomposites increase dramatically with the addition of CNTs within a certain weight fraction, and they reach an optimum when the CNT concentration is about 2 wt.%. Dikshit and Engle [116] used molecular dynamics (MD) modeling to investigate the mechanical characteristics of epoxy DGEBA with and without CNT reinforcement. The results showed that the pure epoxy matrix was less rigid than the CNT enhanced epoxy nanocomposite. The Young's modulus of a CNT-enhanced epoxy nanocomposite was calculated to be 13.27 Gpa and 2.429 GPa without reinforcement. On the other hand, Gao et al. [117] have developed a chemical processing technique that enables continuous spinning of polyamide 6/SWNT fibers by in situ polymerization of caprolactam in the presence of SWNT. Figure 10a–c depict the schematic and pictures of the fiber spinneret setup respectively, of the spinneret arrangement and the composite fiber.

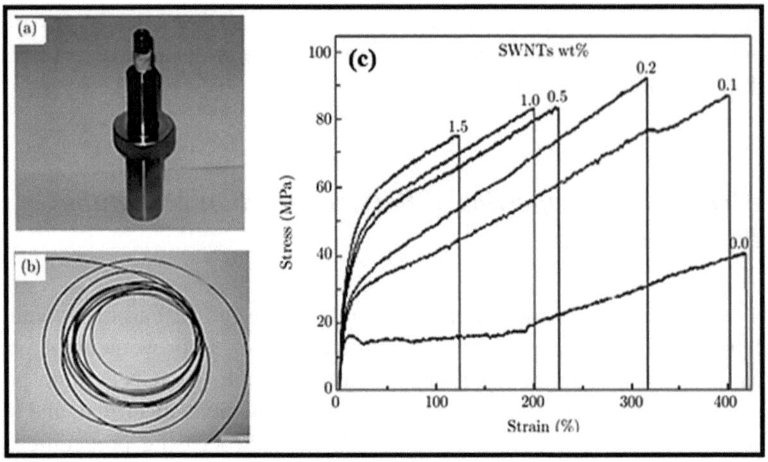

Fig. 10 **a** Photograph of spinneret setup **b** Photograph of the composite fiber and the Stress–strain profiles of nylon 6/SWNT composite fibers at different SWNT loadings. The curves are labeled with the percentage of SWNTs in the polymer matrix [117]

Neat polyamide 6 (PA6) and its composite fibers stress–strain profiles are depicted in Fig. 10(c) The incorporation of SWNT into the PA6 matrix improves the tensile strength and Young's modulus of SWNT-reinforced composite fibers. Tensile modulus and tensile strength rose approximately by 2.7 and 1.9 times, respectively, by incorporating 1.5 wt% SWNTs into PA6, indicating that the nylon fibers became harder and more resistant to deformation.

7.2.2 Compressive Modulus

Sharma et al. [118] reported in their study the development of carbon nanotubes grown fibers with epoxy matrix. They showed that the modification of the fiber/matrix contact is thus predicted to affect composite characteristics. The compressive strengths of these composites were measured, and they showed a significant improvement of 43 and 94% in the longitudinal and transverse compressive strengths, respectively, when compared to carbon fiber composites that went through a similar thermal cycle but did not grow carbon nanotubes. Ramezani et al. [119] have studied the effect of CNT characteristics on the mechanical properties of CNT-reinforced cementitious materials. They revealed that CNT length was shown to have no effect on compressive strength, however longer CNTs outperformed in terms of flexural strength. Moreover, CNTs with small diameters improved compressive strength but decreased the flexural strength. In general, CNTs with average lengths and diameters of 10–20 m and 20–325 nm, respectively, significantly contributed to greater mechanical performance. They also identified that for flexural and compressive strengths, appropriate upper limits of 0.15 c-wt% and 0.20 c-wt% were determined for CNT concentration.

Apart from the mentioned characteristics, CNTs have another property that can affect the compressive behavior of CNT based composites. Mishra et al. [120] studied the effect of SWCNTs with two types of defects on the compressive strength of polypropylene (PP) composite. It was observed that with the increase in percent defect the mechanical properties of the SWCNT-PP composite were decreased. The effects of both Stone–Wales (SW) and vacancy defects were examined. It was observed that the vacancy defect affected the SWCNT-PP nanocomposites more than the SW defect. Li et al. [121] prepared a MWCNT/nickel-coated carbonized loofah fibers/polyether ether ketone composite and showed a superior compressive performance that reached a value of 145.6 Mpa.

7.2.3 Stress Relaxation

Zu et al. [122] performed an investigation of the different aspects influencing the tensile stress relaxation behavior of CNT fibers. They found that in both pure and composite fiber, higher starting strain level, lower strain rate, and larger gauge length resulted in a faster rate of stress decay. The principal cause of time-dependent deformation of CNT-based continuous fibers is sliding among the many short CNTs, which are largely kept together by van der Waals interactions.

In order to improve the interfacial interaction between Mutiwalled carbon nanotubes (MWCNTs) and epoxy matrix, Feng et al. [123] conducted a modification by polydopamine (PDA) coating. Figure 11 shows the evolution of the relaxation modulus at 180 °C for the epoxy composites synthesized with different content of MWCNTs and MWCNTs@PDA hybrids. Clearly, stress relaxation occurred at 180 °C for all samples, suggesting that the epoxy network can flow despite the presence of high-loading fillers.

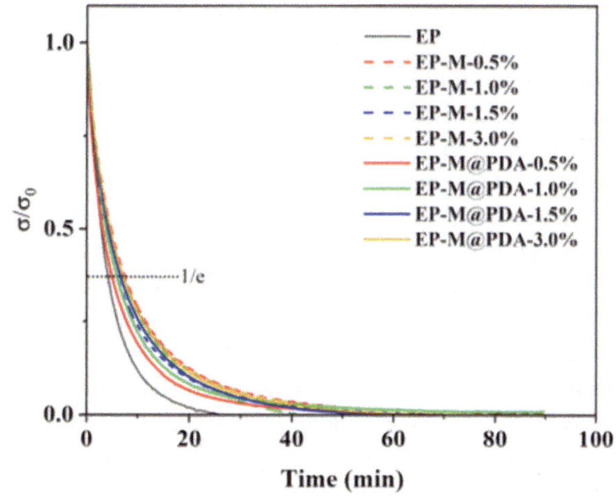

Fig. 11 Stress relaxation curves of epoxy vitrimers at 180 °C [123]

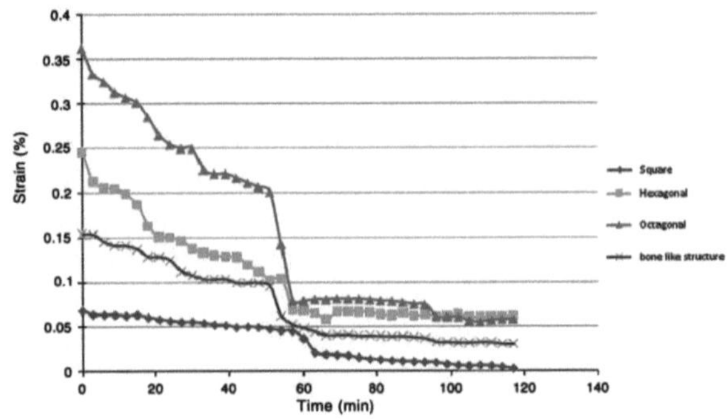

Fig. 12 Creep behavior of PCL/CNT monolayer with different topologies (square, hexagonal, octagonal) [124]

In some studies, the stress relaxation behavior is influenced by the material geometry. Belmonte et al. [124] prepared different 3D bone like scaffold with polycaprolactone–carbon nanotube composites, and the results showed clearly that the creep response is highly dependent on the geometry of the scaffolds (Fig. 12).

7.2.4 Dynamic Viscoelastic Properties

Montazeri et al. [125] have added various weight fractions of multi-walled carbon nanotubes (MWNTs) to the epoxy as resin-forcement to fabricate MWNT/epoxy nanocomposite. The viscoelastic characteristics were determined by the dynamic mechanical thermal analysis and the results indicated that addition of nanotubes to epoxy had significant effect on the viscoelastic properties (Fig. 13). The optimal proportion for maximum modulus was obtained by a 0.5 wt.% MWNT. The storage modulus fluctuates as the percentage of nanotubes increases. At ambient temperature, the glassy storage modulus increases as the nanotube concentration increases. At room temperature, higher contents produce a decrease in modulus. The inclusion of nanotubes increases the elastic characteristics of the epoxy system at high temperatures in the rubbery tray. Because of the large surface area (300 m^2/g) the behavior may be described in terms of an interaction between the MWNT and the epoxy, which limits the potential mobility of the polymer network and therefore increases its thermal stability.

In another study, Suhr et al. [126] performed direct shear tests on epoxy thin films with dense packing of multiwalled carbon nanotube fillers and found substantial viscoelastic behavior with up to a 1,400% increase in loss factor (damping ratio) over the baseline epoxy. Some authors used computational simulations to investigate the viscoelastic properties of SWCNT. In this area, Huang et al. [127] used to evaluate the finite element method (FEM) to analyze the viscoelastic properties of reinforced polypropylene nanocomposites at different temperatures (20, 40, 60, and 80 °C). The changes were quicker as the temperature increases and the relaxation time decreased. Similarly, if the relaxation time is long enough, the fluid-like property of the composite approached the solid-like property.

Fig. 13 Storage modulus for epoxy and nanocomposites with different weight percent at 10 Hz frequency [125]

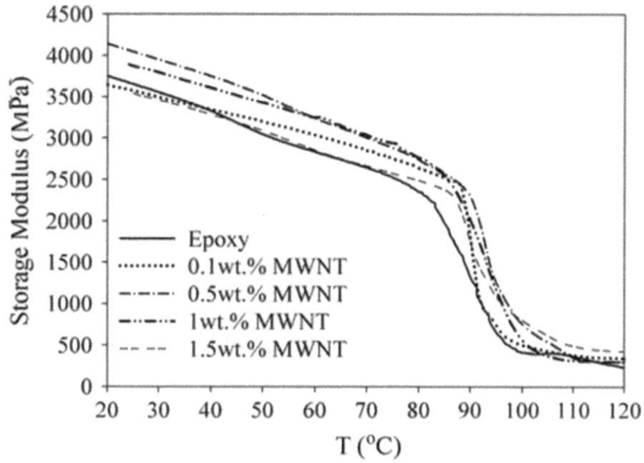

7.3 Polymer/Carbon Nanotubes Nanocomposites Application

Because of their superior properties, carbon-based polymer composites have gained in popularity [128, 129]. The improved properties of carbon nanotubes (CNTs) justify the numerous applications of these polymer nanocomposites. Furthermore, the great compatibility of carbon nanotubes with polymer matrices has increased their potential for use in a broad range of specialized applications, including electronics [130], automotive [131], textiles [132], aerospace [133], sports equipment [134], energy storage devices [135], and filters [136]. Because of their extraordinary stiffness and strength, high fracture resistance, low density, and high energy absorption, polymer nanocomposites reinforced with CNTs have also been regarded as an ideal choice for the construction of ballistic armour materials [137]. In the following sections some examples of the CNT polymer composite in which the mechanical strength is required are detailed.

7.3.1 Photovoltaics

Carbon nanotubes (CNTs) are characterized by their excellent electron transport characteristics and higher electron affinity compared to existing polymer-based light-harvesting donors. These properties make them good candidates for their usage in bulk heterojunction (BHJ)-type hybrid solar cells. These materials have gained a lot of attention for photovoltaic (PV) device applications, thanks to their intriguing features such as high dielectric constant, high charge mobility, and low cost [129].

The first use of this type of material was first initiated by Ago et al. [138] which suggested the replacement of Indium tin oxide (ITO) electrode in PV by MWCNTS polymer-based composite. CNTs can be either applied on the active layer or the electrode in photovoltaic cells. As transparent anodes in polymer solar cells, SWCNTs [139] and MWCNTs [140], have been explored. These materials' features, including their high optical transmittance [141], make them especially desirable for use as transparent conducting electrodes (TCEs) in PV and display applications.

In order to develop SWCNT electrodes with very high transparency and low sheet resistance, considerably increasing the proportion of metallic SWCNTs is a successful strategy, with a major improvement in the PV performance of related devices [142]. However, it has recently been demonstrated that high-quality and highly smooth SWCNT electrodes deposited through ultrasonic spray may effectively replace both (ITO) and/or poly(3,4-ethylenedioxythiophene): poly(styrenesulfonate) (PEDOT:PSS) [143]. The poly(3-hexylthiophene). (P3HT) and [6]-phenyl C61-butyric acid methylester (PCBM) solar cells without ITO or PEDOT had a power conversion efficiency (PCE) of 3.37%, compared to 3.51% and 4.13% for ITO/PEDOT: PSS and SWNT/PEDOT: PSS solar cells, respectively.

On the other hand, the incorporation of SWCNTs into the active layer had an effect on charge separation and transport in the solar device. Schuettfort et al. [144] described a highly pure nanohybrid structure composed of a SWNT covered with a monolayer of P3HT, with an efficiency of 0.57%. In bulk heterojunction P3HT: PCBM devices, acid functionalized SWCNTs achieve efficiencies as high as 1.4%, compared to reference devices with 1% efficiency [145].

7.3.2 Water Treatment

The incorporation of CNTs into polymeric membranes for water treatment has been proposed as a good technique for reducing membrane breakdown and fouling [146]. Shawky et al. [147] have synthesized (MWCNTs)/aromatic polyamide (PA) nanocomposite membranes for water desalination. The authors suggested the formation of a network structure which is responsible of the lower permeability and higher salt rejection. Moreover, the membrane system could eliminate large molecule such as Humic acid which its elimination increased from 54 to 90% when MWCNT loading increased from 0 to 10 mg/g. These CNT composite membranes can also be applied for oil water separation generated from oil industries, domestic sewage, transportation…etc. In this context, Gu et al. [148] have synthesized a superhydrophobic CNT-polystyrene composite membrane for oil–water separation (Fig. 14). This membrane showed a very high rejection capacity of 99.94%. Furthermore, the addition of CNT showed an improvement of the mechanical properties such as tensile strength, and Young's modulus due to the combination of the performances of traditional membrane materials with those of CNTs.

In recent years, the removal of hazardous components in water or wastewater has gained increasing attention because of the tremendous harm they cause to the environment and human health. To investigate the capabilities of pharmaceutical and personal care products removal, a CNTs-PVDF composite membrane [149] was prepared and employed in the removal of triclosan (TCS), acetaminophen (AAP), and ibuprofen (IBU). Toxic metal ions are also considered to be very dangerous for the environment and humans. Thus, the removal of these harmful ions can be

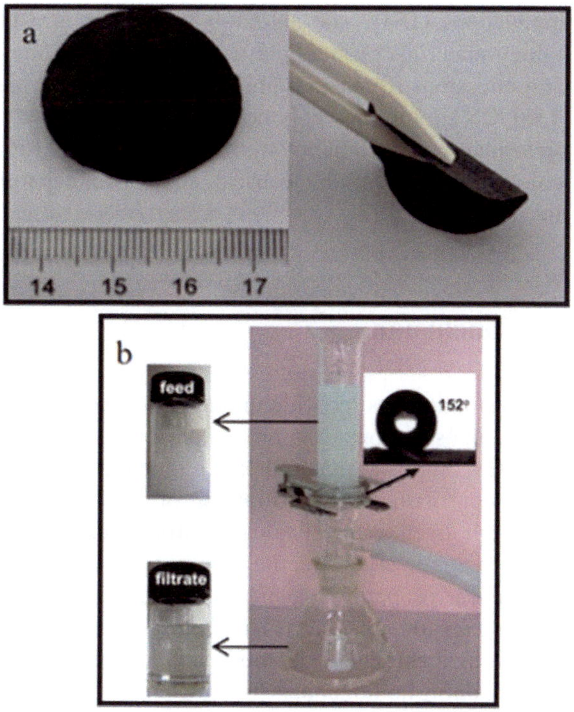

Fig. 14 Oil–water separation for superhydrophobic CNTs-polystyrene composite membrane: photographs of **a** membranes and **b** oil–water separation [148]

performed with CNT-based composite membranes. According to this, Parham et al. [150] have achieved 100% elimination efficiency of heavy metal ions.

7.3.3 Automotive, Military, and Aeronautical

The aerospace and automotive industries need materials with extremely high strength and durability to be integrated as components in their related equipment. Thanks to their outstanding performances, CNTs have been widely evaluated in their incorporation in polymer composites. The usage of CNT/epoxy nanocomposites in the aviation and aerospace industries is clear in the present day, as seen by its employment in the space shuttle [151] and sophisticated commercial aircraft [152]. CNT/polymer composites with broad frequency ranges for absorbing values surpassing 5 dB, such as polyethylene terephthalate (PET), polyethylene (PE) and polypropylene (PP) have prospective uses as radar absorption materials and can be widely employed in commercial and military applications [153].

For de-icing applications in the aircraft sector, the coating was discovered to have low surface resistance and high flexibility. The flame-retardant qualities of 10-dihydro-9-oxa-10 phosphaphenanthrene-10-oxide-based phosphorus tetraglycidyl epoxy nanocomposites were evaluated and found to be appropriate for aerospace applications [154]. The automobile industry is a big consumer of modern technology materials such as plastic. As a result of the goal for light weight and low CO_2 emissions from automobiles, automotive polymers are also garnering attention [155]. CNT reinforced polymer nanocomposites are expected to greatly improve the performance of current automotive technologies [156] due to their superior mechanical, chemical, thermal, electrical, and barrier capabilities, as well as their effect on fire retardancy [157].

8 Conclusion

Carbon nanotubes provide unique prospects for the advancement of polymer nanocomposites toward improved performance and multifunctionality. When CNT-polymer composite materials are aligned, they exhibit a notably high aspect ratio, which contributes to the formation of a wide polymer-filler interface, resulting in very high levels of mechanical reinforcement. In this chapter, CNTs as fillers and their synergetic effects on the different properties of polymer composites especially mechanical properties have been discussed. According to the different reviewed studies, considerable efforts have been conducted to enhance the quality of CNTs and processing techniques. The most challenging issue in achieving CNTs' full potential is maintaining homogenous dispersion of CNTs so that the maximal filler surface area is accessible for charge transport between the filler and matrix. The functionalization of nanotubes provides an easy way to increase dispersion and stress transmission between CNTs and polymer matrix, although more progress is needed in this area to conserve the intrinsic features of CNTs.

CNT is an excellent filler for polymer composites, however certain difficulties must be overcome before the outstanding characteristics of CNT in polymer nanocomposite can be completely reached.

References

1. Colangelo F, Farina I, Travaglioni M, Salzano C, Cioffi R, Petrillo A (2021) Innovative materials in Italy for eco-friendly and sustainable buildings. Materials 14:2048. https://doi.org/10.3390/ma14082048
2. Magee CL (2010) The role of materials innovation in overall technological development. JOM 62:20–24. https://doi.org/10.1007/s11837-010-0043-5
3. Zhang J, Mi Y (2022) Application of new energy composites in sports facilities and fitness equipment. J Nanomater 2022:1–11. https://doi.org/10.1155/2022/7712859
4. Tsai F-C, Shu C-M, Tsai L-C, Ma N, Wen Y, Wen S, Yang Y-K, Zhou W, Xiao H-W, Shu Y-C, Jiang T (2011) Carbon nanotube industrial applications. In: Marulanda JM (ed) Carbon nanotubes applications on electron devices. InTech
5. Gohardani O, Elola MC, Elizetxea C (2014) Potential and prospective implementation of carbon nanotubes on next generation aircraft and space vehicles: a review of current and

expected applications in aerospace sciences. Prog Aerosp Sci 70:42–68. https://doi.org/10.1016/j.paerosci.2014.05.002
6. Anzar N, Hasan R, Tyagi M, Yadav N, Narang J (2020) Carbon nanotube—a review on synthesis, properties and plethora of applications in the field of biomedical science. Sens Int 1:100003. https://doi.org/10.1016/j.sintl.2020.100003
7. Eatemadi A, Daraee H, Karimkhanloo H, Kouhi M, Zarghami N, Akbarzadeh A, Abasi M, Hanifehpour Y, Joo SW (2014) Carbon nanotubes: properties, synthesis, purification, and medical applications. Nanoscale Res Lett 9:393. https://doi.org/10.1186/1556-276X-9-393
8. Daneshvar F, Chen H, Noh K, Sue H-J (2021) Critical challenges and advances in the carbon nanotube–metal interface for next-generation electronics. Nanoscale Adv 3:942–962. https://doi.org/10.1039/D0NA00822B
9. Porwal M (2017) An overview on carbon nanotubes. MOJBB 3. https://doi.org/10.15406/mojbb.2017.03.00045
10. Radushkevich LV, Lukyanovich VM, The structure of carbon produced at thermal decomposition of carbon monoxide on an iron catalyst. Russ J Phys Chem 26:88–95
11. Bacon R (1960) Growth, structure, and properties of graphite whiskers. J Appl Phys 31:283–290. https://doi.org/10.1063/1.1735559
12. Oberlin A, Endo M, Koyama T (1976) Filamentous growth of carbon through benzene decomposition. J Cryst Growth 32:335–349. https://doi.org/10.1016/0022-0248(76)90115-9
13. Iijima S (1991) Helical microtubules of graphitic carbon. Nature 354:56–58
14. Iijima S, Ichihashi T (1993) Single shell carbon nanotubes of 1 nm diameter. Nature 363:603–605
15. Journet C, Maser WK, Bernier P, Loiseau A, de la Chapelle ML, Lefrant S, Deniard P, Lee R, Fischer JE (1997) Large-scale production of single-walled carbon nanotubes by the electric-arc technique. Nature 388:756–758.https://doi.org/10.1038/41972
16. Balasubramanian K, Burghard M (2005) Chemically functionalized carbon nanotubes. Small 1:180–192. https://doi.org/10.1002/smll.200400118
17. Beg S, Rahman M, Jain A, Saini S, Hasnain MS, Swain S, Imam S, Kazmi I, Akhter S (2018) Emergence in the functionalized carbon nanotubes as smart nanocarriers for drug delivery applications. In: Fullerens, graphenes and nanotubes. Elsevier, pp 105–133
18. Drakonakis VM, Aureli M, Doumanidis CC, Seferis JC (2015) Modulus–density negative correlation for CNT-reinforced polymer nanocomposites: modeling and experiments. Compos B Eng 70:175–183. https://doi.org/10.1016/j.compositesb.2014.10.037
19. Journet C, Picher M, Jourdain V (2012) Carbon nanotube synthesis: from large-scale production to atom-by-atom growth. Nanotechnology 23:142001. https://doi.org/10.1088/0957-4484/23/14/142001
20. José-Yacamán M, Miki-Yoshida M, Rendón L, Santiesteban JG (1993) Catalytic growth of carbon microtubules with fullerene structure. Appl Phys Lett 62:657–659.https://doi.org/10.1063/1.108857
21. Journet C, Maser WK, Bernier P, Loiseau A, Chapelle ML, Lefrant S, Deniard P, Lee R, Fisher JE (1997) Large-scale production of single-walled carbon nanotubes by the electric-arc technique. Nature 388:756–758
22. Chrzanowska J, Hoffman J, Małolepszy A, Mazurkiewicz M, Kowalewski TA, Szymanski Z, Stobinski L (2015) Synthesis of carbon nanotubes by the laser ablation method: effect of laser wavelength: synthesis of CNTs by the laser ablation method. Phys Status Solidi B 252:1860–1867. https://doi.org/10.1002/pssb.201451614
23. Kuo TF, Chi CC, Lin IN (2001) Synthesis of carbon nanotubes by laser ablation of graphites at room temperature. Jpn J Appl Phys 40:7147–7150. https://doi.org/10.1143/JJAP.40.7147
24. Chibante LPF, Thess A, Alford JM, Diener MD, Smalley RE (1993) Solar generation of the fullerenes. J Phys Chem 97:8696–8700. https://doi.org/10.1021/j100136a007
25. Laplaze D, Bernier P, Journet C, Vié V, Flamant G, Lebrun M (1997) Carbon sublimation using a solar furnace. Synth Met 86:2295–2296. https://doi.org/10.1016/S0379-6779(97)81133-2

26. De Boer G, Arepalli S, Holmes W, Nikolaev P, Range C, Scott C (2001) Detection of nickel atom by laser induced fluorescence during carbon nanotube formation in a laser produced plume. J Appl Phys 89:5760–5768.https://doi.org/10.1063/1.1365441
27. Keszler AM, Kováts É, Bódis E, Károly Z, Szépvölgyi J (2022) Effect of metallic and non-metallic additives on the synthesis of fullerenes in thermal plasma. Condensed Matter 7:44. https://doi.org/10.3390/condmat7030044
28. Karthikeyan S, Mahalingam P, Karthik M, Large scale synthesis of carbon nanotubes
29. Rashid HU, Yu K, Umar MN, Anjum N, Khan K, Ahmad N, Jan MT, Catalyst role in chemical vapor deposition (CVD) process: a review
30. Laurent Ch, Flahaut E, Peigney A (2010) The weight and density of carbon nanotubes versus the number of walls and diameter. Carbon 48:2994–2996.https://doi.org/10.1016/j.carbon.2010.04.010
31. Berrada N, Desforges A, Bellouard C, Flahaut E, Gleize J, Ghanbaja J, Vigolo B (2019) Protecting carbon nanotubes from oxidation for selective carbon impurity elimination. J Phys Chem C 123:14725–14733. https://doi.org/10.1021/acs.jpcc.8b12554
32. Xiao Y, Ahmed Z, Ma Z, Zhou C, Zhang L, Chan M (2019) Low temperature synthesis of high-density carbon nanotubes on insulating substrate. Nanomaterials 9:473. https://doi.org/10.3390/nano9030473
33. Murali S, Perumal S (2018) Synthesis of nanotubes under carbon environment at low temperature using hydrothermal method. Optik 162:81–85. https://doi.org/10.1016/j.ijleo.2018.01.106
34. Liu B, Jia D, Zhou Y, Feng H, Meng Q (2007) Low temperature synthesis of amorphous carbon nanotubes in air. Carbon 45:1710–1713. https://doi.org/10.1016/j.carbon.2007.04.014
35. Charoenpakdee J, Suntijitrungruang O, Boonchui S (2020) Chirality effects on an electron transport in single-walled carbon nanotube. Sci Rep 10:18949. https://doi.org/10.1038/s41598-020-76047-9
36. Lekawa-Raus A, Patmore J, Kurzepa L, Bulmer J, Koziol K (2014) Electrical properties of carbon nanotube based fibers and their future use in electrical wiring. Adv Funct Mater 24:3661–3682.https://doi.org/10.1002/adfm.201303716
37. Jaspal P, Sharma P (2016) Performance comparison of carbon nanotubes with copper and aluminium as winding material in transformer. Ind J Sci Technol 9. https://doi.org/10.17485/ijst/2016/v9i40/100681
38. Zhang R, Xie H, Zhang Y, Zhang Q, Jin Y, Li P, Qian W, Wei F (2013) The reason for the low density of horizontally aligned ultralong carbon nanotube arrays. Carbon 52:232–238. https://doi.org/10.1016/j.carbon.2012.09.025
39. Collins PG, Zettl A (1996) A simple and robust electron beam source from carbon nanotubes. Appl Phys Lett 69:1969–1971. https://doi.org/10.1063/1.117638
40. Ruoff RS, Qian D, Liu WK (2003) Mechanical properties of carbon nanotubes: theoretical predictions and experimental measurements. C R Phys 4:993–1008. https://doi.org/10.1016/j.crhy.2003.08.001
41. Deng L, Eichhorn SJ, Kao C-C, Young RJ (2011) The effective Young's modulus of carbon nanotubes in composites. ACS Appl Mater Interfaces 3:433–440. https://doi.org/10.1021/am1010145
42. Tingaev MI, Belenkov EA (2017) Hybrid sp ^{2} +sp ^{3} carbon phases created from carbon nanotubes. J Phys: Conf Ser 917:032013. https://doi.org/10.1088/1742-6596/917/3/032013
43. Tracey MMJ, Ebbessen TW, Gibson JM (1996) Exceptionally high Young's modulus observed for individual carbon nanotubes. Nature 381:678–680
44. Krishnan A, Dujardin E, Ebbesen TW, Yianilos PN, Treacy MMJ (1998) Young's modulus of single-walled nanotubes. Phys Rev B 58:14013–14019. https://doi.org/10.1103/PhysRevB.58.14013
45. Lourie O, Wagner HD (1998) Evaluation of Young's modulus of carbon nanotubes by micro-Raman spectroscopy. J Mater Res 13:2418–2422. https://doi.org/10.1557/JMR.1998.0336
46. Lu JP (1997) Elastic properties of carbon nanotubes and nanoropes. Phys Rev Lett 79

47. Xiong Z, Zhong L, Wang H, Li X (2021) Structural defects, mechanical behaviors, and properties of two-dimensional materials. Materials 14:1192. https://doi.org/10.3390/ma14051192
48. Khandoker N, Hawkins SC, Ibrahim R, Huynh CP, Deng F (2011) Tensile strength of spinnable multiwall carbon nanotubes. Procedia Eng 10:2572–2578. https://doi.org/10.1016/j.proeng.2011.04.424
49. Zhao T, Zhang S, Guo Y, Wang Q (2016) Mechanical properties of carbon nanotube fibers at extreme temperatures. Nanoscale 8:233–242. https://doi.org/10.1039/C5NR04472C
50. Kim P, Shi L, Majumdar A, McEuen PL (2001) Thermal transport measurements of individual multiwalled nanotubes. Phys Rev Lett 87:215502. https://doi.org/10.1103/PhysRevLett.87.215502
51. Lindsay L, Broido DA, Mingo N (2009) Lattice thermal conductivity of single-walled carbon nanotubes: beyond the relaxation time approximation and phonon-phonon scattering selection rules. Phys Rev B 80:125407. https://doi.org/10.1103/PhysRevB.80.125407
52. Pop E, Mann D, Wang Q, Goodson K, Dai H (2006) Thermal conductance of an individual single-wall carbon nanotube above room temperature. Nano Lett 6:96–100. https://doi.org/10.1021/nl052145f
53. Hone J, Whitney M, Piskoti C, Zettl A, Thermal conductivity of single-walled carbon nanotubes
54. Grujicic M, Cao G, Gersten B (2004) Atomic-scale computations of the lattice contribution to thermal conductivity of single-walled carbon nanotubes. Mater Sci Eng, B 107:204–216. https://doi.org/10.1016/j.mseb.2003.11.012
55. Ma P-C, Siddiqui NA, Marom G, Kim J-K (2010) Dispersion and functionalization of carbon nanotubes for polymer-based nanocomposites: a review. Compos A Appl Sci Manuf 41:1345–1367. https://doi.org/10.1016/j.compositesa.2010.07.003
56. Xie X, Mai Y, Zhou X (2005) Dispersion and alignment of carbon nanotubes in polymer matrix: a review. Mater Sci Eng R Rep 49:89–112. https://doi.org/10.1016/j.mser.2005.04.002
57. Coleman JN, Khan U, Gun'ko YK (2006) Mechanical reinforcement of polymers using carbon nanotubes. Adv Mater 18:689–706. https://doi.org/10.1002/adma.200501851
58. Qian H, Greenhalgh ES, Shaffer MSP, Bismarck A (2010) Carbon nanotube-based hierarchical composites: a review. J Mater Chem 20:4751. https://doi.org/10.1039/c000041h
59. Bal S, Samal SS (2007) Carbon nanotube reinforced polymer composites—a state of the art. Bull Mater Sci 30:379–386. https://doi.org/10.1007/s12034-007-0061-2
60. Niyogi S, Hamon MA, Hu H, Zhao B, Bhowmik P, Sen R, Itkis ME, Haddon RC (2002) Chemistry of single-walled carbon nanotubes. Acc Chem Res 35:1105–1113. https://doi.org/10.1021/ar010155r
61. Hirsch A, Vostrowsky O (2005) Functionalization of carbon nanotubes. In: Schlüter AD (ed) Functional molecular nanostructures. Springer, Berlin Heidelberg, Berlin, Heidelberg, pp 193–237
62. Kuznetsova A, Popova I, Yates JT, Bronikowski MJ, Huffman CB, Liu J, Smalley RE, Hwu HH, Chen JG (2001) Oxygen-containing functional groups on single-wall carbon nanotubes: NEXAFS and vibrational spectroscopic studies. J Am Chem Soc 123:10699–10704. https://doi.org/10.1021/ja011021b
63. Nagasawa S, Yudasaka M, Hirahara K, Ichihashi T, Iijima S (2000) Effect of oxidation on single-wall carbon nanotubes. Chem Phys Lett 328:374–380
64. Mickelson ET, Chiang IW, Zimmerman JL, Boul PJ, Lozano J, Liu J, Smalley RE, Hauge RH, Margrave JL (1999) Solvation of fluorinated single-wall carbon nanotubes in alcohol solvents. J Phys Chem B 103:4318–4322. https://doi.org/10.1021/jp9845524
65. Pehrsson PE, Zhao W, Baldwin JW, Song C, Liu J, Kooi S, Zheng B (2003) Thermal fluorination and annealing of single-wall carbon nanotubes. J Phys Chem B 107:5690–5695. https://doi.org/10.1021/jp027233s
66. Ray SC, Pao CW, Tsai HM, Chiou JW, Pong WF, Chen CW, Tsai M-H, Papakonstantinou P, Chen LC, Chen KH (2007) A comparative study of the electronic structures of oxygen- and chlorine-treated nitrogenated carbon nanotubes by x-ray absorption and scanning photoelectron microscopy. Appl Phys Lett 91:202102. https://doi.org/10.1063/1.2807275

67. Chen YK, Green MLH, Griffin JL, Hammer J, Lago RM, Tsang SC (1996) Purification and opening of carbon nanotubes via bromination. Adv Mater 8:1012–1015.https://doi.org/10.1002/adma.19960081216
68. Chen J, Hamon MA, Hu H, Chen Y, Rao AM, Eklund PC, Haddon RC (1998) Solution properties of single-walled carbon nanotubes. Science 282:95–98. https://doi.org/10.1126/science.282.5386.95
69. Brunetti FG, Herrero MA, de Muñoz JM, Giordani S, Díaz-Ortiz A, Filippone S, Ruaro G, Meneghetti M, Prato M, Vázquez E (2007) Reversible microwave-assisted cycloaddition of aziridines to carbon nanotubes. J Am Chem Soc 129:14580–14581.https://doi.org/10.1021/ja077927k
70. Sangolkar AA, Pawar R (2020) Prediction of the [4 + 2]- and [5 + 4]-cycloaddition reactions in zig-zag carbon nanotubes *via* an ambimodal transition state: density functional theory calculations. RSC Adv 10:11111–11120. https://doi.org/10.1039/C9RA10252C
71. Syrgiannis Z, Melchionna M, Prato M (2014) Covalent carbon nanotube functionalization. In: Kobayashi S, Müllen K (eds) Encyclopedia of polymeric nanomaterials. Springer, Berlin Heidelberg, Berlin, Heidelberg, pp 1–8
72. Rinzler AG, Liu J, Dai H, Nikolaev P, Huffman CB, Rodríguez-Macías FJ, Boul PJ, Lu AH, Heymann D, Colbert DT, Lee RS, Fischer JE, Rao AM, Eklund PC, Smalley RE (1998) Large-scale purification of single-wall carbon nanotubes: process, product, and characterization. Appl Phys A Mater Sci Process 67:29–37. https://doi.org/10.1007/s003390050734
73. Martin-Martinez M, Machado BF, Serp P, Morales-Torres S, Silva AMT, Figueiredo JL, Faria JL, Gomes HT (2020) Carbon nanotubes as catalysts for wet peroxide oxidation: the effect of surface chemistry. Catal Today 357:332–340.https://doi.org/10.1016/j.cattod.2019.03.014
74. Thi Mai Hoa L (2018) Characterization of multi-walled carbon nanotubes functionalized by a mixture of HNO3/H2SO4. Diamond and Related Materials 89:43–51.https://doi.org/10.1016/j.diamond.2018.08.008
75. Xie J, Ahmad MN, Bai H, Li H, Yang W (2010) Concentration and temperature controlled oxidation and cutting of single-walled carbon nanotubes by ammonium persulfate. Sci China Chem 53:2026–2032. https://doi.org/10.1007/s11426-010-4061-5
76. Malikov EY, Akperov OH, Muradov MB, Eyvazova GM, Maharramov AM, Kukovecz Á, Kónya Z (2014) Oxidation of multiwalled carbon nanotubes using different oxidation agents like nitric acid and potassium permanganate. News of Baku Univ 4:49–58
77. Li C, Wang D, Liang T, Wang X, Wu J, Hu X, Liang J (2004) Oxidation of multiwalled carbon nanotubes by air: benefits for electric double layer capacitors. Powder Technol 142:175–179. https://doi.org/10.1016/j.powtec.2004.04.037
78. Simmons JM, Nichols BM, Baker SE, Marcus MS, Castellini OM, Lee C-S, Hamers RJ, Eriksson MA (2006) Effect of ozone oxidation on single-walled carbon nanotubes. J Phys Chem B 110:7113–7118. https://doi.org/10.1021/jp0548422
79. Klink S, Ventosa E, Xia W, La Mantia F, Muhler M, Schuhmann W (2012) Tailoring of CNT surface oxygen groups by gas-phase oxidation and its implications for lithium ion batteries. Electrochem Commun 15:10–13.https://doi.org/10.1016/j.elecom.2011.11.012
80. Li C, Zhao A, Xia W, Liang C, Muhler M (2012) Quantitative studies on the oxygen and nitrogen functionalization of carbon nanotubes performed in the gas phase. J Phys Chem C 116:20930–20936. https://doi.org/10.1021/jp306866q
81. Hirsch A (2002) Functionalization of single-walled carbon nanotubes. Angew Chem Int Ed 41:1853. https://doi.org/10.1002/1521-3773(20020603)41:11%3c1853::AID-ANIE1853%3e3.0.CO;2-N
82. Pathak AK, Borah M, Gupta A, Yokozeki T, Dhakate SR (2016) Improved mechanical properties of carbon fiber/graphene oxide-epoxy hybrid composites. Compos Sci Technol 135:28–38. https://doi.org/10.1016/j.compscitech.2016.09.007
83. Lu X, Hiremath N, Hong K, Evora MC, Ranson VH, Naskar AK, Bhat GS, Kang N-G, Mays JW (2017) Improving mechanical properties of carbon nanotube fibers through simultaneous solid-state cycloaddition and crosslinking. Nanotechnology 28:145603. https://doi.org/10.1088/1361-6528/aa6223

84. Jun LY, Mubarak NM, Yee MJ, Yon LS, Bing CH, Khalid M, Abdullah EC (2018) An overview of functionalised carbon nanomaterial for organic pollutant removal. J Ind Eng Chem 67:175–186. https://doi.org/10.1016/j.jiec.2018.06.028
85. Kharlamova MV, Paukov M, Burdanova MG (2022) Nanotube functionalization: investigation. Methods and Demonstrated Appl. Mater 15:5386. https://doi.org/10.3390/ma15155386
86. Chen RJ, Bangsaruntip S, Drouvalakis KA, Wong Shi Kam N, Shim M, Li Y, Kim W, Utz PJ, Dai H (2003) Noncovalent functionalization of carbon nanotubes for highly specific electronic biosensors. Proc Natl Acad Sci USA 100:4984–4989.https://doi.org/10.1073/pnas.0837064100
87. Pang X, Imin P, Zhitomirsky I, Adronov A (2010) Amperometric detection of glucose using a conjugated polyelectrolyte complex with single-walled carbon nanotubes. Macromolecules 43:10376–10381. https://doi.org/10.1021/ma101862b
88. Vaisman L, Marom G, Wagner HD (2006) Dispersions of surface-modified carbon nanotubes in water-soluble and water-insoluble polymers. Adv Funct Mater 16:357–363. https://doi.org/10.1002/adfm.200500142
89. Holzinger M, Baur J, Haddad R, Wang X, Cosnier S (2011) Multiple functionalization of single-walled carbon nanotubes by dip coating. Chem Commun 47:2450–2452. https://doi.org/10.1039/C0CC03928D
90. Zhang J, Lee J-K, Wu Y, Murray RW (2003) Photoluminescence and electronic interaction of anthracene derivatives adsorbed on sidewalls of single-walled carbon nanotubes. Nano Lett 3:403–407. https://doi.org/10.1021/nl025952c
91. Goff AL, Moggia F, Debou N, Jegou P, Artero V, Fontecave M, Jousselme B, Palacin S (2010) Facile and tunable functionalization of carbon nanotube electrodes with ferrocene by covalent coupling and π-stacking interactions and their relevance to glucose bio-sensing. J Electroanal Chem 641:57–63. https://doi.org/10.1016/j.jelechem.2010.01.014
92. Zhan K, Liu H, Zhang H, Chen Y, Ni H, Wu M, Sun D, Chen Y (2014) A facile method for the immobilization of myoglobin on multi-walled carbon nanotubes: poly(methacrylic acid-co-acrylamide) nanocomposite and its application for direct bio-detection of H2O2. J Electroanal Chem 724:80–86. https://doi.org/10.1016/j.jelechem.2014.04.012
93. Basiuk Golovataya-Dzhymbeev EV, Rybak-Akimova EV, Basiuk VA, Acosta-Najarro D, Saniger JM (2002) Adsorption modification of single-walled carbon nanotubes with tetraaza-annulene macrocyclic complexes. Nano Lett 2:1249–1252.https://doi.org/10.1021/nl025791c
94. Byrne MT, Gun'ko YK (2010) Recent advances in research on carbon nanotube-polymer composites. Adv Mater 22:1672–1688.https://doi.org/10.1002/adma.200901545
95. Huang YY, Ahir SV, Terentjev EM (2006) Dispersion rheology of carbon nanotubes in a polymer matrix. Phys Rev B 73:125422. https://doi.org/10.1103/PhysRevB.73.125422
96. Al-Saleh MH, Al-Anid HK, Hussain YA (2013) CNT/ABS nanocomposites by solution processing: proper dispersion and selective localization for low percolation threshold. Compos Part A: Appl Sci Manuf 46:53–59.https://doi.org/10.1016/j.compositesa.2012.10.010
97. Du F, Fischer JE, Winey KI (2003) Coagulation method for preparing single-walled carbon nanotube/poly(methyl methacrylate) composites and their modulus, electrical conductivity, and thermal stability. J Polym Sci B Polym Phys 41:3333–3338. https://doi.org/10.1002/polb.10701
98. Kasaliwal GR, Göldel A, Pötschke P, Heinrich G (2011) Influences of polymer matrix melt viscosity and molecular weight on MWCNT agglomerate dispersion. Polymer 52:1027–1036. https://doi.org/10.1016/j.polymer.2011.01.007
99. Alig I, Pötschke P, Lellinger D, Skipa T, Pegel S, Kasaliwal GR, Villmow T (2012) Establishment, morphology and properties of carbon nanotube networks in polymer melts. Polymer 53:4–28. https://doi.org/10.1016/j.polymer.2011.10.063
100. Mezghani K, Farooqui M, Furquan S, Atieh M (2011) Influence of carbon nanotube (CNT) on the mechanical properties of LLDPE/CNT nanocomposite fibers. Mater Lett 65:3633–3635. https://doi.org/10.1016/j.matlet.2011.08.002

101. Kaminsky W (2018) Polyolefin-nanocomposites with special properties by in-situ polymerization. Front Chem Sci Eng 12:555–563. https://doi.org/10.1007/s11705-018-1715-x
102. Coleman JN, Khan U, Blau WJ, Gun'ko YK (2006) Small but strong: a review of the mechanical properties of carbon nanotube–polymer composites. Carbon 44:1624–1652.https://doi.org/10.1016/j.carbon.2006.02.038
103. Zamani Keteklahijani Y, Shayesteh Zeraati A, Sharif F, Roberts EPL, Sundararaj U (2020) In situ chemical polymerization of conducting polymer nanocomposites: effect of DNA-functionalized carbon nanotubes and nitrogen-doped graphene as catalytic molecular templates. Chem Eng J 389:124500. https://doi.org/10.1016/j.cej.2020.124500
104. Wang C, Xu J, Yang J, Qian Y, Liu H (2017) In-situ polymerization and multifunctional properties of surface-modified multiwalled carbon nanotube-reinforced polyimide nanocomposites. High Perform Polym 29:797–807. https://doi.org/10.1177/0954008316657862
105. Wu C-S (2009) Synthesis and characterization of poly(trimethylene terephthalate) nanocomposites incorporating multi-walled carbon nanotubes. J Appl Polym Sci 114:1633–1642. https://doi.org/10.1002/app.30754
106. Trujillo M, Arnal ML, Müller AJ, Laredo E, Bredeau St, Bonduel D, Dubois PH (2007) Thermal and morphological characterization of nanocomposites prepared by in-situ polymerization of high-density polyethylene on carbon nanotubes. Macromolecules 40:6268–6276.https://doi.org/10.1021/ma071025m
107. Kwon S-M, Kim H-S, Myung SJ, Jin H-J (2008) Poly(methyl methacrylate)/multiwalled carbon nanotube microspheres fabricated via in-situ dispersion polymerization. J Polym Sci B Polym Phys 46:182–189. https://doi.org/10.1002/polb.21353
108. Song W-H, Ni Q-P, Zheng Z, Tian L-Y, Wang X-L (2009) The preparation of biodegradable polyurethane/carbon nanotube composite based on *in situ* cross-linking. Polym Adv Technol 20:327–331. https://doi.org/10.1002/pat.1271
109. Jiang Q, Wang X, Zhu Y, Hui D, Qiu Y (2014) Mechanical, electrical and thermal properties of aligned carbon nanotube/polyimide composites. Compos B Eng 56:408–412. https://doi.org/10.1016/j.compositesb.2013.08.064
110. Sankar N, Reddy MN, Prasad RK (2016) Carbon nanotubes dispersed polymer nanocomposites: mechanical, electrical, thermal properties and surface morphology. Bull Mater Sci 39:47–55. https://doi.org/10.1007/s12034-015-1117-3
111. Watt MR, Gerhardt RA (2020) Effect of processing on the properties and morphology of MWCNT-polymer networks. Mater Res Express 7:015075. https://doi.org/10.1088/2053-1591/ab67fd
112. Manoj Kumar R, Sharma SK, Manoj Kumar BV, Lahiri D (2015) Effects of carbon nanotube aspect ratio on strengthening and tribological behavior of ultra high molecular weight polyethylene composite. Compos Part A: Appl Sci Manuf 76:62–72.https://doi.org/10.1016/j.compositesa.2015.05.007
113. Nayak SY, Shenoy S, Hameed Sultan MT, Kini CR, Seth A, Prabhu S, Safri SNA (2021) Effect of CNT-based resin modification on the mechanical properties of polymer composites. Front Mater 7:609010. https://doi.org/10.3389/fmats.2020.609010
114. Zhao W, Li T, Li Y, O'Brien DJ, Terrones M, Wei B, Suhr J, Lucas Lu X (2018) Mechanical properties of nanocomposites reinforced by carbon nanotube sponges. J Materiomics 4:157–164. https://doi.org/10.1016/j.jmat.2018.01.006
115. Liu M, Younes H, Hong H, Peterson GP (2019) Polymer nanocomposites with improved mechanical and thermal properties by magnetically aligned carbon nanotubes. Polymer 166:81–87. https://doi.org/10.1016/j.polymer.2019.01.031
116. Dikshit MK, Engle PE (2018) Investigation of mechanical properties of CNT reinforced epoxy nanocomposite: a molecular dynamic simulations. Mater Phys Dyn 37:7–15
117. Gao J, Itkis ME, Yu A, Bekyarova E, Zhao B, Haddon RC (2005) Continuous spinning of a single-walled carbon nanotube−nylon composite fiber. J Am Chem Soc 127:3847–3854. https://doi.org/10.1021/ja0446193
118. Sharma SP, Lakkad SC (2010) Compressive strength of carbon nanotubes grown on carbon fiber reinforced epoxy matrix multi-scale hybrid composites. Surf Coat Technol 205:350–355. https://doi.org/10.1016/j.surfcoat.2010.06.055

119. Ramezani M, Kim YH, Sun Z (2020) Mechanical properties of carbon-nanotube-reinforced cementitious materials: database and statistical analysis. Mag Concr Res 72:1047–1071. https://doi.org/10.1680/jmacr.19.00093
120. Mishra B, Sharma S (2022) Compressive behavior of carbon nanotube reinforced polypropylene composites under high strain rate: insights from molecular dynamics. Mech, Comput, Appl: An Int J 13:19–40. https://doi.org/10.21203/rs.3.rs-890632/v1
121. Li S, Li J, Ma N, Liu D, Sui G (2019) Super-compression-resistant multiwalled carbon nanotube/nickel-coated carbonized loofah fiber/polyether ether ketone composite with excellent electromagnetic shielding performance. ACS Sustain Chem Eng 7:13970–13980. https://doi.org/10.1021/acssuschemeng.9b02447
122. Zu M, Li Q, Zhu Y, Zhu Y, Wang G, Byun J-H, Chou T-W (2013) Stress relaxation in carbon nanotube-based fibers for load-bearing applications. Carbon 52:347–355. https://doi.org/10.1016/j.carbon.2012.09.036
123. Feng Y, Nie Z, Deng P, Luo L, Hu X, Su J, Li H, Fan X, Qi S (2022) An Effective approach to improve the thermal conductivity, strength, and stress relaxation of carbon nanotubes/epoxy composites based on vitrimer chemistry. IJMS 23:8833. https://doi.org/10.3390/ijms23168833
124. Mattioli-Belmonte M, Vozzi G, Whulanza Y, Seggiani M, Fantauzzi V, Orsini G, Ahluwalia A (2012) Tuning polycaprolactone–carbon nanotube composites for bone tissue engineering scaffolds. Mater Sci Eng: C 32:152–159.https://doi.org/10.1016/j.msec.2011.10.010
125. Montazeri A, Pourshamsian K, Riazian M (2012) Viscoelastic properties and determination of free volume fraction of multi-walled carbon nanotube/epoxy composite using dynamic mechanical thermal analysis. Mater Des 1980–2015(36):408–414. https://doi.org/10.1016/j.matdes.2011.11.038
126. Suhr J, Koratkar N, Keblinski P, Ajayan P (2005) Viscoelasticity in carbon nanotube composites. Nature Mater 4:134–137. https://doi.org/10.1038/nmat1293
127. Huang J, Rodrigue D, Dong L (2021) Effect of temperature on the viscoelastic properties of carbon nanotube reinforced polypropylene composites. Adv Mater Sci Eng 2021:1–12. https://doi.org/10.1155/2021/6630408
128. Tarfaoui M, Lafdi K, El Moumen A (2016) Mechanical properties of carbon nanotubes based polymer composites. Compos B Eng 103:113–121. https://doi.org/10.1016/j.compositesb.2016.08.016
129. Khan W, Sharma R, Saini P (2016) Carbon nanotube-based polymer composites: synthesis, properties and applications. In: Berber MR, Hafez IH (eds) Carbon nanotubes—current progress of their polymer composites. InTech
130. Huang Q, Wang J, Li C, Zhu J, Wang W, Huang Y, Zhang Y, Jiao H, Zhang S, Meng H, Zhang M, Wang X (2022) Intrinsically flexible all-carbon-nanotube electronics enabled by a hybrid organic–inorganic gate dielectric. npj Flex Electron 6:61. https://doi.org/10.1038/s41528-022-00190-8
131. Krishnan A, Shandilya S, HS B, Gupta P (2020) A review on applications of carbon nanotubes in automobiles. Int J Mech Eng Technol (IJMET) 11. https://doi.org/10.34218/IJMET.11.1.2020.018
132. Shahidi S, Moazzenchi B (2018) Carbon nanotube and its applications in textile industry—a review. The J Textile Inst 109:1653–1666. https://doi.org/10.1080/00405000.2018.1437114
133. Guzman de Villoria R, Ydrefors L, Hallander P, Ishiguro K, Nordin P, Wardle B (2012) Aligned carbon nanotube reinforcement of aerospace carbon fiber composites: substructural strength evaluation for aerostructure applications. In: 53rd AIAA/ASME/ASCE/AHS/ASC structures, structural dynamics and materials conference <b> 20th AIAA/ASME/AHS adaptive structures conference
 14th AIAA. American institute of aeronautics and astronautics. Honolulu, Hawaii
134. Fu B, Gao J (2020) Study on the application of graphene polymer material in sports equipment. J Phys: Conf Ser 1676:012044. https://doi.org/10.1088/1742-6596/1676/1/012044
135. Sun L, Wang X, Wang Y, Zhang Q (2017) Roles of carbon nanotubes in novel energy storage devices. Carbon 122:462–474. https://doi.org/10.1016/j.carbon.2017.07.006

136. Srivastava A, Srivastava ON, Talapatra S, Vajtai R, Ajayan PM (2004) Carbon nanotube filters. Nat Mater 3:610–614. https://doi.org/10.1038/nmat1192
137. Mylvaganam K, Zhang LC (2007) Ballistic resistance capacity of carbon nanotubes. Nanotechnology 18:475701. https://doi.org/10.1088/0957-4484/18/47/475701
138. Ago H, Petritsch K, Shaffer MSP, Windle AH, Friend RH (1999) Composites of carbon nanotubes and conjugated polymers for photovoltaic devices. Adv Mater 11:1281–1285. https://doi.org/10.1002/(SICI)1521-4095(199910)11:15%3c1281::AID-ADMA1281%3e3.0.CO;2-6
139. Jeon I, Matsuo Y, Maruyama S, Single-walled carbon nanotubes in solar cells
140. Feng Y, Ju X, Feng W, Zhang H, Cheng Y, Liu J, Fujii A, Ozaki M, Yoshino K (2009) Organic solar cells using few-walled carbon nanotubes electrode controlled by the balance between sheet resistance and the transparency. Appl Phys Lett 94:123302. https://doi.org/10.1063/1.3103557
141. Hu L, Hecht DS, Grüner G (2010) Carbon nanotube thin films: fabrication, properties, and applications. Chem Rev 110:5790–5844. https://doi.org/10.1021/cr9002962
142. Tyler TP, Brock RE, Karmel HJ, Marks TJ, Hersam MC (2011) Electronically monodisperse single-walled carbon nanotube thin films as transparent conducting anodes in organic photovoltaic devices. Adv Energy Mater 1:785–791. https://doi.org/10.1002/aenm.201100274
143. Barnes TM, Bergeson JD, Tenent RC, Larsen BA, Teeter G, Jones KM, Blackburn JL, van de Lagemaat J (2010) Carbon nanotube network electrodes enabling efficient organic solar cells without a hole transport layer. Appl Phys Lett 96:243309. https://doi.org/10.1063/1.3453445
144. Schuettfort T, Snaith HJ, Nish A, Nicholas RJ (2010) Synthesis and spectroscopic characterization of solution processable highly ordered polythiophene–carbon nanotube nanohybrid structures. Nanotechnology 21:025201. https://doi.org/10.1088/0957-4484/21/2/025201
145. Kymakis E, Kornilios N, Koudoumas E (2008) Carbon nanotube doping of P3HT: PCBM photovoltaic devices. J Phys D: Appl Phys 41:165110. https://doi.org/10.1088/0022-3727/41/16/165110
146. Brunet L, Lyon DY, Zodrow K, Rouch J-C, Caussat B, Serp P, Remigy J-C, Wiesner MR, Alvarez PJJ (2008) Properties of membranes containing semi-dispersed carbon nanotubes. Environ Eng Sci 25:565–576. https://doi.org/10.1089/ees.2007.0076
147. Shawky HA, Chae S-R, Lin S, Wiesner MR (2011) Synthesis and characterization of a carbon nanotube/polymer nanocomposite membrane for water treatment. Desalination 272:46–50. https://doi.org/10.1016/j.desal.2010.12.051
148. Gu J, Xiao P, Chen J, Liu F, Huang Y, Li G, Zhang J, Chen T (2014) Robust preparation of superhydrophobic polymer/carbon nanotube hybrid membranes for highly effective removal of oils and separation of water-in-oil emulsions. J Mater Chem A 2:15268. https://doi.org/10.1039/C4TA01603C
149. Wang Y, Zhu J, Huang H, Cho H-H (2015) Carbon nanotube composite membranes for microfiltration of pharmaceuticals and personal care products: capabilities and potential mechanisms. J Membr Sci 479:165–174. https://doi.org/10.1016/j.memsci.2015.01.034
150. Parham H, Bates S, Xia Y, Zhu Y (2013) A highly efficient and versatile carbon nanotube/ceramic composite filter. Carbon 54:215–223. https://doi.org/10.1016/j.carbon.2012.11.032
151. Watson KA, Connell JW (2006) Polymer and carbon nanotube composites for space applications. In: Carbon nanotechnology. Elsevier, pp 677–698
152. O'Donnell S, Sprong K, Haltli B (2004) Potential impact of carbon nanotube reinforced polymer composite on commercial aircraft performance and economics. In: AIAA 4th aviation technology, integration and operations (ATIO) forum. American Institute of Aeronautics and Astronautics, Chicago, Illinois
153. Fan Z, Luo G, Zhang Z, Zhou L, Wei F (2006) Electromagnetic and microwave absorbing properties of multi-walled carbon nanotubes/polymer composites. Mater Sci Eng, B 132:85–89. https://doi.org/10.1016/j.mseb.2006.02.045

154. Shree Meenakshi K, Pradeep Jaya Sudhan E, Ananda Kumar S, Umapathy MJ (2011) Development and characterization of novel DOPO based phosphorus tetraglycidyl epoxy nanocomposites for aerospace applications. Prog Organic Coat 72:402–409.https://doi.org/10.1016/j.porgcoat.2011.05.013
155. Lule ZC, Kim J (2020) Thermally conductive polybutylene succinate composite filled with Si-O-N-C functionalized silicon carbide fabricated via low-speed melt extrusion. Eur Polymer J 134:109849. https://doi.org/10.1016/j.eurpolymj.2020.109849
156. Naskar AK, Keum JK, Boeman RG (2016) Polymer matrix nanocomposites for automotive structural components. Nat Nanotech 11:1026–1030. https://doi.org/10.1038/nnano.2016.262
157. Grm V, Zavec D, Dražić G (2020) A carbon-nanotubes-based heating fabric composite for automotive applications. Mater Technol 54:761–768. https://doi.org/10.17222/mit.2019.238

Mechanical Characterization of Graphene Nanoparticles

Azzam Ahmed

Abstract Graphene nanoparticles (GNPs) are a new class of nanoparticles with remarkable properties which make them attractive for a variety of applications. Their unique properties, such as high strength, high electrical conductivity, and lightweight, making them promising for use in a variety of fields including electronics, energy storage, and biomedical applications. In order to fully exploit their potential, however, the mechanical properties of GNPs must be characterized. The mechanical characterization of GNPs has been an active area of research in recent years. High Young's modulus, high hardness, and high yield strength make GNPs attractive for a variety of applications, and further research into their mechanical characterization is likely to lead to even more potential applications.

Keywords Graphene · Graphene oxide (GO) · Reduced graphene oxide (rGO) · Mechanical characterization · Graphene nanoparticles (GNP)

1 Introduction

Graphene is the thinnest material in the world and has immediately attracted worldwide attention. Due to the particularity of the graphene material structure, it has high electrical conductivity, high thermal conductivity, high strength, and high transparency, etc. It is a material with great development prospects in the future, so it has triggered the development of this kind of material. The structure, performance, preparation of materials, and some research directions of preparation and characterization of composite materials, involving disciplines including physics, chemistry, biology, materials science, optics, etc. The excellent performance provides a wide

A. Ahmed (✉)
Department of the Textile Engineering, College of Engineering and Technology of Industries, Sudan University of Science and Technology, Khartoum, Sudan
e-mail: azzam.ko@hotmail.com

Safat College of Science and Technology, Khartoum, Sudan

© The Author(s), under exclusive license to Springer Nature Singapore Pte Ltd. 2023
H. Abdellaoui et al. (eds.), *Mechanics of Nanomaterials and Polymer Nanocomposites*, Smart Nanomaterials Technology,
https://doi.org/10.1007/978-981-99-2352-6_3

range of ideas for its application in many fields. This chapter focuses on the Mechanical characterization of graphene nanoparticles and then graphene samples could be characterized by Optical microscopy, electron microscopy, AFM, and Raman spectroscopy. This chapter summarizes the most recent findings on the mechanical characterization of graphene. In 2004, Andre K. Geim of the University of Manchester in the United Kingdom prepared graphene. Geim and his colleagues stumbled across a new, easy way to do it. They forcibly separated the graphite into smaller fragments, peeled off the thinner graphite flakes from the fragments, then stuck a special plastic tape to the sides of the flakes, peeled the tape, and the flakes split in two. Repeating this process over and over again yielded thinner and thinner graphite flakes, some of which consisted of only one single layer of carbon atoms, in the final they made graphene [1]. The initiation of graphene has caused a research boom all over the world. Not only is it the thinnest known material, it's also very strong and rigid; as a single substance, it transfers electrons at room temperature faster than any known conductor [2–5]. Graphene has a very special structure at the atomic scale, which must be described by relativistic quantum physics [6–8]. The structure of graphene is very stable. So far, researchers have not found any missing carbon atoms in graphene [9]. The connection between each carbon atom in graphene is very flexible. When an external mechanical force is applied, the surface of the carbon atom bends and deforms, so that the carbon atoms do not need to be rearranged to adapt to the external force, and the structure remains stable. This stable lattice structure gives carbon atoms excellent electrical conductivity. Electrons in graphene move through their orbits without being scattered by lattice defects or the introduction of foreign atoms [10, 11]. Because the interatomic force is very strong, at room temperature, even if the surrounding carbon atoms collide, the electrons in graphene are disturbed very little. The biggest feature of graphene is that the movement speed of electrons in it reaches 1/300 of the speed of light, far exceeding the movement speed of electrons in general conductors [12–16]. This makes the electrons in graphene, or more accurately, the "electric charge carriers", behave very similarly to relativistic neutrinos. Graphene is a two-dimensional carbon nanomaterial, which is a new material with a hexagonal honeycomb lattice structure composed of carbon atoms with sp2 hybrid orbitals [17–19]. Graphene has excellent optical, electrical, and mechanical properties, and has important application prospects in materials science, micro-nano processing, energy, biomedicine, and drug delivery, and is considered to be a revolutionary material in the future [20–22]. Graphene-based nanoparticles have gained great interest from scientists as lubricant additives due to their remarkable physical and chemical properties. With the increasing use of vegetable oils as biodegradable lubricants, especially in areas with high environmental impact, many studies have focused on using graphene-based nanosheets as additives for biodegradable lubricants, thereby making them Outperforms commercial mineral oils [23, 24]. There are several common characterization methods for graphene, mainly including SEM [25–27], TEM, AFM, FT-IR [28–32], Raman, PL, UV-Vis, NH_3-TPD, XRD, XPS, molecular dynamics (AIMD) simulation and EPR [33–35], etc. Although the three methods of micromechanical exfoliation, chemical vapor deposition, and epitaxial growth can obtain high-quality graphene, they are not conducive to large-scale production due to factors such as harsh reaction conditions, high cost, and low yield [36]. In comparison, the graphene oxide (GO) solution reduction method (Reduction of Graphene Oxide Solution) is easy to operate and has a large output. At the same time, the product form of graphene sol also brings convenience to the further processing and molding of materials [37, 38]. At present, the reduction methods of GO mainly include: chemical reduction method (the reducing agent used includes hydrazine, dimethylhydrazine, hydroquinone, sodium borohydride, sulfur-containing compounds, etc.), thermal reduction method and ultraviolet-assisted reduction method [39–41]. In 2006, Ruoff et al. [42]. used hydrazine as a reducing agent for the first time to prepare single-layer graphene by chemical reduction of GO solution, but the prepared graphene still contained a very small number of oxygen-containing groups, and its conjugated structure also had certain inconsistencies like integrity, so it is generally called Chemically Reduced Graphene Oxide (CRG). GO is generally prepared by using the Hummers method to intercalate graphite oxide to obtain graphite oxide, and then exfoliate (cleavage) it in water and other solvents by mechanical action such as ultrasonic vibration [43]. In this process, due to the strong interaction between water molecules and the oxygen-containing functional groups (carboxyl, epoxy and hydroxyl) formed during the oxidation of graphite, they can be easily inserted into the gap between graphite oxide layers, and at the same time it can be cleaved by swelling under the action of ultrasound. Graphene nanosheets were batch-fabricated by reducing GO with hydroquinone in aqueous solution [44]. Preparation of graphite by chemical reduction of GO aqueous solution and mixture of GO water and N, N-dimethylacetamide by a series of sulfur-containing compounds (sodium bisulfite, sulfur dioxide, thionyl chloride, sodium thiosulfate, and sodium sulfide) alkenes, and found that sodium bisulfite and thionyl chloride were comparable in reducing power to hydrazine hydrate [45, 46]. There is a problem in the preparation of graphene by the reduction of pristine GO aqueous solution, that is, in the solution, the gradual reduction of oxygen-containing groups will cause the weakening of the hydrophilicity of the reduced GO flakes, thereby rapidly agglomerating. Therefore, it is necessary to add polymers or surfactants to modify CRG during the reduction process to prevent π-π stacking. By reducing the mixed aqueous solution of GO and polystyrene sulfonate sodium salt, a black aqueous solution of CRG coated with the polymer was obtained. GO coated with sodium dodecylbenzene sulfonate was reduced by hydrazine, and then chemically modified by aryl diazonium salt to obtain CRG coated with sodium dodecylbenzene sulfonate, which could be dispersed in dimethyl formamide, N, N-dimethylacetamide and N-methyl pyrrolidone at a concentration of 1 mg/mL. Related studies on the modification of CRG suspensions using small organic molecules and nanoparticles have been reported. The thermal reduction of GO is another method used to prepare graphene. Graphene was prepared by microwave-assisted rapid thermal reduction of GO in a mixture of water and N, N-dimethylacetamide. The reduction of GO was completed within a few minutes (1–10 min), while the degree of product reduction depended on the duration of microwave treatment. Graphene can also be prepared by UV-assisted reduction. A black suspension of TiO_2-coated chemically modified graphene was obtained by irradiating an ethanol solution containing a TiO_2/GO

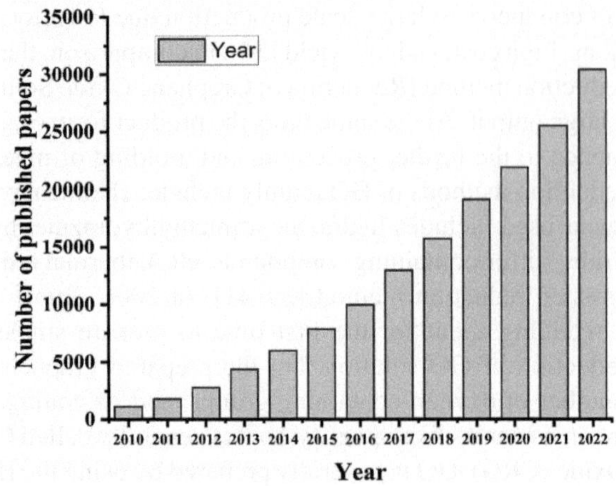

Fig. 1 The number of mechanical characterization graphene research publications by year, Elsevier. *Source* Web of Science

mixture, where TiO_2 is not only a photocatalyst but also a dispersant. Graphene was successfully prepared by photocatalytic reduction of the mixture of GO and TiO_2 nanocomposites by UV irradiation. In summary, there are many methods to prepare graphene by reducing GO solution. The biggest advantages of these methods are simple experimental conditions, low cost and mass production. At the same time, the original material is cheap graphite, so it is expected to realize graphene large-scale production of alkenes.

Since graphene is just a single layer of carbon atoms connected in a hexagonal pattern, it is also extremely thin and lightweight and an attractive material for nanotechnology applications. On the other hand, the increasing number of graphene characterization publications in journals, as shown in Fig. 1 shows the importance of graphene after innovation.

The mechanical properties of graphene are outstanding in terms of high strength and high modulus. The strength of graphene reaches 130 GPa, which is about 100 times that of ordinary steel materials. At the same time, its tensile strength is as high as 125 GPa, its elastic modulus is 1.1 TPa, and it is light in weight. The theoretical value of its specific surface area is 2630 m^2/g, and its Young's modulus is 1.0 TPa. The hardness of graphene is higher than that of diamond with a Mohs hardness of 10, but it also has good toughness, can be bent, and has excellent ductility [47].

2 Classifications and Characteristics of Graphene

Graphene, a two-dimensional form of carbon, is one of the most promising materials of the twenty-first century. This remarkable material has a wide range of potential applications in the fields of electronics, energy storage, and even medicine. Research into graphene has been ongoing since its discovery in 2004, and its unique properties have been extensively studied. In this review paper, we will discuss the classifications and characteristics of graphene. Classifications of graphene can be divided into two main categories: monolayer and multilayer. Monolayer graphene is a single layer of carbon atoms arranged in a two-dimensional lattice structure. This form of graphene is extremely thin and light, and it has been shown to possess excellent electrical, optical, and mechanical properties. Multilayer graphene consists of multiple layers of carbon atoms arranged in a stacked, three-dimensional structure. This form of graphene is slightly thicker than monolayer graphene, and it can be tuned to have different electrical and optical properties. The remarkable characteristics of graphene are largely due to its atomic structure. Graphene has a very strong bond between the carbon atoms, which gives it a high degree of stability and strength. It also has a very low density and can be stretched up to 20% of its original length. Graphene has an extremely high electron mobility, which makes it highly conductive. It is also very lightweight, and it has a high thermal conductivity. Graphene is mainly divided into: single-layer graphene, double-layer graphene and multi-layer graphene (Figs. 2, 3, 4 and 5). Single-layer graphene: refers to a two-dimensional carbon material composed of a layer of carbon atoms that are periodically densely packed in a benzene ring structure (i.e., a hexagonal honeycomb structure). Double-layer graphene: refers to a two-dimensional carbon material composed of two layers of carbon atoms that are periodically stacked closely in a benzene ring structure in different stacking ways. Multi-layer graphene: refers to a two-dimensional carbon material composed of 3–10 layers of carbon atoms that are periodically densely packed in a benzene ring structure and stacked in different stacking ways.

Graphene has excellent mechanical properties, electrical and thermal conductivity, large specific surface area and excellent adsorption properties, good strength, flexibility, and other properties, and has a wide range of application values in the field of polymer nanocomposite materials [50]. It is the material with the highest thermal conductivity so far and has very good thermal conductivity, so it is widely used in the new heating industry, making the heating process more comfortable and convenient. Graphene is used in the manufacture of some high-tech products, such as display screens, touch screens, and sensors that you often encounter, all of which use graphene. What exactly is the reason for graphene's popularity? The main reason

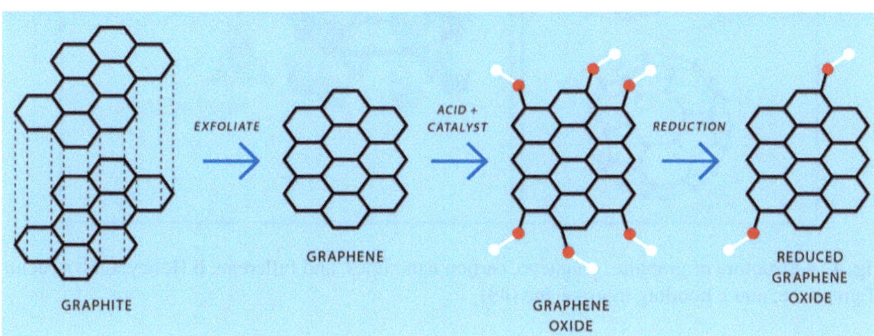

Fig. 2 Schematic diagram of the family of graphene [48]

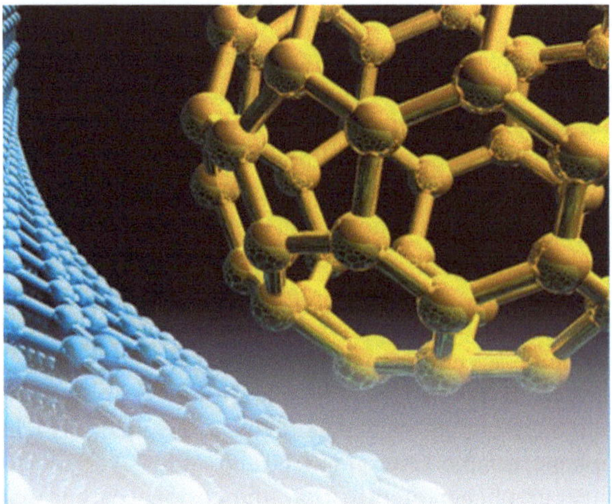

Fig. 3 3D Schematic diagram of graphene structure

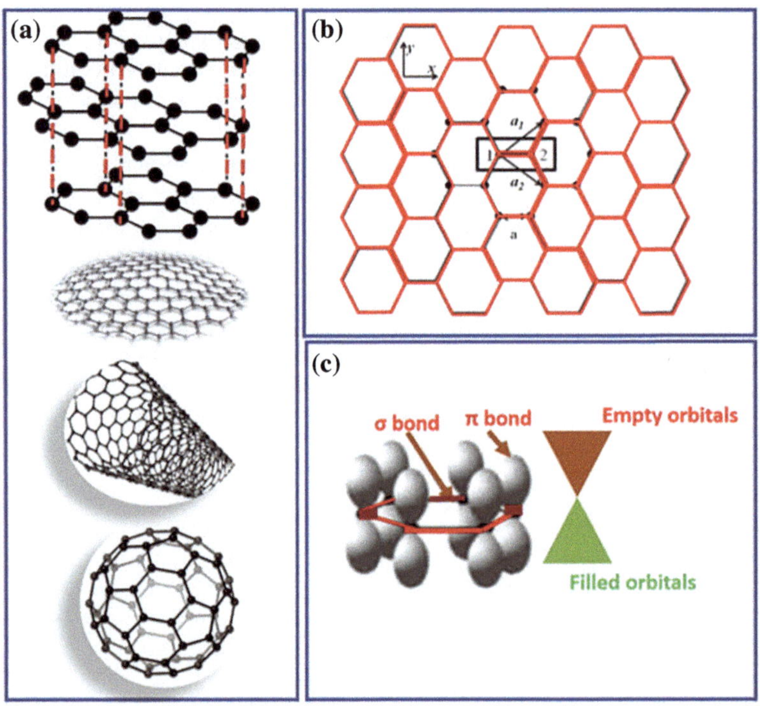

Fig. 4 **a** Structure of graphite, graphene, carbon nanotubes, and fullerene, **b** Honeycomb structure of graphene, and **c** bonding in graphene [49]

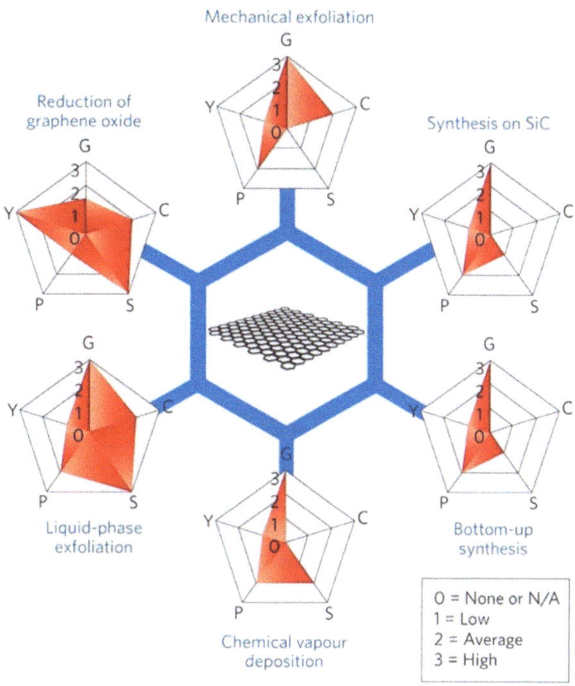

Fig. 5 The key characteristics of the most common graphene production methods in a scale of 0–3; (G) refers to the graphene quality, (C) refers to the cost of production (a low value corresponds to high cost of production), (S) refers to the scalability, (P) refers to the purity and (Y) refers to the yield of each preparation route [52]

is that this material has strong electrical conductivity, and the toughness of graphene is 200 times higher than that of steel plate [51].

The production of graphene has various process, conditions, and raw materials, Fig. 4, it can be seen that each method presents different characteristics in terms of the output and for this reason, the selection of the method should be performed each time, based on the application for which the graphene will be used.

2.1 Advantages and Disadvantages of Graphene

Graphene hazards are potential damage and irritation to the skin and eyes, as well as inhalation and ingestion. The jagged edges of graphene nanoparticles are so sharp and strong that they can easily penetrate the cell membranes of human skin and immune cells [53, 54]. Graphene does pose a potentially serious hazard to humans and other animals. Graphene is thinner than paper, harder than diamond, and more conductive than copper. Its potential harm is divided into two levels, one is the damage to tissue cells [55]. Brown's elite team of microbiologists, technical engineers and raw materials biologists have studied the potential toxic side effects of graphene materials on tissues and cells. They emphasized that the jagged edges of graphene-based gold nanoparticles are so sharp and strong that they can pierce the human skin

Table 1 Main advantages and disadvantages of graphene [57–61]

Name	Advantages	Disadvantages
Graphene	High electrical and thermal and conductivities	Hydrophobicity
	High control on functionalization	High cost
		Difficult workability
		Small production
GO	Water dispersibility	Lower electrical and thermal conductivity
	Polar functionalization	Surface random functionalization
	Low cost	Poor control on post-preparation functionalization
	Easy workability	
rGO	High electrical and thermal conductivity	Hydrophobicity
	Good control on functionalization	Difficult workability
	Less expensive than neat graphene	Properties related to production methodology used

and the cytoplasm of cellular immunity [56]. Graphite is manufactured by chemical or mechanical exfoliation separating thin carbon layers to form a dry powder that can be inhaled or exposed. If these dry powders (also called material fragments) are inhaled into the body and come into contact with human cells, then these material fragments are likely to cut through human cells and be absorbed by human cells. It can be clearly seen that the possible products and substances that will appear in the process of studying graphene will bring more possible hazards than graphene itself. The main advantages and disadvantages of graphene are presented in Table 1.

3 Mechanical Characterization Methods

Graphene is a new carbonaceous material that is densely packed into a two-dimensional honeycomb lattice structure by a single layer of carbon atoms. It has excellent semiconductor properties such as large specific surface area, high carrier mobility, and high thermal conductivity, making graphene gradually become a research hotspot. For each material, its testing and characterization techniques are indispensable in the preparation and quality inspection of materials. This chapter introduces several commonly used characterization techniques for analyzing and characterizing the structure of graphene materials. The characterization of graphene is mainly divided into images and maps. The images are mainly based on optical microscopy, transmission electron microscopy (TEM), scanning electron microscopy (SEM) and atomic force microscopy (AFM), while the maps are mainly based on Raman spectroscopy. (Raman), infrared spectroscopy (IR), X-ray photoelectron spectroscopy (XPS) and ultraviolet spectroscopy (UV) as representatives. Among them, TEM, SEM, Raman, AFM and optical microscopy are generally used to determine the number of layers of graphene and mechanical characterization, while IR, XPS and UV can characterize the structure of graphene and monitor the synthesis process of graphene.

3.1 Optical Microscopy

Optical microscopy is an effective method for quickly and easily characterizing the number of graphene layers [62]. Geim et al. found that if the single-layer graphene is attached to a SiO_2 layer Si wafer with a certain thickness (300 nm) on the surface, it can be observed under an optical microscope. This is due to the interference between the single-layer graphite layer and the substrate on the light and a certain contrast, so the single-layer graphene can be distinguished under an optical microscope. Studies by Roddaro et al. [63]. have shown that graphene is visible under an optical microscope because of the interface between its air-graphite layer-SiO_2 layer. Blake et al. [64]. proposed the use of narrow-band filters so that graphene can be observed on SiO_2 layers of any thickness, and can even be observed on other films such as Si_3N_4 and PM-MA (polymethyl methacrylate) films. Observing graphene with an optical microscope makes it possible to further accurately characterize the number of graphene layers, which lays the foundation for the controlled preparation and physical properties of graphene. Reports of optical pictures of the thin-layer graphite are also controversial, in some cases the single-layer graphene is not visible, while other experiments show that simple optical methods are visible. In recent years, researchers have proposed a variety of methods to improve the image contrast of graphene, such as narrow-band illumination and selection of suitable substrates. Reflectance and contrast spectroscopy have also been used to identify graphene layers. It is primarily used to determine the number of layers in graphene by using the contrast difference between the graphene layers and the substrate. Size and shape of the graphene structure is also obtained from this technique with different graphene layers of graphene on the SiO_2/Si substrate [65]. Optical microscopy and Raman spectra have been used to obtain graphene with different layers on the SiO_2/Si substrate as shown in Fig. 6.

3.2 Scanning Electron Microscope (SEM)

SEM is one of the effective tools to characterize the morphology of graphene. Copper is used as the substrate to grow graphene by CVD method. Carbon is almost insoluble in copper, and graphene grows in the surface adsorption mode [67]. Figure 7, shows a large amount of graphene flakes with a curled morphology consisting of a thin wrinkled paper like structure, typical for graphene sheets deposited on a solid support.

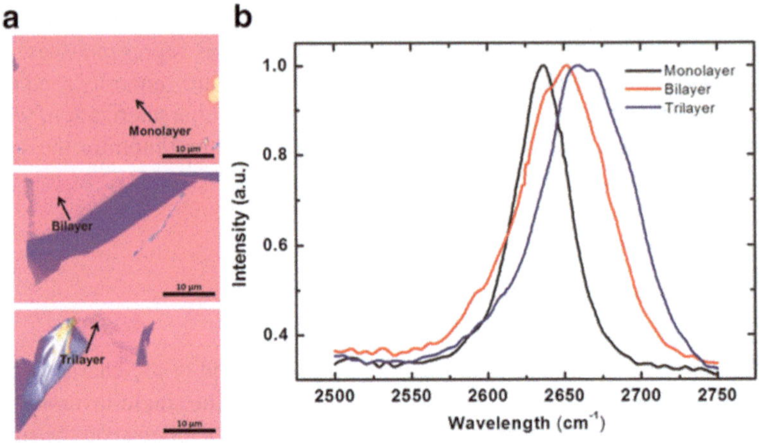

Fig. 6 Graphene with different layers on SiO_2/Si substrate. **a** Optical image and **b** Raman spectra of graphene with different layers on SiO_2/Si substrate, which are measured and collected using 632.8-nm laser [66]

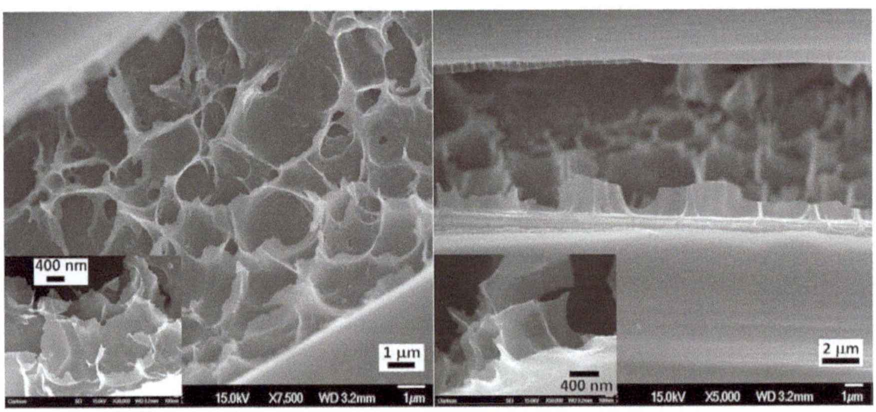

Fig. 7 Scanning electron microscope (SEM) images of the graphene nanosheets [68]

3.3 *Transmission Electron Microscope (TEM)*

The TEM method can observe the microscopic topography of the graphene surface, and can measure the clear suspended graphene structure and atomic-scale details. At the same time, single-layer and multi-layer graphene can be identified by electron diffraction patterns [69]. TEM images show that graphene is relatively thin. The number of layers can be determined from the selected area electron diffraction (SEAD) profile. The regular hexagonal pattern has a uniform distribution, and the diffraction spots of the second inner layer are brighter than that of the innermost layer, which indicates that the typical feature of bilayer graphene sheets is to have good crystallinity [70]. Recently, transmission electron microscopy has attracted the attention of researchers as a structural characterization tool for suspended graphene, and TEM is capable of measuring low-resolution images and atomic-scale details. Furthermore, high-resolution and spherical aberration-corrected transmission electron microscopy studies of single-layer defect sites and adsorbed light atoms demonstrate that suspended graphene can serve as an ideal supporting membrane. Rapid developments in the fields of both electron microscopy and graphene will continue to provide a rich ground for future insights [71].

3.4 Atomic Force Microscope (AFM)

AFM can be used to obtain information on the lateral size, area and thickness of graphene, but generally it can only be used to distinguish the single-layer or double-layer graphene. Of course, it is an atomic force microscope (AFM) looking at the height map, a single layer of graphene is less than 1 nm. It should be said that AFM is the most convenient means to characterize graphene materials. Of course, care should be taken to distinguish dust, salts, and graphene molecules during AFM characterization. The number of layers can be obtained by measuring the thickness of a single-layer sheet, and it can be calculated according to [72]. Figure 8, shows atomic force microscopy images of the suspended graphene drum skin before and after optical forging. Bottom: analogue presentation of how a material can become stiffer when it is corrugated. Graphene is an ultrathin material characterized by its ultrasmall bending modulus, superflimsiness. Now the researchers at the Nanoscience Center of the University of Jyväskylä have demonstrated how an experimental technique called optical forging can make graphene ultrastiff, increasing its stiffness by several orders of magnitude [73].

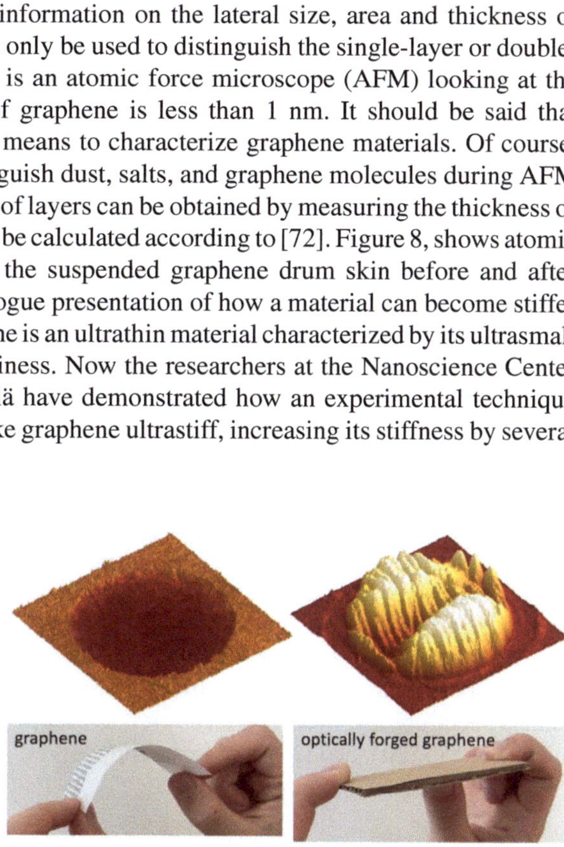

Fig. 8 Atomic force microscopy images of the suspended graphene drum skin before and after optical forging [73]

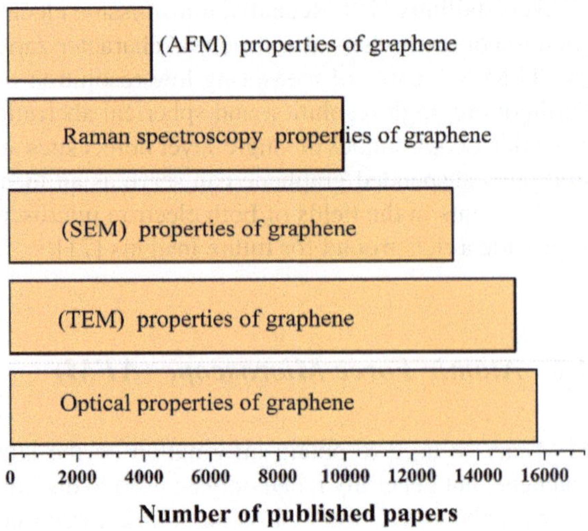

Fig. 9 The number of mechanical characterization method for graphene publications in 2022, Elsevier. *Source* Web of Science

3.5 Raman Spectrum (Raman)

The Raman method is based on the analysis of the Raman scattering effect when light passes through the sample. It can analyze the number of layers, defects, crystal structure, phonon Characterization of energy bands, etc., is an important means of testing and analyzing graphene materials. There are also articles showing that Raman spectroscopy can also be used to test the mechanical properties and thermal conductivity of graphene by detecting phonon frequency and other methods. It can be seen that this method is a very important characterization method for graphene research. It provides great help for the research of graphene materials by characterizing the structure of graphene layers, defects, etc., as well as mechanical properties and thermal conductivity. A new series of nanofluids based on graphene dispersed in 2-hydroxyethylammonium lactate (ML) ionic liquid was developed, Raman spectra showed a strong interaction between ML and graphene [74]. Some of the common characterization methods for graphene focusing on identify number of layers and structure morphology of graphene have been used and published as seen in Fig. 9. All the properties mentioned in the graph including Atomic Force Microscope (AFM), Raman spectroscopy, Scanning electron microscope (SEM), Transmission electron microscope (TEM), and Optical Microscopy.

4 Tensile Strength and Young's Modulus of Graphene

Graphene is the strongest and hardest crystalline structure of all materials. Its tensile strength and elastic modulus are 125 GPa and 1.1 TPa, respectively. The strength limit of graphene is $42N/m^2$. The strength of ideal graphene is about 100 times that of ordinary steel, and a graphene sheet with an area of $1m^2$ can bear a mass of 4 kg.

Graphene can be used as a typical two-dimensional reinforcement material and has a potential application value in the field of composite materials. The measurement of young modulus of the graphene has been used a lot of techniques such as theoretical and experimental to obtain it. Table 2, presents some of the methods that are used for calculation of the graphene young modulus.

The researchers found that the graphene sample particles could withstand a maximum pressure of about 2.9 micronewtons per 100 nm of distance before they began to crumble. According to scientists' calculations, this result is equivalent to applying a pressure of 55 Newtons to break a 1-m-long graphene [83, 84]. If physicists were able to produce graphene as thick as an ordinary food plastic bag (about 100 nm thick), it would take almost 20,000 Newtons of pressure to tear it apart. In other words, if the bag were made of graphene, it would be able to hold an item weighing about two tons.

Table 2 Young modulus of graphene measurements

Mechanical property	Graphene and its derivatives	Measurement method and its Value	Refs.
Young's modulus	Single layer graphene sheet	Functional tight binding (DFTB) and Density function theory (DFT), 995.14 GPa	Memarian et al. [75]
Young's modulus	Monolayer graphene	Atomic force microscope, 1.0 ± 0.1 TPa	Lee et al. [47]
Young's modulus	Single layer graphene sheet	Different effective thickness, 0.974 TPa	Lu [76]
Young's modulus	Single layer graphene sheet	Molecular dynamics simulation, 1.238 TPa	Jin and Yuan [77]
Young's modulus	Single layer graphene sheet	Tight binding molecular dynamics, 1.24 TPa	Hernandez et al. [78]
Young's modulus	Single layer graphene sheet	Molecular structural mechanics model, 1.06 TPa	Xiao et al. [79], Wu et al. [80]
Young's modulus	Single layer graphene sheet	Atomic force microscopy, 1.03 TPa	Cui et al. [81]
Young's modulus	Graphene/polymer composites	Hierarchical multi-scale modeling approach, 7.02 GPa	Rafiee and Eskandariyun [82]

5 Summary

At present, the graphene industry has become one of the most promising emerging industries, and it is likely to bring a new round of innovation to people's society in the future. Nowadays, there are all kinds of new products in our life, and the products developed in different fields are different, but many products cannot be used without the support of raw materials. Graphene is now known as black gold, and some scientists predict that graphene will become the very interested of materials in the future. In case of mechanical characterization, scanning electron microscopy, transmission electron microscopy and atomic force microscopy can all observe and analyze the surface morphology of graphene samples. At the same time, scanning electron microscopy can analyze the composition of the sample, and transmission electron microscopy can observe the lattice structure of the sample. The atomic force microscope can also test the thickness of the sample. Fourier transform infrared spectrometer; Raman spectrometer and X-ray diffractometer can test and analyze the insufficient reduction of oxygen-containing functional groups and lattice defects of graphene samples. Raman spectrometer can also calculate the number of sample layers and analyze the types of lattice defects and concentration. as Also, the ability to test mechanical and thermal properties, is an important means of graphene research. There are many analysis and test methods for graphene, but there are problems such as overlapping functions, imperfect evaluation methods and inconsistent performance test methods, which need to be gradually improved in future research work. With the continuous progress of graphene research, the characterization methods of graphene are becoming more and more abundant. However, the thickness of graphene is generally only a few atomic layers, and differences in crystal defects, surface adsorbed substances, and preparation methods will cause differences in characterization results. Whether it is optical microscopy, SEM and TEM atomic force microscopy, Raman spectroscopy, IR spectroscopy, or XPS spectroscopy or UV–Vis spectroscopy, although graphene can be characterized to a certain extent, there are certain limitations. It is necessary to select an appropriate characterization method according to the needs, and compare and confirm the results obtained with each other to obtain accurate information about graphene. In conclusion, graphene is an incredibly versatile material with a wide range of potential applications. Its unique properties make it a promising material for the twenty-first century, and research into its potential is ongoing. It is classified into two main categories: monolayer and multilayer, and it has a variety of electrical, optical, and mechanical properties. It can also be used in a variety of medical applications. Graphene has the potential to revolutionize the way we look at materials and their applications.

References

1. Novoselov KS, Geim AK, Morozov SV, Jiang D-E, Zhang Y, Dubonos SV, et al (2004) Electric field effect in atomically thin carbon films. Science 306(5696):666–669
2. Geim AK (2012) Graphene prehistory. Phys Scr 2012(T146):014003
3. Ghany NAA, Elsherif SA, Handal HT (2017) Revolution of graphene for different applications: state-of-the-art. Surf Interfaces 9:93–106
4. Randviir EP, Brownson DA, Banks CE (2014) A decade of graphene research: production, applications and outlook. Mater Today 17(9):426–432
5. Si C, Sun Z, Liu F (2016) Strain engineering of graphene: a review. Nanoscale 8(6):3207–3217
6. Katsnelson MI (2007) Graphene: carbon in two dimensions. Mater Today 10(1–2):20–27
7. Katsnelson M, Novoselov K (2007) Graphene: new bridge between condensed matter physics and quantum electrodynamics. Solid State Commun 143(1–2):3–13
8. Geim AK, Kim P (2008) Carbon wonderland. Sci Am 298(4):90–97
9. Mao S, Pu H, Chen J (2012) Graphene oxide and its reduction: modeling and experimental progress. RSC Adv 2(7):2643–2662
10. Yu W, Sisi L, Haiyan Y, Jie L (2020) Progress in the functional modification of graphene/graphene oxide: a review. RSC Adv 10(26):15328–15345
11. Banhart F, Kotakoski J, Krasheninnikov AV (2011) Structural defects in graphene. ACS Nano 5(1):26–41
12. Trikkaliotis DG, Christoforidis AK, Mitropoulos AC, Kyzas GZ (2021) Graphene oxide synthesis, properties and characterization techniques: a comprehensive review. ChemEngineering 5(3):64
13. Zhang T (2022) Lattice and energy band. Springer, Graphene, pp 55–70
14. Yu-Qian Z, Yu-Chen L, Shi-Xi Z (2012) Nature of the dynamics of graphene. business, economics, financial sciences, and management: Springer, pp 75–82
15. Tian W, Liu X, Yu W (2018) Research progress of gas sensor based on graphene and its derivatives: a review. Appl Sci 8(7):1118
16. Weiss NO, Zhou H, Liao L, Liu Y, Jiang S, Huang Y et al (2012) Graphene: an emerging electronic material. Adv Mater 24(43):5782–5825
17. Xu M, Liang T, Shi M, Chen H (2013) Graphene-like two-dimensional materials. Chem Rev 113(5):3766–3798
18. CeNeR R, AeK S, KeS S, Graphene GA (2009) The new two-dimensional nanomaterial. Angew Chem Int Ed 48(42):7752–7777
19. Du Q-S, Tang P-D, Huang H-L, Du F-L, Huang K, Xie N-Z et al (2017) A new type of two-dimensional carbon crystal prepared from 1, 3, 5-trihydroxybenzene. Sci Rep 7(1):1–11
20. Chen X, Zhang S, Han W, Wu Z, Chen Y, Wang S (2018) A review on application of graphene-based microfluidics. J Chem Technol Biotechnol 93(12):3353–3363
21. Huang Z, Shao G, Li L (2022) Micro/nano functional devices fabricated by additive manufacturing. Prog Mater Sci 101020
22. Feng J (2022) Preparation and application of graphene. Acad J Sci Technol 4(1):59–62
23. Chouhan A, Mungse HP, Khatri OP (2020) Surface chemistry of graphene and graphene oxide: a versatile route for their dispersion and tribological applications. Adv Coll Interface Sci 283:102215
24. Sun J, Du S (2019) Application of graphene derivatives and their nanocomposites in tribology and lubrication: a review. RSC Adv 9(69):40642–40661
25. Xie J, Spallas JP (2012) Different contrast mechanisms in SEM imaging of graphene. Agil Technol
26. Kim SM, Hsu A, Lee Y-H, Dresselhaus M, Palacios T, Kim KK et al (2013) The effect of copper pre-cleaning on graphene synthesis. Nanotechnology 24(36):365602
27. Paulchamy B, Arthi G, Lignesh B (2015) A simple approach to stepwise synthesis of graphene oxide nanomaterial. J Nanomed Nanotechnol. 6(1):1

28. Guo S, Wen D, Zhai Y, Dong S, Wang E (2011) Ionic liquid–graphene hybrid nanosheets as an enhanced material for electrochemical determination of trinitrotoluene. Biosens Bioelectron 26(8):3475–3481
29. Wang H, Yuan X, Wu Y, Huang H, Zeng G, Liu Y et al (2013) Adsorption characteristics and behaviors of graphene oxide for Zn (II) removal from aqueous solution. Appl Surf Sci 279:432–440
30. Zhang X, Li K, Li H, Lu J, Fu Q, Chu Y (2014) Graphene nanosheets synthesis via chemical reduction of graphene oxide using sodium acetate trihydrate solution. Synth Met 193:132–138
31. Yuan W, Li B, Li L (2011) A green synthetic approach to graphene nanosheets for hydrogen adsorption. Appl Surf Sci 257(23):10183–10187
32. Pham TA, Kim JS, Kim JS, Jeong YT (2011) One-step reduction of graphene oxide with L-glutathione. Colloids Surf, A 384(1–3):543–548
33. Zhang X, Sun J, Tang K, Wang H, Chen T, Jiang K et al (2022) Ultralow detection limit and ultrafast response/recovery of the H2 gas sensor based on Pd-doped rGO/ZnO-SnO2 from hydrothermal synthesis. Microsyst Nanoeng 8(1):1–12
34. Gupta A, Sakthivel T, Seal S (2015) Recent development in 2D materials beyond graphene. Prog Mater Sci 73:44–126
35. Lim E, Hong X, Tan MK, Yu H, Wu H, Hung YM (2022) Distinctive evaporation characteristics of water and ethanol on graphene nanostructured surfaces. Int J Heat Mass Transf 183:122174
36. Yu J, Li J, Zhang W, Chang H (2015) Synthesis of high quality two-dimensional materials via chemical vapor deposition. Chem Sci 6(12):6705–6716
37. Chabot V, Higgins D, Yu A, Xiao X, Chen Z, Zhang J (2014) A review of graphene and graphene oxide sponge: material synthesis and applications to energy and the environment. Energy Environ Sci 7(5):1564–1596
38. Bi H, Yin K, Xie X, Zhou Y, Wan N, Xu F et al (2012) Low temperature casting of graphene with high compressive strength. Adv Mater 24(37):5124–5129
39. Gao W (2015) The chemistry of graphene oxide. Springer, Graphene oxide, pp 61–95
40. Zhao G, Wen T, Chen C, Wang X (2012) Synthesis of graphene-based nanomaterials and their application in energy-related and environmental-related areas. RSC Adv 2(25):9286–9303
41. Chamoli P, Srivastava T, Tyagi A, Raina K, Kar KK (2020) Urea and cow urine-based green approach to fabricate graphene-based transparent conductive films with high conductivity and transparency. Mater Chem Phys 242:122465
42. Moon IK, Lee J, Ruoff RS, Lee H (2010) Reduced graphene oxide by chemical graphitization. Nat Commun 1(1):1–6
43. Cravotto G, Cintas P (2010) Sonication-assisted fabrication and post-synthetic modifications of graphene-like materials. Chem–A Eur J 16(18):5246–5259
44. Nufer S, Brunton A, Dalton AB (2019) Applications of high surface area carbon nanomaterials, University of Sussex
45. Cao N, Zhang Y (2015) Study of reduced graphene oxide preparation by Hummers' method and related characterization. J Nanomater 2015
46. Song P, Zhang X, Sun M, Cui X, Lin Y (2012) Synthesis of graphene nanosheets via oxalic acid-induced chemical reduction of exfoliated graphite oxide. RSC Adv 2(3):1168–1173
47. Lee C, Wei X, Kysar JW, Hone J (2008) Measurement of the elastic properties and intrinsic strength of monolayer graphene. Science 321(5887):385–388
48. Diamante L (2021) https://graphene-flagship.eu/graphene/news/graphene-101-all-about-graphene-and-its-many-forms/
49. Pareek A, Venkata Mohan S (2019) Chapter 1.4—Graphene and its applications in microbial electrochemical technology. In: Mohan SV, Varjani S, Pandey A (eds) Microbial electrochemical technology. Elsevier, pp 75–97
50. Zhu Y, Murali S, Cai W, Li X, Suk JW, Potts JR et al (2010) Graphene and graphene oxide: synthesis, properties, and applications. Adv Mater 22(35):3906–3924
51. Hwang J, Yoon T, Jin SH, Lee J, Kim TS, Hong SH et al (2013) Enhanced mechanical properties of graphene/copper nanocomposites using a molecular-level mixing process. Adv Mater 25(46):6724–6729

52. Raccichini R, Varzi A, Passerini S, Scrosati B (2015) The role of graphene for electrochemical energy storage. Nat Mater 14(3):271–279
53. Chen L, Li J, Chen Z, Gu Z, Yan L, Zhao F et al (2020) Toxicological evaluation of graphene-family nanomaterials. J Nanosci Nanotechnol 20(4):1993–2006
54. Phull JS, Kaur H, Randhawa DK, Kaur S, Kaur R, Sharma R et al (2022) Investigation of bulk and electronic properties of dimethyl disulphide adsorption on boron doped zigzag graphene nanoribbon: DFT study. Harbin Gongye Daxue Xuebao/J Harbin Inst Technol 54(10):190–195
55. Li J, Zeng H, Zeng Z, Zeng Y, Xie T (2021) Promising graphene-based nanomaterials and their biomedical applications and potential risks: a comprehensive review. ACS Biomater Sci Eng 7(12):5363–5396
56. Iqbal HM, Bilal M, Nguyen TA (2021) Nano-bioremediation: fundamentals and applications. Elsevier
57. Bulat PV, Lebedev SP (2014) Overview of the graphene film technologies. Life Sci J 11(7s):198–202
58. Magne TM, de Oliveira Vieira T, Alencar LMR, Junior FFM, Gemini-Piperni S, Carneiro SV, et al (2022) Graphene and its derivatives: understanding the main chemical and medicinal chemistry roles for biomedical applications. J Nanostruct Chem 12(5):693–727
59. Whitener KE Jr, Sheehan PE (2014) Graphene synthesis. Diamond and Related Mater 46:25–34
60. Amiri A, Naraghi M, Ahmadi G, Soleymaniha M, Shanbedi M (2018) A review on liquid-phase exfoliation for scalable production of pure graphene, wrinkled, crumpled and functionalized graphene and challenges. FlatChem. 8:40–71
61. Prolongo S, Moriche R, Jiménez-Suárez A, Sánchez M, Ureña A (2014) Advantages and disadvantages of the addition of graphene nanoplatelets to epoxy resins. Eur Polym J 61:206–214
62. Yang J, Yao H (2020) Automated identification and characterization of two-dimensional materials via machine learning-based processing of optical microscope images. Extreme Mech Lett 39:100771
63. Michalik J, Roddaro S, Casado L, Ibarra M, De Teresa J (2011) Quantification and minimization of disorder caused by focused electron beam induced deposition of cobalt on graphene. Microelectron Eng 88(8):2063–2065
64. Rashidian Vaziri MR (2020) High-contrast optical microscopy of graphene sheets. Microscopy Res Technique 83(9):1132–1140
65. Adetayo A, Runsewe D (2019) Synthesis and fabrication of graphene and graphene oxide: a review. Open J Compos Mater 9(02):207
66. Huang C-W, Lin H-Y, Huang C-H, Shiue R-J, Wang W-H, Liu C-Y, et al (2012) Layer-dependent morphologies of silver on n-layer graphene. Nanoscale Res Lett 7(1):1–6
67. Yang F, Liu Y, Wu W, Chen W, Gao L, Sun J (2012) A facile method to observe graphene growth on copper foil. Nanotechnology 23(47):475705
68. Koushanpour A, Guz N, Gamella M, Katz E (2016) Graphene-functionalized 3D-carbon fiber electrodes-preparation and electrochemical characterization. Electroanalysis 28(9):1943–1946
69. Bangert U, Gass M, Zan R, Pan CT (2011) Scanning transmission electron microscopy and spectroscopy of suspended graphene. Physics and Applications of Graphene-Experiments: IntechOpen, pp 377–408
70. Ansari S, Giannelis EP (2009) Functionalized graphene sheet—Poly (vinylidene fluoride) conductive nanocomposites. J Polym Sci, Part B: Polym Phys 47(9):888–897
71. Huang PY, Meyer JC, Muller DA (2012) From atoms to grains: transmission electron microscopy of graphene. MRS Bull 37(12):1214–1221
72. Ismail A, Goh PS, Tee JC, Sanip SM, Aziz M (2008) A review of purification techniques for carbon nanotubes. NANO 3(03):127–143
73. M. Superflimsy graphene turned ultrastiff by optical forging (2021). https://phys.org/news/2021-05-superflimsy-graphene-ultrastiffby-optical-forging.html
74. Manuel Martínez-Rubio P, Dolores Avilés M, Arias-Pardilla J, José Carrión-Vilches F, Sanes J, Dolores Bermúdez M, et al (2022) Physicochemical characterisation of graphene-ammonium lactate ionic liquid nanofluid. J Mol Liquids 367:120446

75. Memarian F, Fereidoon A, Ganji MD (2015) Graphene Young's modulus: molecular mechanics and DFT treatments. Superlattices Microstruct 85:348–356
76. Lu JP (1997) Elastic properties of carbon nanotubes and nanoropes. Phys Rev Lett 79(7):1297
77. Jin Y, Yuan F (2003) Simulation of elastic properties of single-walled carbon nanotubes. Compos Sci Technol 63(11):1507–1515
78. Hernandez E, Goze C, Bernier P, Rubio A (1998) Elastic properties of C and B x C y N z composite nanotubes. Phys Rev Lett 80(20):4502
79. Xiao J, Gama B, Gillespie J Jr (2005) An analytical molecular structural mechanics model for the mechanical properties of carbon nanotubes. Int J Solids Struct 42(11–12):3075–3092
80. Wu Y, Zhang X, Leung A, Zhong W (2006) An energy-equivalent model on studying the mechanical properties of single-walled carbon nanotubes. Thin-walled Struct 44(6):667–676
81. Cui Y, Wang G, Wang W, Cui X, Dong W, Wang C et al (2022) Trade-off between interface stiffening and Young's modulus weakening in graphene/PMMA nanocomposites. Compos Sci Technol 225:109483
82. Rafiee R, Eskandariyun A (2020) Predicting Young's modulus of agglomerated graphene/polymer using multi-scale modeling. Compos Struct 245:112324
83. Gao C, Feng P, Peng S, Shuai C (2017) Carbon nanotube, graphene and boron nitride nanotube reinforced bioactive ceramics for bone repair. Acta Biomater 61:1–20
84. Ferrari AC, Bonaccorso F, Fal'Ko V, Novoselov KS, Roche S, Bøggild P et al (2015) Science and technology roadmap for graphene, related two-dimensional crystals, and hybrid systems. Nanoscale 7(11):4598–4810

Nanostructured Metals: Optical, Electrical, and Mechanical Properties

B. Arulmurugan, G. Kausalya Sasikumar, and L. Rajeshkumar

Abstract Polymer nanocomposites are naturally an attractive research area owing to their versatile applications. To enhance the polymer nanocomposites' strength, researchers prepare to hybridize with metal and its alloys. So, in this article, the focus is specifically on the interaction between the metal and polymer nanocomposites are discussed. The physicochemical and plasmonic properties, along with their potential applications of metal nanostructures, have piqued the interest of researchers in recent years. Many researchers have spent time and energy perfecting methods for the precise production of metal nanostructures with specific functional properties. Recent advances in the production and implementation of metal nanostructures are the primary focus of this chapter.

Keywords Physicochemical and plasmonic properties · Metal nanostructure applications · Production of metal nanostructures · Metal nanostructure-based nanocomposites

1 Introduction

Metals with nanostructures seem to be polycrystals with microstructural attributes on a magnitude below 100 nm. This may take the form of high-angle grain boundaries low-angle grain boundaries, and sub-grain domain or mosaic frameworks. Metals with dimensions between 1 and 100 nm are called metal nanoparticles. Catalysis, electronics, optoelectronics, sensing, and other fields have found widespread use for metal nanostructures in recent years. For these uses, it would be ideal if metal nanostructures could be designed and fabricated straightforwardly on functional

B. Arulmurugan
Department of Mechanical Engineering, KPR Institute of Engineering and Technology, Coimbatore, Tamilnadu 641407, India

G. K. Sasikumar · L. Rajeshkumar (✉)
Centre for Research and Development, KPR Institute of Engineering and Technology, Coimbatore, Tamilnadu 641407, India
e-mail: lrkln27@gmail.com

© The Author(s), under exclusive license to Springer Nature Singapore Pte Ltd. 2023
H. Abdellaoui et al. (eds.), *Mechanics of Nanomaterials and Polymer Nanocomposites*, Smart Nanomaterials Technology,
https://doi.org/10.1007/978-981-99-2352-6_4

substrates, where their cost, dimensions, and shape, could be precisely controlled [1–5]. It is common knowledge that metal nanomaterials' structure, size, composition, and shape directly affect their properties and functions. As a result, the last two decades have seen extensive research into the controlled fabrication of metal nanocrystals for various purposes. Compared to other nanostructures like nanowires, nanoparticles, nanosheets, nanoplates, and nanorods, they also demonstrated some interesting properties of their own [6–9].

2 Applications

In addition to their remarkable fundamental features, metal NPs have long attracted attention owing to their potential applications in optical waveguides, catalysis, surface-enhanced Raman scattering (SERS), and sensors. The detection of proteins, DNA, and RNA using photovoltaic devices, has all attracted attention recently. Metal nanoparticles are also showing promise as a potential treatment for cancer and other disorders [10–12]. New technical uses of metal nanomaterials include not only the traditional field of catalysis but also sensing, energy conversion, nanophotonics, biological detection, nanoelectronics, therapy, and imaging [13, 14] (Fig. 1).

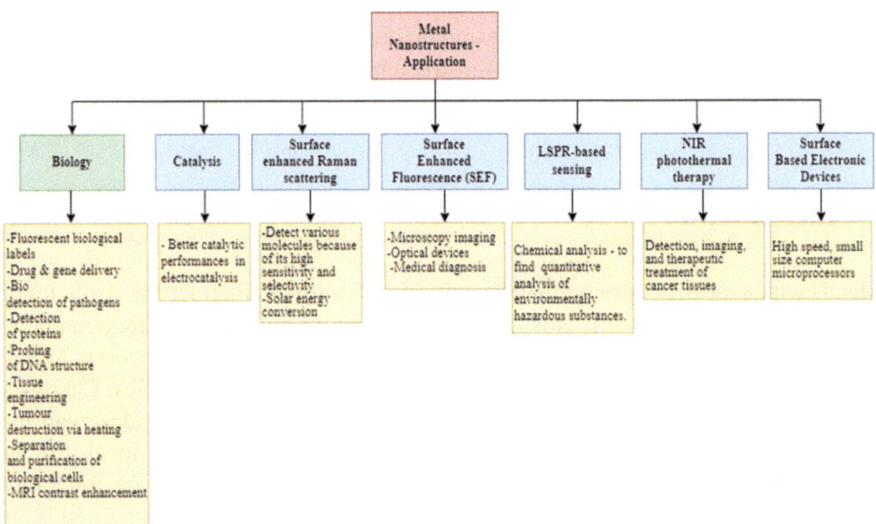

Fig. 1 Applications of metal nanostructures

2.1 Catalysis

High surface areas, nanoplates, and sheets, specifically those made from noble metals, are exceptional candidates for an extensive range of catalytic applications. Some nanoplates and sheets made of noble metals have shown superior catalytic capabilities in electrocatalysis compared to commercially available catalysts [15, 16].

2.2 SERS

Because of its excellent sensitivity and selectivity, SERS is one of the most powerful methods for detecting different compounds. Ag and Au nanoparticles, in particular, are the most effective SERS substrates. With its remarkable sensitivity and selectivity, SERS has found widespread application as a surface-sensitive potent tool for detecting an extensive range of compounds. Substrates for SERS that use metal plasmonic nanostructures, such as Ag and Au, are particularly effective. SERS activities, which can enhance electromagnetic field (E-field) and thus cause "hot spots", were typically kept in plasmonic nanostructures having sharp edges. Recent work by many researchers highlights the significance of metal nanostructure sharpness in SERS by demonstrating, for instance, that the signal intensity of 4-mercapto-pyridine in triangular-shaped and hexagonal-shaped Pd nanoplate is 4.31 and 3.43 times that of Pd cuboctahedra. SERS signals could be amplified through precise manipulation of nanostructured sharp edges; however, maintaining this sharpness presents a challenge. Bimetallic nanostructures like Pd–Ag core–shell had been produced for SERS applications and displayed very steady signals to overcome this issue [15].

2.3 SEF—Surface Enhanced Fluorescence

The fluorescence method has numerous uses, including microscope imaging, optical devices, and medical diagnosis. Scientists are very interested in increasing the fluorescence sensitivities of fluorophore surfaces; nevertheless, the challenge becomes more intriguing when the scope is narrowed to a single molecule. The SEF method is employed for this function. It is predicated on resonant molecules emitting in the vicinity of emitters. The near-field connection among excited state fluorophores with surface modes is crucial to the success of SEF. In particular, surfaces having a plasmonic nanostructure and localized surface plasmons are productive SEF substrates. Similar to how SERS investigations are affected by molecular distance towards plasmonic nanostructures, SEF experiments are affected by molecular range from nanoparticles.

2.4 LSPR-Based Sensing

Metal nanoplates and nanosheets show promise for refractive index sensing since their LSPR is highly sensitive to the individual dielectric constant of their surroundings. For this reason, Ag nanoplates were effectively utilized to detect various anions with a sensitivity of 1×10^{-6} m was attained [16].

2.5 NIR Photothermal Therapy

Metal nanostructures have found a novel and potentially fruitful use in photothermal monitoring and treatment for cancer. To kill cancer cells, photothermal therapy uses light to convert into heat, and photothermal imaging uses this heat to create an image. Highly efficient and rapid (1–2 ps) conversion of light energy to heat energy makes metal nanostructures superior to heat conductors and photothermal exchangers. Therefore, nanoparticles show promise for use in cancer photothermal imaging and treatment. Indeed, multiple pilot investigations have demonstrated that metal nanostructures can be used for detecting, imaging, and treating cancer tissues. A gold and otherwise silver nanoparticle might undergo a significant temperature change when in resonance, including an excitation source. At first, nanoparticles were introduced to tumors either by exposure to them or injected directly into them. Although this approach has the potential to be successful, it needs more precision [3, 17]. If the particles land somewhere the exciting light should reach, the cells there will be killed. One strategy for improving the efficiency of particle distribution is to enhance the properties of the particles with various kinds of medicines that target cancer. Firm LSPR absorption peaks in the NIR zone, great photothermal conversion performances, outstanding biocompatibilities, along with photothermal reliabilities make nanoplates and nanosheets potentially useful in NIR photothermal therapy [18].

Strong LSPR absorption, particularly within near-infrared (NIR) range, photothermal stability, and biocompatibility make metal nanostructures a potential frontrunner for photothermal therapy [19–22]. For instance, NIR laser light was used to illuminate palladium (Pd) nanosheets in such an aqueous medium, and the resulting NIR photothermal effects at different temperatures were then measured. After 10 min of illumination with a near-infrared laser, a solution containing Pd nanosheets rose from 28 °C to 48.7 °C. In comparison, NIR laser irradiation of a 1 ml water solution devoid of Pd nanosheets produced a 0.5 C increase in temperature. When this method was used on liver cancer cells, over one hundred percent of the cancer cells were eliminated after being exposed to a laser for only 5 minutes. Pd nanosheets showed superior photothermal stability compared to Au and Ag after being exposed to a NIR laser for 30 min. Pd nanosheets coated with a small layer of Ag were found to be more stable under light and heat. Since infrared light may possess excellent tissue penetration and minimal optical absorption through blood and soft tissues, it is encouraging that reports of Au nanoprisms exhibiting outstanding photothermal stability with a more extended NIR laser irradiation period have been published [23, 24].

2.6 *Surface-Based Electronic Devices*

Molecular absorption is the fundamental concept of molecular sensors. The plasmon emission's peak, location, and strength are all affected by the molecules' ability to absorb them. The development of fast, compact computer microprocessors may benefit from optical integration and electrical circuiting. Surface plasmons, in contrast, are widely recognized as crucial for data transmission in computer processors. To better sustain such high frequencies (100 THz), plasmonic nanostructures are preferable to traditional wires (10 GHz). When applied to creating practical electronics like plasmonstors, plasmonic nanostructures stand out due to their unique ability to operate at high frequencies. Electronic devices depending on plasmonic circuits are often developed using lithographic processes. The use of lithographic methods is also beneficial in resolving the issues of location, geometry, and orientation. The thickness of monolayers on colloid films can be measured using surface plasmons because of their sensitivity to the propagation medium. Numerous businesses, including Biacore, have produced instruments depending on this idea. In 2009, a group of Korean researchers found a way to increase the effectiveness of organic light-emitting diodes [25–27].

3 Nanostructured Metals

In the latest days, nanostructures have given engineers unprecedented control over materials with dimensions between 1 and 100 nm. They have allowed scientists to obtain a more profound knowledge of the connections among origin, size, composition, and shape. This newfound understanding has served as a link between the worlds of university nanotechnology research and commercialized goods. The size, composition, dimensions, and origin of nanomaterials and nanostructures are all critical factors in classifying these phenomena. An enhancement in the worth of each category is possible through the foresightful prediction of the unique characteristics of nanostructures. Nanostructures are artificial structures with a size between 1 and 100 nm [28, 29]. They can be fabricated from a metal oxide, metal, composite, carbon, organic, or inorganic material. Inorganic nanostructures are typically defined as those depending on a metal or metal oxide. Metal NPs like lead or Au, metal oxide NPs including titanium dioxide, and semiconductors including ceramics and silicon can all be created using these nanostructures.

Synthesized metal nanostructures (nanoparticles) can be created using either constructive or destructive means. Nanoparticles made of metals have interested researchers for more than a century and are now widely used in the life and physical sciences. Nanoparticles (NPs) can be manufactured from virtually any metal. Nanoparticle synthesis typically employs the metals cobalt (Co), copper (Cu), lead (Pb), gold (Au), zinc (Zn), aluminum (Al), cadmium (Cd), iron (Fe), and silver (Ag), cylindrical or spherical, crystalline or amorphous, and as little as 10–100 nm, inorganic metal NPs have a wide range of desirable characteristics [8]. The primary motivations for their synthesis are the improved performance and reactivity of nanostructures depending on metal oxides. To alter the features and characteristics of metal-based NPs, oxide-based nanostructures are created. For instance, iron (Fe) NPs are instantly oxidized to iron oxide (Fe_2O_3) in the presence of oxygen, even at ambient temperature. Zinc oxide (ZnO), iron oxide (Fe_2O_3), silicon dioxide (SiO_2), cerium oxide (CeO_2), titanium dioxide (TiO_2), aluminum oxide (Al_2O_3), and agnetite (Fe_3O_4), are typical metal oxides used in nanostructure production (CeO_2). Due to their unique features, these nanoparticles are expected to significantly influence numerous industries, including but not limited to aerospace, medicine, (photo) catalysis, and electronics [30–32].

4 Synthesis Routes of Metal Nanostructures

Synthesis of metal nanoparticles typically involves reducing metal ions or salts using suitable falling reagents. Au nanoparticles rose to prominence among the earliest identified nanomaterials due to their simple fabrication. When a reduction agent is added to a gold salt solution or heated in a glass matrix, the solution's color changes from yellow to a deep crimson or burgundy, signifying the quick conversion from dissolved ($Au3^+$ ions) salt to an accumulation of tiny gold metal particles [33–36].

4.1 Top-Down Route

Graphite and bulk crystals of Transition-metal di-chalcogenides (TMDs) are two examples of layered materials in which the layers stack with each other through weak van der Waals forces. Thus, top-down methods, including mechanical exfoliation, solvent-assisted exfoliation, and chemical or electrochemical Li intercalation and exfoliation, can be used to formulate graphene and 1—or few-layer nanosheets directly of TMDs. On the other hand, metallic atoms bond to one another in a powerful way. Consequently, it is exceedingly challenging to formulate nanosheets and nanoplates by straightway exfoliating their bulk crystal [37, 38].

4.2 Bottom-Up Route

Many bottom-up synthesis techniques are developed for creating metal nanoplates and sheets. Self-assembly of nanoparticles, seeded growth, photochemical synthesis, 2D template-assisted synthesis, polyol reduction ultrasonic methods, hydrothermal and solvothermal methods, halide ion confined growth, carbon monoxide, biological synthesis, and surfactant-mediated synthesis are all examples of bottom-up techniques that begin with metal salts or small metal nanoparticles. Top-down methods are developed to make metal nanoplates and sheets [39], including nanoimprint lithography, electron beam nanolithography, and nanocubes, hole-mask colloidal lithography, in addition to bottom-up synthesis. There has been a success in synthesizing silver nanospheres nanowires through a polymer-mediated polyol process. To create each morphology, it was necessary to adjust the silver precursor and polymer concentrations and cover specific facets with the polymer. Besides adjusting their size, the properties of metal nanoparticles can also be fine-tuned by manipulating their shapes [40, 41].

5 Characterization Techniques of Nanostructured Metals

Nanomaterial study relies heavily on elucidating their structures. Given that most nanostructures are too small to be seen with an optical microscope, it is essential to employ suitable methods for accurately characterizing their internal and external architecture and surface at the molecular or atomic level. This is vital for discovering their basic properties and technical and functional performance in technological contexts. X-ray absorption fine structure (EXAFS), Scanning Electron Microscopy (SEM), Energy dispersive X-ray spectroscopy (EDX), scanning tunneling microscopy (STM), X-ray absorption spectroscopy (XAS), extended atomic force microscopy (AFM), transmission electron microscopy (TEM), X-ray absorption near edge structure (XANES), and X-ray diffraction (XRD), can all be (dynamic light scattering) [42–46]. Various methods vary in their sensitivity to surfaces. Practices vary in terms of whether or not they require knowledge of a particular element. Knowing what kind of data is needed to make an informed decision about the method. Nanostructured Metals with Tuneable Optical Properties and Applications [47, 48].

Microscopy techniques such as electron microscopy (EM) and XRD can be used to learn more about nanomaterial sizes, shapes, and morphologies. Surface and topological characteristics can be studied with scanning probe microscopy (SPM), which includes atomic force microscopy (AFM). EM is a powerful technique for characterizing the nanoscale structure of various materials. Images are formed by the interaction of accelerated electrons with an electron-transparent sample in transmission EM (TEM). High-resolution TEM allows for resolution on the sub-nanometer scale (HRTEM). To identify the expressed faces, one can observe the lattice fringe spacing in crystalline specimens at this resolution [49]. Secondary electrons have been emitted from the sample's surface, which is beneficial because they can reveal more details about the material. Secondary electrons have characteristic peaks that can be used in electron energy loss spectrometry (EELS) and X-ray energy dispersive spectroscopy (XEDS) to provide quantitative elemental distribution. The method of electron diffraction (ED) can be applied with TEM to help figure out crystal structure. Diffraction patterns produced by electrons passing via crystalline or semi-crystalline solids reveal details about the crystal structure and the distances between lattice repeats. Since the resolution of optical microscopy is restricted to roughly half the wavelength of light, it is not typically used to characterize the structure of metal nanoparticles [50–52].

Nonetheless, methods including scattering near-field scanning optical microscopy (SNSM) and near-field scanning optical microscopy (NSOM) have been developed and are in use to circumvent this limitation (s-NSOM). Utilizing the difference in phase between the incident beam and the field scattered mostly by particles, this method creates an image of the particles as a ring-shaped interference pattern. These techniques allow for the visualization of particles with sizes below 100 nm. However, the resolution of this optical method is too low to help characterize the structure of nanomaterials, so its application is restricted. Popular surface imaging methods with nanometer resolution include SEM and AFM. AFM belongs to a group of procedures known as scanning probe microscopy (SPM). Techniques like these involve scanning a sharp tip across a sample's surface and sensing the resulting surface-tip interaction [53]. In scanning tunneling microscopy (STM), a three-dimensional photograph of the specimen is generated by monitoring the current flowing through the microscope's tip. AFM is more flexible because it can also take pictures of samples that don't conduct electricity. It does this by moving an atomic structure's sharp tip on a flexible cantilever anywhere along the material's surface and estimating the force upon that tip from the surface. The use of a piezoelectric stage allows for atomic-level precision. Using this method has many benefits, such as requiring little to no specimen preparation but rather coatings and being usable in environments that don't vacuum or are wet. However, it has some drawbacks, such as a low scanning speed and a small inspection area. SEM, like TEM, makes use of the short De-Broglie wavelength to create an image, but instead of detecting transmitted electrons, it detects electrons that have been scattered from the surface of the material. Although it does not produce accurate 3D maps as AFM does, it does deliver precise shape and size details and exterior topologies based on the feature size. Secondary electrons, just like TEMs, can be used to reveal more about a material's elemental makeup. Additional data about the sample can be gleaned from X-ray methods, including XRD, XAS, and X-ray photoelectron spectroscopy (XPS). XRD is a technique for determining the crystal phase and material type by measuring the distance between crystallographic planes. More so, the Debye-Scherrer equation can extrapolate details more about crystallite grain sizes (less than 100 nm) first from the complete width at half the maximum of reflections. In addition to XRD, XPS can quantify the material's elemental makeup, and several XAS variants can investigate the material's specific geometric and electronic properties [54–56].

Few research works used SEM, AFM, and absorption spectroscopy to visualize and characterize the nanoisland films created by heat treatment fine gold films at high heat. Experimental research was conducted on thin gold films with an adequate thickness of 2, 5, and 7 nm after annealing at 500, 700, and 900 °C, respectively. Absorption measurements corroborated theoretical calculations showing that gold islands have been roughly spherical. From 2 to 5 nm, the film thickness, and thus the islands, grew from 15 to 40 nm in size, as affirmed by SEM images. Getting prepared localized surface plasmon resonance substrates using gold nanoisland films has been proposed as an improvement over using gold colloid films [57]. Some other researchers synthesized the cobalt nanostructures by placing an AAO template in a sol and vibrating it at a high frequency for about 10 min before annealing it at 648 K again at 873 K in a hydrogen atmosphere total of 1 h. XRD, SEM, and TEM studies all revealed that the cobalt nanostructures present were highly textured and had a low L/D ratio. EPR experiments in an orthogonal direction to the (002) crystallographic plane at room temperature revealed an isotropic signal [58].

6 Properties of Metal Nanostructures

6.1 Optical Properties

Localized surface plasmon resonance (LSPR) is a cohesive modulation of the conduction electrons that commonly happens in the visible to the near-UV area of the spectrum. It is this property of metal nanostructures that has inspired a number of these implementations. Absorption but rather scattering cross sections of metal nanoparticles might be enormously large since several electrons contribute to the LSPR. Many of these implementations rely on strong plasmon resonance and its sensitivity to particle surroundings and inter-particle linkages. The plasmon resonance can also examine the particles and learn more about their characteristics [59].

The shape of metal nanoparticles has a profound influence on their optical properties. There have been many recent advances in our ability to regulate the expansion of nanoparticles. Recent advances in nanotechnology have disproved mainly the long-held theory that a material's characteristics have been determined solely by its makeup. See here to see how changing a material's nanostructure can result in various changes to its attributes despite maintaining the same composition. Quantum effects manifest themselves as material constituent electrons' energy levels are profoundly impacted by confinement due to a nanoparticle's size decrease. Plasmons, which are collective electron oscillations, are strongly amplified by captivity, significantly affecting the material's optical and electrical characteristics. Because those plasmons propagate, their properties can change depending on the medium they encounter outside the nanostructure. The sensing potential of this effect is enormous. There are now many substrate and overlayer options for metal layers and nanostructures, but rather nanoparticles, which significantly expands our ability to fine-tune the materials' properties [60, 61].

Researchers are looking into enhancing electromagnetic fields via localized plasmons throughout island-like 2D structures but rather nanoparticles, which can be excited by incident light alone. This has implications for the development of more effective Raman spectroscopy methods. Surface plasmon resonance (SPR) and low-frequency surface plasmon resonance (LSPR) sensing are cutting-edge techniques that rely on noble metal nanostructures because of their inherent properties as ideal substrates. Groups specializing in developing the SERS technique have conducted extensive research into the impact of electromagnetic field augmentation of localized plasmons. Noble metal nanoparticles with enhanced reactivity have proven helpful in conventional gas sensing techniques. The efficiency of organic solar cells is another area where nanostructured precious metals show tremendous potential for progression [62].

Metal nanoparticles (NPs) including gold (Au), silver (Ag), copper (Cu), and so on, have attracted a lot of attention from engineers and scientists in contemporary years owing to their unique physicochemical properties, nanometer size, and surface plasmon behavior. Few of these metal NPs show a wide variety of applications with weak interaction, despite their widespread use within catalysis, electronics for circuit development, and as self-assembled novel nanostructure materials. Plasmonics refers to the study of how photons in a field interact with nanostructures made of metal. Metal nanostructures are a source to transform light into a localized electric field in metals, a phenomenon known as a localized surface plasmon. This property defines the plasmonic nanostructure materials. Tiny metal nanostructures, size, and shape can be precisely controlled and colocalized, improving the efficiency of incident light. Applications include plasmon-assisted photo lithography, nanoscale lasing, super lenses, single-molecule spectroscopy, SERS, plasmon-enhanced fluorescence, light harvesting, quantum computing, augmentation of non-linear optical signals, photo-catalysis, biochemical sensing, therein probably solar to chemical energy conversion with plasmonic metal nanoscopic nanostructures are attracting increasing attention to nanoplasmonic research [63].

Large-scale chemical preparation of Au and Ag nanostructures for plasmonic implementations has been performed in the last few decades; because of its distinct and convenient physicochemical characteristics for the nexgen of plasmonic know-hows, Ag has become an excellent option in plasmonics despite being 50 times less expensive than Au. It is important to remember that plasmonics only works when there is an incident light source, as the creation of surface plasmon requires a pairing between electromagnetic activation and metal nanostructure. Nanoplasmonics is one of a kind because it bridges the gap between the nano- and microscales by employing nanostructures ranging in size from 10 s to hundreds of nanometers. Plasmonics seeks to understand the manipulation of light at the nanoscale utilizing metallic structures and, as such, represents a novel area of research within the broader scope of nanotechnology [64]. With the intense ligh—matter interaction, nanostructured metal displays optical characteristics like scattering, diffraction, and surface plasmon unavailable in bulk metals. Because of recent developments in lithography technology, a delicate and complex metal nanostructure that was earlier impossible to achieve is now within reach [65].

6.2 *Electrical Properties*

The enhancement of system efficiency in OSCs with metal nanostructures has been due to the optical and electrical impact caused by the nanostructures. It has been reported by multiple groups that changes in the electrical characteristics of the photoactive layer and CTLs possess significant implications for OSC performance [66]. Very unusual electrical properties are indeed exhibited by the silver nanostructures here on the gallium arsenide surface (GAS). Silver accumulations on the GAS must have a high enough mass density to shunt the substrate conductivity entirely. When the film is subjected to high temperatures during annealing, it breaks down into individual nanoparticles with a much lower conductivity than the substrate. Most intriguing are the properties of low-temperature-annealed island films, which can be halted there at the percolation verge. Films in this manner have a high resistance at low voltages but switch to a low-impedance state at higher voltages, with 7 V as the transition voltage. Whenever the voltage is lowered, the low resistance remains unchanged. Due to this, the film retains information about the voltage used to create it [67, 68].

The plasmonic impacts of metal nanostructures were widely exploited to boost the productivity of organic solar cells (OSCs). Some researchers discovered increased active layer absorption due to plasmonic-optical effects like localized surface plasmon resonances, scattering effect, proliferating surface plasmon resonances, etc., which are well established. In addition, it has recently been reported that incorporating plasmonic metal nanostructures into various OSC layers improves performance by a similar amount due to the improved electrical characteristics of OSCs by plasmonic-electrical effects. Alongside experimental and theoretical results, the mechanisms of every plasmonic-electrical effect, such as the redirected flow of charge carriers, the lowered extraction barrier of the carrier transport layer, and the redistributed generation of excitons in the active layer, might be investigated. Plasmonic-electrical effects have been shown to increase the OSCs' performance, and the potential integration of plasmonic-optical and electrical consequences holds excellent promise for further enhancing OSCs' performance [69, 70].

Some researchers looked into the electrical characteristics of metal–insulator–insulator–metal nanostructures with bi-layers of Ta_2O_5/Al_2O_3 and Nb_2O_5/Al_2O_3 as rectifiers. Al was used as the metal contact for the ultra-thin (1–6 nm) insulator layers deposited using atomic-layer accumulation or rf magnetron sputtering. To determine the band gap and optical characteristics of restricted band gap insulator layers, spectroscopic ellipsometry was used, and atomic force microscopy was used to assess the surface roughness of metal contacts. From the current–voltage characteristics, we can see that the device exhibits excellent low-voltage large-signal along with small-signal nonlinearities, including an imbalance of 18 at 0.35 V, a change rate of nonlinearity of 7.5 V^{-1}, and high sensitivity of 9 A/W at 0.2 V. Resonant tunneling explains the dramatic current boost observed at 2 V in a Ta_2O_5/Al_2O_3 machine [71].

6.3 Mechanical Properties

There are three significant obstacles that nanocrystalline materials face when trying to achieve ductility. The first is processing artifacts, the second is tension force instability, and the third is fracture nucleation or transmission instability. Material characteristics at the macroscopic and microscopic scales can be quite different for the same material when the substance is in nanostructured form, as opposed to macroscopic bulk form. Mechanically, Au/Ag nanoboxes are intriguing as well. Cubic nanoboxes assisted on such a silicon substrate were mechanically probed to evaluate their hardness and flow stress. The nanoboxes' tip-to-tip length is 205 nm, and their wall thickness is 24 nm. Nanoboxes have been shown to exhibit linear hardening at small deformations and parabolic hardening at large deformations. Results demonstrate that the slender Au nanobox has greater strength after linear and parabolic toughening than the solid Au nanosphere (200 nm) or the 400 nm thick, thin film. Multiple electronic wearables have been designed for long periods spent outdoors [72, 73].

Measurements of touch sensitivity and thermal and mechanical stability of agile touchscreen panels (TSPs) are crucial for these wearable devices. Their high touch sensitivities are due, in part, to the high value of their dielectric constants (k). Research into transparent and flexible cover layers with high k and excellent mechanical and thermal reliability is crucial. Herein, a novel method is reported for fabricating cellulose nanofiber (CNF) films that are both elastic and see-through. To significantly raise k, these films are embedded with ultra-long metal nanofibers, which also function as nanofillers. The influences of the nanostructures along with their contents of these fillers here on the optical and dielectric characteristics of the films were also assessed through the manipulation of their dimensions and aspect ratios. Incredibly aligned, ultra-long nanofibers might be broken apart using a stretching method, allowing for precise regulation of nanofiber length. These high-k films with embedded nanofibers have been not only mechanically but also thermally stable. They outperform commercially available transparent plastics regarding Young's modulus, tensile strength, and thermal expansion. High-k carbon nanofiber (CNF) film for smartphones can produce highly sensitive TSPs, suggesting that it could play a crucial role in developing cutting-edge handheld electronics [74–76].

The mechanical characteristics of nanowires can now be quantified thanks to a new method developed by Wu et al. Nanowire bending beneath lateral load from an AFM edge is the basis of this technique. Using this method, they evaluated the mechanical characteristics of the polyol-produced multi-twinned silver nanowires and discovered that the nanowires are much more durable than bulk silver [77]. They found, surprisingly, that the plastic distortion at collapse is less than forty percent of the elastic fracture. Unusual strength and brittle fracture, they said, are the result of the nanowires' grain alignment and grain boundary organization. To achieve this uniform hardening, the primary slip guidance therein nanowire grain intertwines only with twinning boundary lines that run the length of the wire. The polyol method results in silver nanowires with a remarkable structure consisting of highly oriented nanoscale grains, making them ideal candidates for strengthening through microstructure modification. The twinning boundaries in this nanowire form constrain the movement of dislocation motion and ensure that they intersect with all possible slip systems. Grain boundary-hardened materials are made these nanowires efficient [78, 79].

7 Conclusion

The current chapter summarises the metal nanostructure, assessing its application in various ways, including its acts as SEF, catalysis, LSPR-based sensing, SERS, NIR photothermal therapy, and Surface-Based Electronic Devices (SBED). The other thing to be noted is the classification method of a metal nanostructure. The production techniques of metal nanostructures and their structural characterization. Finally, it discussed the various characteristics of metal nanostructures with different modalities, including optical, electrical, and mechanical perspectives.

References

1. Ruffino F, Grimaldi MG (2019) Nanostructuration of thin metal films by pulsed laser irradiations: a review. Nanomaterials 9(8):1133
2. Saravanakumar A, Rajeshkumar L, Balaji D, Jithin Karunan MP (2020) Prediction of wear characteristics of AA2219-Gr matrix composites using GRNN and Taguchi-based approach. Arab J Sci Eng 45(11):9549–9557.https://doi.org/10.1007/s13369-020-04817-8
3. Bhuvaneswari V, Rajeshkumar L, Saravanakumar R, Balaji D (2022) Synthesis and characterization of bioceramics reinforced aluminium matrix composites. Archives of Metal Mater 67(4):1217–1226. https://doi.org/10.24425/amm.2022.141045
4. Bhuvaneswari V, Rajeshkumar L, Ross KNS (2021) Influence of bioceramic reinforcement on tribological behaviour of aluminium alloy metal matrix composites: experimental study and analysis. J Market Res 15:2802–2819. https://doi.org/10.1016/j.jmrt.2021.09.090
5. Wang YM, Ma E, Valiev RZ, Zhu YT (2004) Tough nanostructured metals at cryogenic temperatures. Adv Mater 16(4):328–331
6. Xu JJ, Zhang WC, Guo YW, Chen XY, Zhang YN (2022) Metal nanoparticles as a promising technology in targeted cancer treatment. Drug Delivery 29(1):664–678
7. Ramesh M, Rajeshkumar L, Bhoopathi R (2021) Carbon substrates: a review on fabrication, properties and applications. Carbon Lett 31(4):557–580. https://doi.org/10.1007/s42823-021-00264-z
8. Deepa C, Rajeshkumar L, Ramesh M (2022) Preparation, synthesis, properties and characterization of graphene-based 2D nano-materials for biosensors and bioelectronics. J Market Res 19:2657–2694. https://doi.org/10.1016/j.jmrt.2022.06.023

9. Bhuvaneswari V, Synthesis and Applications of Metal-Based Nanomaterials. In: Sud D, Singla AK, Gupta MK (eds) Nanomaterials in manufacturing processes. CRC Press, London, pp 189–218. https://doi.org/10.1201/9781003154884
10. Amirthagadeswaran KS (2019) Corrosion and wear behaviour of nano Al_2O_3 reinforced copper metal matrix composites synthesized by high energy ball milling. Part Sci Technol 38(2):228–235. https://doi.org/10.1080/02726351.2018.1526834
11. Yang B, Jin S, Guo S, Park Y, Chen L, Zhao B, Jung YM (2019) Recent development of SERS technology: semiconductor-based study. ACS Omega 4(23):20101–20108
12. Rajeshkumar L, Suriyanarayanan R, Hari KS, Babu SV, Bhuvaneswari V, Karunan MJ (2020) Influence of boron carbide addition on particle size of copper zinc alloys synthesized by powder metallurgy. IOP Conf Ser: Mater Sci Eng 954(1):012008
13. Qazi UY, Javaid R (2016) A review on metal nanostructures: preparation methods and their potential applications. Adv Nanoparticles 5(01):27
14. Khan I, Saeed K, Khan I (2019) Nanoparticles: properties, applications and toxicities. Arab J Chem 12(7):908–931
15. Devarajan B, Saravanakumar R, Sivalingam S, Bhuvaneswari V, Karimi F, Rajeshkumar L (2021) Catalyst derived from wastes for biofuel production: a critical review and patent landscape analysis. Appl Nanosci 12:3677–3701. https://doi.org/10.1007/s13204-021-01948-8
16. Schwartzberg AM, Zhang JZ (2008) Novel optical properties and emerging applications of metal nanostructures. The J Phys Chem C 112(28):10323–10337
17. Radhika N, Sasikumar J, Arulmozhivarman J (2020) Tribo-mechanical behaviour of Ti-based particulate reinforced as-cast and heat treated A359 composites. SILICON 12(11):2769–2782
18. Nasrollahzadeh M, Issaabadi Z, Sajjadi M, Sajadi SM, Atarod M (2019) Types of nanostructures. Interface Sci Technol 28:29–80
19. Klębowski B, Depciuch J, Parlińska-Wojtan M, Baran J (2018) Applications of noble metal-based nanoparticles in medicine. Int J Mol Sci 19(12):4031
20. Arulmozhivarman J, Radhika N, Mohanraj T, Vairavel M (2019) The effect of bio-cutting fluids on surface roughness during end milling of A359 aluminium alloy. Int J Mech Prod Eng Res Dev 9:987–996
21. Mahendran R, Rajkumar P, Nirmal L, Karthikeyan S, Rajeshkumar L (2021) Effect of deep cryogenic treatment on tool life of multilayer coated carbide inserts by shoulder milling of EN8 steel. J Braz Soc Mech Sci Eng 43(8):1–11. https://doi.org/10.1007/s40430-021-03100-7
22. Wiley B, Sun Y, Mayers B, Xia Y (2005) Shape-controlled synthesis of metal nanostructures: the case of silver. Chem–A Eur J 11(2):454–463
23. Ortiz N, Skrabalak SE (2014) On the dual roles of ligands in the synthesis of colloidal metal nanostructures. Langmuir 30(23):6649–6659
24. Saravanakumar A, Rajeshkumar L, Sisindri Reddy G, et al (2023) Multi-response optimization of turning parameters for AZ91D magnesium alloy using gray-based Taguchi approach. In: Pradhan P, Pattanayak B, Das HC, Mahanta P (eds) Recent advances in mechanical engineering. lecture notes in mechanical engineering. Springer, Singapore, pp 389–397. https://doi.org/10.1007/978-981-16-9057-0_41
25. Neretina S, Hughes RA, Gilroy KD, Hajfathalian M (2016) Noble metal nanostructure synthesis at the liquid–substrate interface: new structures, new insights, and new possibilities. Acc Chem Res 49(10):2243–2250
26. Saravanakumar A, Suresh Babu J, Harikrishna A, Rajeshkumar L, Sathiyamoorthy V (2022) Optimization of cutting zone temperature in machining of magnesium alloy using Taguchi method. In: Chaurasiya PK, Singh A, Verma TN, Rajak U (eds) Technology innovation in mechanical engineering. Lecture notes in mechanical engineering. Springer, Singapore, pp. 459–468. https://doi.org/10.1007/978-981-16-7909-4_42
27. Ross NS, Ananth M, Jafferson JM, Rajeshkumar L, Kumar MS (2022) Performance assessment of vegetable oil–based MQL in milling of additively manufactured AlSi10Mg for sustainable production. Biomass Convers Biorefinery. https://doi.org/10.1007/s13399-022-02967-3
28. Pinna N, Niederberger M (2008) Surfactant-free nonaqueous synthesis of metal oxide nanostructures. Angew Chem Int Ed 47(29):5292–5304

29. Newhouse RJ, Zhang JZ (2012) Optical properties and applications of shape-controlled metal nanostructures. Reviews in Plasmonics, Springer, New York, pp 205–238
30. Wepasnick KA, Smith BA, Bitter JL, Howard D (2010) Chemical and structural characterization of carbon nanotube surfaces. Anal Bioanal Chem 396(3):1003–1014
31. Ramesh M, Marimuthu K, Karuppuswamy P, Rajeshkumar L (2021) Microstructure and properties of YSZ-Al2O3 functional ceramic thermal barrier coatings for military applications. Boletín de la Sociedad Española de Cerámica y Vidrio 61(6):641–652. https://doi.org/10.1016/j.bsecv.2021.06.004
32. Seerangan RN, Putta P, Srinivasan N, Venkatesh B (2022) Investigation on formability of tailored AA7075 thin sheets by friction stir processes (No. 2022–28–0349). SAE Technical Paper
33. Syed A, Al Saedi MH, Bahkali AH, Elgorgan AM, Kharat M, Pai K, Pichtel J, Ahmad A (2022) αAu2S nanoparticles: fungal-mediated synthesis, structural characterization and bioassay. Green Chem Lett Rev 15(1):61–70
34. Rajenthirakumar D, Srinivasan N, Sridhar R (2022) Development of a micro-forming system for micro-extrusion process of micro-pin in AZ80 alloy. In: Machines, mechanism and robotics, Springer, Singapore, pp 1227–1229
35. Rajenthirakumar D, Srinivasan N, Sridhar R (2019) Numerical simulation of micro forming of bio-absorbable AZ80 magnesium alloy. In: Advances in computational methods in manufacturing. Springer, Singapore, pp 3–11
36. Filippo E, Serra A, Buccolieri A, Manno D (2010) Green synthesis of silver nanoparticles with sucrose and maltose: morphological and structural characterization. J Non-Cryst Solids 356(6–8):344–350
37. Nashner MS, Frenkel AI, Adler DL, Shapley JR, Nuzzo RG (1997) Structural characterization of carbon-supported platinum– ruthenium nanoparticles from the molecular cluster precursor PtRu5C (CO) 16. J Am Chem Soc 119(33):7760–7771
38. Saravanakumar A, Sreenivas P, Rajeshkumar L (2022) Optimization of drilling process parameters for self-lubricants reinforced aluminium metal matrix composites. Mater Today: Proc 52:1461–1465. https://doi.org/10.1016/j.matpr.2021.11.196
39. Kumar S, Kumar S, Jain S, Verma NK (2012) Magnetic and structural characterization of transition metal co-doped CdS nanoparticles. Appl Nanosci 2(2):127–131
40. Gupta G, Tanaka D, Ito Y, Shibata D, Shimojo M, Furuya K, Mitsui K, Kajikawa K (2008) Absorption spectroscopy of gold nanoisland films: optical and structural characterization. Nanotechnology 20(2):025703
41. Rajenthirakumar D, Srinivasan N, Sridhar R (2018) Experimental investigations on micro scaled plastic deformation of bio-absorbable magnesium AZ80 alloy. Procedia Manufacturing 15:1635–1640
42. Calin G, Mitrea G, Marcu C, Burlui A, Trusca R, Vasile E, Iacomi F (2019) Synthesis and structural characterization of cobalt nanostructures in AAO membranes. Rev Chim 70(9):3444–3447
43. Kvítek O, Siegel J, Hnatowicz V, Švorčík V (2013) Noble metal nanostructures influence of structure and environment on their optical properties. J Nanomater 2013:15. https://doi.org/10.1155/2013/743684
44. Srinivasan N, Rajenthirakumar D, Sridhar R (2019) Micro-scaled plastic deformation behavior of biodegradable AZ80 magnesium alloy: experimental and numerical investigation. The Int J Adv Manuf Technol 102(9):3531–3541
45. Arasu PT, Dhanasekaran R, Kumar PS, Srinivasan N (2013) Effect of hardness and microstructure on En 353 steel by heat treatment. Int J Eng Sci 2:1–5
46. Kannan T, Arulmurugan B, Rajeshkumar L (2022) Performance analysis of AA3103 and AA6063 dissimilar weld joints by friction stir welding. Songklanakarin J Sci Technol 44(2):316–322
47. Hartland GV (2011) Optical studies of dynamics in noble metal nanostructures. Chem Rev 111(6):3858–3887

48. Kreibig U, Gartz M, Hilger A, Hövel H, Quinten M, Wagner D, Ditlbacher H (2006) A short survey of optical properties of metal nanostructures. In: Kassing R, Petkov P, Kulisch W, Popov C (eds) Functional properties of nanostructured materials. Springer, Singapore, pp 75–110
49. Goldys EM, Calander N, Drozdowicz-Tomsia K (2011) Extreme sensitivity of the optical properties of metal nanostructures to minor variations in geometry is due to highly localized electromagnetic field modes. The J Phys Chem C 115(3):676–682
50. Ovchinnikov V (2011) Effect of adhesion layer on morphology and optical properties of self-organized metal nanostructures. In: International conference on quantum, nano and micro technologies, pp 6–11
51. Ross NS, Srinivasan N, Amutha P, Gupta MK, Korkmaz ME (2022) Thermo-physical, tribological and machining characteristics of Hastelloy C276 under sustainable cooling/lubrication conditions. J Manuf Process 80:397–413
52. Srinivasan N, Rajenthirakumar D, Sridhar R, Amutha P (2020) Tribomechanical performance of MgO–ZnO nanoparticles as lubricating additives in the microextrusion process. Proc Inst Mech Eng C J Mech Eng Sci 234(22):4543–4553
53. Kang J (2022) Bottom-up processed metal nanostructures for optoelectronic devices, The Korea Advanced Institute of Science and Technology
54. Gladskikh PV, Gladskikh IA, Toropov NA, Baranov MA, Vartanyan TA (2015) Correlation between structural, optical, and electrical properties of self-assembled plasmonic nanostructures on the GaAs surface. J Nanopart Res 17(11):1–6
55. Rajeshkumar LN, Amirthagadeswaran KS (2019) Variations in the properties of copper-alumina nanocomposites synthesized by mechanical alloying. Mater Technol 53:57–63. https://doi.org/10.17222/mit.2018.122
56. Choy WC, Ren X (2015) Plasmon-electrical effects on organic solar cells by incorporation of metal nanostructures. IEEE J Sel Top Quantum Electron 22(1):1–9
57. Nguyen HT, Tran TT, Bhatt V, Kumar M, Song J, Yun JH (2021) Enhancement of Schottky junction silicon solar cell with CdSe/ZnS quantum dots decorated metal nanostructures. Appl Sci 12(1):83
58. Hao E, Zou S, Schatz GC (2004) Optical properties of one-dimensional metal nanostructures. MRS Online Proceedings Library (OPL), p 818
59. Weerakkody AD, Sedghi N, Mitrovic IZ et al (2015) Enhanced low voltage nonlinearity in resonant tunneling metal–insulator–insulator–metal nanostructures. Microelectron Eng 147:298–301
60. Koch CC (2003) Optimization of strength and ductility in nanocrystalline and ultrafine grained metals. Scripta Mater 49(7):657–662
61. Lu X, Chen J, Skrabalak SE, Xia Y (2007) Galvanic replacement reaction: a simple and powerful route to hollow and porous metal nanostructures. Proc Inst Mech Eng, Part N: J Nanoeng Nanosyst 221(1):1–16
62. Ji S, Jang J, Cho E et al (2017) High dielectric performances of flexible and transparent cellulose hybrid films controlled by multidimensional metal nanostructures. Adv Mater 29(24):1700538
63. Yang M, Chen X, Wang Z et al (2021) Zero→ two-dimensional metal nanostructures: an overview on methods of preparation, characterization, properties, and applications. Nanomaterials 11(8):1895
64. Siddiqui H, Singh N, Katiyar D et al (2022) Electrochemical additive manufacturing (ECAM): a new approach to fabricate metal nanostructures. Mater Today: Proc 72:2741–2748
65. Zhao ZJ, Shin SH, Choi DG, Park SH, Jeong JH (2018) Shape-controlled 3D periodic metal nanostructures fabricated via nanowelding. Small 14(6):1703102
66. Elechiguerra JL, Reyes-Gasga J, Yacaman MJ (2006) The role of twinning in shape evolution of anisotropic noble metal nanostructures. J Mater Chem 16(40):3906–3919
67. Ruggeri G, Covolan VL, Bernabò M et al (2013) Metal nanostructures with magnetic and biodegradable properties for medical applications. J Braz Chem Soc 24:191–200
68. Huang X, Tang S, Xiaoliang M, et al (2011) Freestanding palladium nanosheets with plasmonic and catalytic properties. Nat Nanotechnol 6(1):28–32

69. Pelaz B, Grazu V, Ibarra A et al (2012) Tailoring the synthesis and heating ability of gold nanoprisms for bioapplications. Langmuir 28(24):8965–8970
70. Srinivasan N, Rajenthirakumar D, Sridhar R, Niranjan G (2020) Combined influence of size effect and temperature in microscale deformation of 6063 aluminum gear. J Braz Soc Mech Sci Eng 42(6):1–10
71. Xiaoqing H, Tang S, Liu B, Ren B, Zheng N (2011) Enhancing the photothermal stability of plasmonic metal Nanoplates by a core-shell architecture. Adv Mater 23(30):3420–3425
72. Xue C, Mirkin CA (2007) pH-switchable silver nanoprism growth pathways. Angew Chem 119(12):2082–2084
73. Balaji D, Ranga J, Bhuvaneswari V, Arulmurugan B et al (2022) Additive manufacturing for aerospace from inception to certification. J Nanomater 2022:1–18. https://doi.org/10.1155/2022/7226852
74. Shaoheng T, Chen M, Zheng N (2014) Sub-10-nm Pd Nanosheets with renal clearance for efficient near-infrared photothermal cancer therapy. Small 10(15):3139–3144
75. Saravanakumar A, Ravikanth D, Rajeshkumar L, Balaji D, Ramesh M (2021) Tribological behaviour of MoS2 and graphite reinforced aluminium matrix composites. IOP Conf Ser: Mater Sci Eng 1059(1):012021
76. Gupta MK, Etri HE, Korkmaz ME et al (2022) Tribological and surface morphological characteristics of titanium alloys: a review. Archives of Civil and Mech Eng 22(2):1–23
77. Xiaoqing H, Tang S, Yang J (2011) Etching growth under surface confinement: an effective strategy to prepare mesocrystalline Pd nanocorolla. J Am Chem Soc 133(40):15946–15949
78. Ross NS, Sheeba PT, Jebaraj M, Stephen H (2022) Milling performance assessment of Ti-6Al-4V under CO2 cooling utilizing coated AlCrN/TiAlN insert. Mater Manuf Processes 37(3):327–341
79. Zengxia Z, Shi S, Huang Y, Tang S, Chen X (2014) Simultaneous photodynamic and photothermal therapy using photosensitizer-functionalized Pd nanosheets by single continuous wave laser. ACS Appl Mater Interfaces 6(11):8878–8885

Green Nanomaterials: Processing, Characterization and Applications

Melkie Getnet Tadesse and Jörn Felix Lübben

Abstract This chapter presents an introduction to green nanomaterials, synthesized by green approach. Brief methods of green nanomaterials processing, advantages of green processing and their limitations are discussed. The different types of green nanomaterials with their processing methods are explored. This chapter also provides an information on how the green nanomaterials are characterized and advanced methods of characterizing equipments are reviewed. Furthermore, various applications of green nanomaterials are detailed.

Keywords Green nanomaterials · Sustainability · Green chemistry · Characterization

1 Introduction

In recent most recent times, the use and development of functional materials have been dynamically increased. Such advanced materials can be used as high-performance energy storage materials and electroactive tissues [28, 90, 93], carbon dioxide fixation [72], electronic devices [42, 84]. However, such production techniques do not follow the sustainable approach with respect to the environment. Therefore, the world is highly seeking nanomaterials with the same function but their production and application should follow being environmentally friendly. In the era of sustainable chemistry and technology, green nanomaterials make a distinct advancement and play a crucial role. Green nanomaterials brought an advantage to the world such that environmental and human being are on the safest side. Green nanomaterials are the future key role players when we think of sustainability. Greener

M. G. Tadesse (✉) · J. F. Lübben
Sustainable Engineering (STE), Albstadt-Sigmaringen University, 72458 Albstadt, Germany
e-mail: tadesse@hs-albsig.de

M. G. Tadesse
Ethiopian Institute of Textile and Fashion Technology, Bahir Dar University, Bahir Dar 1037, Ethiopia

© The Author(s), under exclusive license to Springer Nature Singapore Pte Ltd. 2023
H. Abdellaoui et al. (eds.), *Mechanics of Nanomaterials and Polymer Nanocomposites*, Smart Nanomaterials Technology,
https://doi.org/10.1007/978-981-99-2352-6_5

approach has more advantageous than that of the conventional manufacturing techniques with respect to environment and human health [75]. Green nanomaterials are considered to be safe to the human life as well as the environment as they are obtained from natural products. Green nanomaterials have diverse applications [52]. Their applications can be varied from small electronics applications to advanced medical applications. Green nanomaterials are mostly developed using nanotechnology technique [81]. On the last few decades, green nanomaterials are increasingly researched, which has attracted most of the scientific sector because of copious applications in almost all areas.

Conventional manufacturing of electronic devices such as coating [92, 94], 3D printing [89], carbonization [90, 91] for various applications produces a lot of emissions to the environment. However, production of engineered nanoparticles with greener approach removes such issues [75]. Nanotechnology has become the most widely used potential technological approach in producing nanomaterials in a pollution-free environment [39]. Plants are the abundant sources for the preparation of green nanomaterials. For instance, Abo Dalam et al. [1] reported the preparation of green nanomaterials from rice waste. Rice waste was found to be a potential candidate to replace toxic chemicals. Agricultural wastes dominate in the preparation of green nanomaterials [37, 73, 78, 104].

In this chapter, the processing of green nanomaterials, their characterization and areas of applications will be discussed in detail.

2 Green Nanomaterial Processing

In the twenty-first century, nanotechnology plays a crucial role in the development of modern nanomaterials, which have diverse applications in various fields. However, most of the raw materials used in the current advanced manufacturing are from nonrentable resources. This has to be replaced with a more sustainable raw material where green nanomaterial processing can be a decisive factor in the future of nanotechnology. Green nanomaterials are the so called green because they are eco-friendly, economically feasible, technically acceptable and have diverse and sustainable applications. In this view, processing of some green nanomaterials will be discussed in this chapter.

2.1 Processing of Nanocellulose—A Green Nanomaterial

A nanocellulose is a nanosized cellulose-based fibre processed by bacteria or resulting from plants and extracted from aqueous processes. The most abundant source of nanocellulose is agricultural waste [15, 54, 74]. These are naturally available at cheaper costs. The general principle of nanocellulose processing can be represented in Fig. 1.

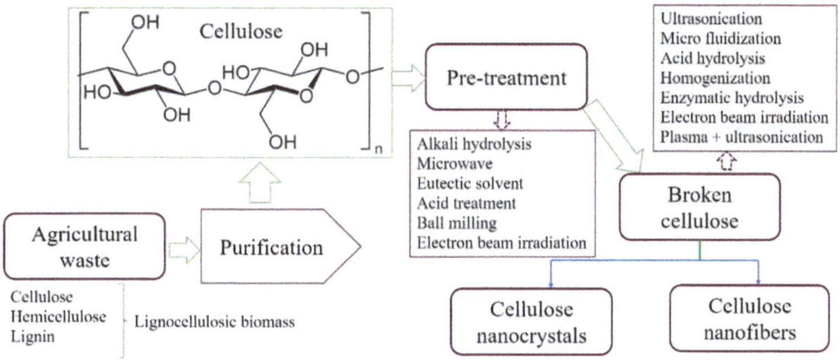

Fig. 1 Chemical and mechanical processing of green cellulose nanomaterials from agricultural (biomass) raw materials.

As can be observed in Fig. 1, various emerging and green technologies have been employed to process green nanomaterials from biomass products (wastes). Not only the final product but also the inside process followed in the processing of green cellulose-based nanomaterials are of environmentally friendly and safe to human life. Therefore, the simple processing methods, biodegradability of the substances used and products, eco-friendly features, recyclability, green chemistry applied and pollution-free process, are the major gainful features of green nanomaterial processing from biomass wastes (Table 1).

Table 1 Summary of the processing techniques in the production of green nanocellulose materials

Source	Pre-treatment	Extraction	Refs.
Tea stalk	H_2O_2 treatment	Acid hydrolysis	Guo et al. [34]
Rice husk and straw	Alkali hydrolysis	Ultrasonication	Rezanezhad et al. [77]
Corn cob	Alkali hydrolysis	Acid hydrolysis, pulp refining, [a]TEMPO)-mediated oxidation	Liu et al. [49]
Apple pomace	Alkali hydrolysis	Acid hydrolysis	Meliko˘glu et al. [58]
Soy hulls	Bleaching	Reactive extrusion	Debiagi et al. [18]
Wheat straw	Alkali hydrolysis	Acid hydrolysis and mechanical	Nehra and Chauhan [66]
Bagasse (orange)	Bleaching, $NaClO_2$	Ultrasound treatment	Mariño et al. [56]
Beer residue	Acid hydrolysis	Ultrasonication	Shahabi-Ghahafarrokhi et al. [82]
Bagasse (sugarcane)	Bleaching, NaCl	Acid hydrolysis	Mandal and Chakrabarty [55]
Citrus waste	Alkali hydrolysis	Enzymatic Hydrolysis	Mariño et al.[56]

[a] 2,2,6,6-tetramethylpiperidine-1-oxyl

Liu et al. [51] reported engineered nanocellulose hydrogel materials for the applications in drug delivery. Less risks of poisonousness, biodegradability, biocompatibility, being recyclability, and tunable surface properties make nanocellulose hydrogels for the perfect choice for such applications. Nanocellulose follows biotechnological production routes. Nanocellulose-based green nanomaterials sowed various applications such as supercapacitor [95], flexible energy storage devices [19], biomedical applications [98] and environmental remediation [83]. Shak K.P.Y. et al. reported different nanocellulose green nanomaterials including cellulose nanofiber (CNF) and cellulose nanocrystal and can be obtained from different sources including algae, bacteria, plants, and tunicates. According to their report, nanocellulose-based green nanomaterials follow various production routes such as mechanical disintegration, chemical reaction and biological reaction. Such processing methods are sustainable in terms of raw materials and the production methods including their application.

Overall, the processing of green nanomaterials from cellulose-based agricultural wastes is relatively inexpensive, biodegradable, environmentally friendly process, cheap raw materials available in the biomass and could be an alternative in the future production of green nanomaterials. In integrated treatment approaches for the production of nanocellulose, synergistic effects were seen. Nanocellulose-based nanocomposites have shown various benefits that make them well-suited for environmental protection.

2.2 *Processing of PLA Composites*

Green nanomaterials having environmentally friendly attributes that are economically feasible and technically fit contributes more a sustainable future. In this context, polylactic acid (PLA) is fully biodegradable abiotically using mechanisms of hydrolysis, thermal decomposition and photodegradation. This shows that the use of PLA could be the best alternative to petroleum-based polymers. PLA plays an important role in the challenges of obtaining green nanomaterials that do not harm the environment as well as the human life. Table 2 summarizes the synthesis of green nanomaterials based on PLA.

Lignin/PLA nanocomposites [32], cellulose/PLA nanocomposites [61], and other green nanomaterials are now becoming common practices in the processing of nanomaterials to replace the conventional materials. More dominantly, bio-based nanomaterials are now in an encouraging development to compete with fossil fuel-based industry. PLA-based composites are now used for the production of packaging materials, and future advancement is required in order to use these green nanomaterials for various applications.

Table 2 Processing of green nanomaterials based on polylactic acid (PLA).

Composites	Processing methods	Properties	Refs.
Plant extract	Sonication	Stable PLA	Kumari et al. [47]
Layered silicate	Extrusion using a co-rotating	Morphology and mechanical properties were improved	Cumkur et al. [17]
Silica	Surface-grafting reaction	Improved toughness and tensile strength	Yan et al. [101]
CNCs	Dispersion methods	An increase in elastic modulus and slight decrease in yield strength	Haque et al. [36]
$Al(OH)_3$	Melt processing	Enhancement in oxygen and water vapour barrier properties	Katiyar et al. [43]
TiO_2	Film formation	Highly transparent	Nakayama and Hayashi [65]
–	Electrospinning	Enhanced mechanical and thermal properties	Wu et al. [100]
Calcium sulphate	Melt-blending	Excellent flame-retardant properties	Marius et al. [57]
Magnetite NPs	Melt-compounding	improved thermal properties	Murariu et al. [63]

2.3 *Processing of Cellulose-Based Composites*

Cellulose-based nanomaterials can be obtained in water as dilute suspensions. Water is the most accessible liquid medium with respect to other polar liquid mediums. Green nanocomposites with fully biodegradable cellulose-based nanomaterials have been prepared [81]. The concept of sustainability and safety lies when a material is obtained naturally and needs safe processing. In this regard, plant fibres play a key role in the manufacturing of composites where a cellulose part act as reinforcements.

Lignin-containing cellulose-based nanomaterials can be easily processed with low energy and chemical consumption [50]. In this regard, lignin-containing cellulose nanomaterials can be prepared several mechanical methods such as grinding, ultrasonication and high-pressure homogenization. Hydrolysis using inorganic and organic acids, alkali treatment, organosolv fractionation, 2,2,6,6-Tetramethylpiperidine-1-oxyradical (TEMPO) oxidation, enzymatic treatment, and deep eutectic solvent treatment have been deeply reviewed as a potential technique to prepare lignin-containing cellulose-based green nanomaterials. These methods are found to be cheap in terms of energy utilization and have no problems with respect to chemical recovery.

Fig. 2 Illustration of the number of scientific publications since 2008 using search terms green nanomaterials (GNM), cellulose-based nanomaterials (CNM) and green nanomaterials based on cellulose (GNMC). Data analysis completed using PubMed search system on 08 November 2022

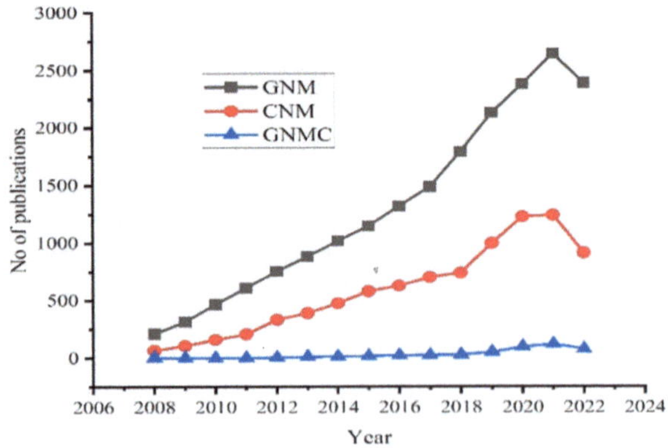

Cellulose-based nanomaterials have been used as a filler in the manufacturing of polymer nanocomposites [21]. These types of fillers can help to improve the mechanical properties of the composites. Furthermore, thermoplastic processing has been reported to process the cellulose-based green nanomaterials [61]. To obtain biodegradable, and eco-friendly materials with improved strength and stiffness, cellulose-based green nanomaterials are irreplaceable. That is why sometimes, some researchers called cellulose-based nanomaterials are new generation materials that can solve the hot global issues: pollution [60]. The use of cellulose-based nanomaterials in green nanomaterials is increasing rapidly since 2008. Figure 2 proves this one.

Zhang et al. [105] reported the processing of green cellulose nanocrystals (CNCs) using acid–CNC serration techniques such as centrifugation, gravity settling and microfiltration. Both techniques are efficiently utilized to process CNCs and both can contribute to the reduction of global warming. Such techniques have the great potential in sustainable manufacturing of green nanomaterials. Deep eutectic solvent has been implemented to prepare CNCs in a greener approach [86].

Not only eutectic solvents but also acids such as formic acid has been reported on the preparation cellulose-based nanomaterials using rapid fractionation [106]. Overall, cellulose-based nanomaterials can be prepared using various principles and methods in greener approach in which the sustainability and healthy processing of nanomaterials for various applications such as healthcare, bio-composites, energy and water purification are realized [24].

3 Green Nanomaterial Characterizations

Due to the fact that eco-friendly, sustainability and green processing, green nanomaterials are now become popular in nanotechnology manufacturing arena. Before applications, green nanomaterials must be characterized. The most common way of characterizing green nanomaterials is depicted to determine its functionality and chemical and physical properties. Particle size, morphology, specific surface area, size distribution, crystallinity and their elemental composition are the main characteristics that determine the behaviour of green nanomaterials [4]. The most frequently used tools help to characterize green nanomaterials are X-ray diffraction (XRD), energy dispersive spectroscopy (EDAX), transmission electron microscope (TEM), UV–visible absorption spectroscopy, Fourier transform infrared (FTIR) spectroscopy, atomic force microscopy (AFM), scanning electron microscopy (SEM), IR infrared (IR) spectroscopy. In this section, some green nanomaterial characterization methods are discussed.

3.1 X-ray Diffraction

Characterization of nanomaterials using X-ray diffraction is becoming popular due to its most efficient and cost-effectiveness solution. It is a new entry in this regard. XRD has been used to characterize the crystalline nature of the nanomaterials. For instance, the report in Ref. [4] showed the characterization of Ag nanoparticles using X-ray diffraction methods. Crystallinity is the most important property of green nanomaterials to determine its functionality, behaviour and usability. XRD makes it simple to determine the size, shape, lattice parameter determination, and phase fraction analysis of the unit cell for any substance [11]. Furthermore, XRD can also use to characterize and help to obtain important information regarding the nanomaterial's crystal structure, made up (its composition), crystalline grain size [88]. The characterization of green nanomaterials is paramountcy important because they are synthesized from several chemical or mechanical routes and in addition, they have big potential in advanced technological solutions for sustainable environment. In this regard, X-ray diffraction plays an important role.

3.2 Energy Dispersive Spectroscopy (EDX)

Energy Dispersive Spectroscopy (EDX) is a method of analysis used to characterize a sample's chemical composition or elemental composition [30]. It depends on the interaction of an X-ray excitation source and a sample. Its ability to characterize materials is largely a result of the fundamental idea that every element has a special atomic structure that allows for a special collection of peaks on its electromagnetic

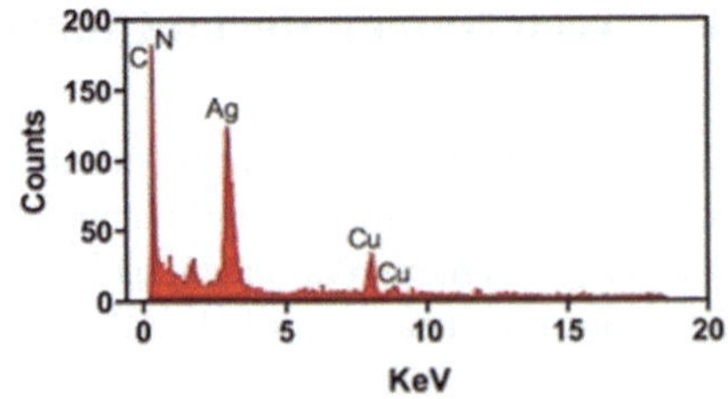

Fig. 3 EDX spectra of AgNPs. Reprinted under Licence number: 5440061494102 with Elsevier publishing company [3]

emission spectrum (which is the main principle of spectroscopy). EDX has been used to determine the characteristic peaks of nanomaterials [3]. Figure 3 shows the EDX patterns of silver nanoparticles (AgNPs).

Several individual particles underwent local EDX microanalysis to determine the chemical composition of AgNPs. Major energy peaks in EDX spectra were found to correspond to Ag, N, C and Cu. Around the primary peak of Ag, many peaks were seen, each of which represented a different valency state of the metal in the particles. Samples placed on C-coated copper grids frequently show peaks that correspond to Cu and C. Generally, EDX is an X-ray technique used to identify the elemental compositions of green nanomaterials. In conclusion, using the EDX, the composition and the quantity of the elements of the green nanomaterials wich are placed near or at the surface of the green nanomaterial sample, can be measured. Therefore, EDX is considered as a useful tool in all works that requires elemental determination.

3.3 *Atomic Force Microscopy (AFM)*

Atomic force microscopy sometimes called scanning microscopy (SFM) is a kind of microscopy implementing very high resolution in the manometer scale using a scanning probe where topographic imaging, force measurement and manipulation are achieved using such machines [29]. AFM is capable of imaging the surface of nanomaterials [41]. The working principles of AFM are shown in Fig. 4.

Green nanomaterials have been characterized using AFM for the identification of the cell wall damage [103]. Furthermore, the morphology of the nanoparticles was studied using AFM [70]. Therefore, AFM is one of the important techniques to characterize green nanomaterials. Depending on the final applications of green nanomaterials, several useful information can be obtained using AFM like other morphological techniques.

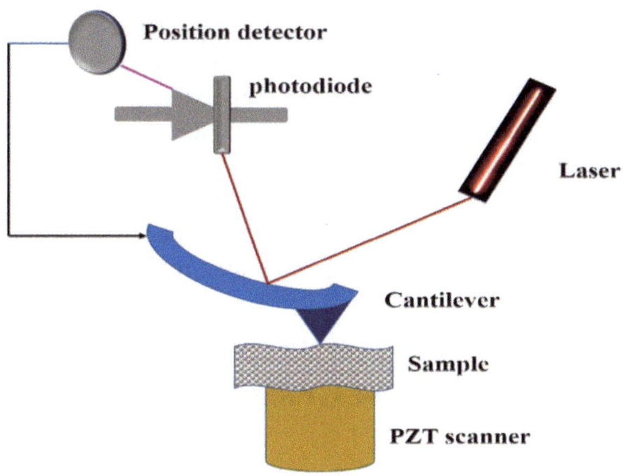

Fig. 4 AFM working principles. Reproduced with permission under licence number: 5440210772360, Elsevier [41]

3.4 Other Characterization Techniques

There are lots of advanced techniques used to characterize green nanomaterials. Among these, scanning electron microscopy has been used by several researchers [70] to observe the topographical imaging of the green nanomaterials. From this imaging, the surface analysis of the green nanomaterials can be made, and very important information can be obtained. Another important technique that has been used to characterize green nanomaterials is transmission electron microscopy (TEM) [70]. TEM has very high resolution of 0.05 nm, high energy resolution of 7 meV, which helped to identify the size, composition, strain and morphology of green nanomaterials [87].

Green nanomaterials can be produced using metal-based nanoparticles. Salam et al. [80] demonstrated the preparation of zinc oxide nanoparticles from Ocimum basilicum L. var. purpurascens Benth. -Lamiaceae leaf extract. Zinc has been also synthesized from brown marine macroalga Sargassum muticum aqueous extract in green approach [6]. This provides an eco-friendly alternative to conventional physical and chemical methods as this approach eliminates the use of toxic chemicals. In addition, zinc oxide catalyst has been prepared in greener approach under microwave irradiation using banana for biodiesel synthesis from fish waste lipid [23]. This makes zinc nanoparticles heavily employed for the preparation of green nanomaterials form using extraction principles. Silver nanoparticles have been synthesized in a greener approach using dried medicinal plant of basil [2]. Another metallic element for the synthesis of green nanomaterial is gold nanoparticle. It has been reported that green synthesized gold nanoparticle dispersed porous carbon composites have been used for the production of electrochemical energy storage. Biocompatible silver and gold nanoparticles from Memecylon umbellatum have been demonstrated for the possible use of the green nanomaterials for chemical sensors [5]. Green synthesis of iron-based nanomaterials for environmental remediation has been reported somewhere else [9]. A strong reason to look into new methods of the synthesis of green nanomaterials such as metallic compounds in the nanoparticle range is growing significantly. In this aspect, the use of green-based nanoparticles over the past many decades and their uses in environmental protection is highly increased. As a result, the synthesis of these green nanomaterials now takes into account the concepts of green chemistry, waste reduction, energy efficiency, safer solvents, and the use of safe and nontoxic precursor materials, giving rise to substantial research in this field. Therefore, extraction of green metallic nanoparticles from plant-based materials is becoming the hot issue especially for those environmentalist researchers.

4 Applications of Green Nanomaterials

Wide-ranging applications are intended for green nanomaterials in various sectors. Several researches are now focusing on the development of green nanomaterials for different applications from conventional to advanced applications areas. Sustainability and environmentally friendly process of green nanomaterials are the main deriving factors in the increasing demand for green nanomaterials in different sectors. Hence, in the future, there could be an industrial revolution regarding green nanomaterials. Green chemistry was effective for nanoparticle synthesis using extracts [31]. In this chapter, an overview of the applications of green nanomaterials in the quest of many industrial applications is discussed.

4.1 Wastewater Treatment and Remediation

The rapid intervention of industry for the economic development brought a great influence on the change in the environment with respect to water and land pollution. The industry at large and the textile and chemical industry more specifically contributes the lion share in today's polluted water and huge amount of hard waste accumulation. On the other hand, the dynamic upgrade in the use of advanced technology also fetched a thoughtful impact on the environmental pollution. Therefore, some mechanisms must be devised to reduce such impact. Most recent studies indicated that green nanomaterials are amongst the mechanisms used to treat the wastewater in greener and more sustainable approach. Chitosan nanoparticle, one of the most studied biocompatible and biodegradable fibrous substance has been employed as an adsorbent for the removal of dyes in the textiles industry [38, 40, 108] and wastewater treatment [20, 67, 102]. In addition, removal of dyes, heavy metals, and pesticides from wastewater have been reported using green nanomaterials [69].

Moreover, green nanomaterials sourced from microorganism, plants, carbohydrates, fats, and proteins are used as potential wastewater remediation treatment [85]. Not only the cost of the synthetic materials but also the problem with the environmental pollution problem, the world has now forced to direct their attention to find green alternatives and in this regard, green nanomaterials are the perfect solution for wastewater treatment and remediation actions.

4.2 Packaging

Food packaging industry requires safe, biocompatible, nontoxicity and bio-based materials. Nanotechnology touches food packaging industry as it touches others. Now, this technology finds its application in green nanomaterials packaging systems. Nanotechnology plays an important role in packaging, keeping for the longer time without losing its content and freshness, processing and testing of food items [13]. Green nanomaterials act as active and intelligent packaging materials [76].

The inclusive choice of green nanomaterials could make its range for food packaging applications. Cellulose nanomaterials [45], biopolymers [10], starch nanocrystals and clay and silicate nanoparticles [97] and eco-friendly polymer composites [62] find their applications towards food packaging. The green nanomaterials used for food packaging must consider antimicrobial properties as food might be attacked from microbes. In addition, green nanomaterials have been used to design eco-designed nanoparticles and nanosystems for sustainable agrifood applications [8].

4.3 Coatings

As with other technologies, coating technology also uses green nanomaterials in different coating industries. The use of coating is enhancing corrosion resistance, wear resistance, surface properties of the materials and hence nanomaterials will increase the quality of coatings when incorporated [27]. Phytic acid-incapacitated green nanoparticles for green anticorrosion coatings have been reported somewhere else [96] and proved that a new way of ecologically responsive anticorrosion coatings.

Balamurugan et al. [7] reported coating of cotton fabric with silver green nanoparticles and claimed the best antimicrobial characteristics out of it. Ultraviolet protective coatings based on ZnO nanoparticles also experimented and have improved UV protection ability due to the green nanomaterials [12].

Other coatings such as self-cleaning coatings [48], anti-COVID coatings [25], hydrophobic coatings [53] and hydrophilic coatings[14] are among those that are already available in the market. In general, green nanomaterials have been utilized in coating industry for various applications such as energy, protective, self-acting, medical, sensors, agriculture and textiles areas.

4.4 Biomedical Treatment

Green nanomaterials more specifically cellulose nanomaterials have been used as biomedical materials because of their excellent properties such as biocompatibility, low toxicity and biodegradability [22]. The use of keratin-based green nanomaterial for the drug delivery carriers has been reported somewhere else [26]. Biomedical application is a sensitive issue and scientific works performed use green nanomaterials for antimicrobial and antioxidant applications [59]. In this regard, natural products are well known for their inherent properties that might use for various applications.

The use of green nanomaterials based on ZnO for biomedical applications has been extensively studied [16, 99]. In addition, green silver and gold nanomaterials have reported somewhere else as potential applications for biomedical treatment [79]. Due to biocompatibility, high surface area to volume ratio and other important properties green nanoparticles (GNPs) gain attention in biomedical applications specially those extracted from bacteria, plant, fungi and algae [44].

In general, green nanomaterials that are developed by means of some green nanotechnologies intended to be used for biomedical treatment as these materials are safe, biocompatible, biodegradable and nontoxic nature. Due to this fact, there are lots of research works undergoing in multidisciplinary approach.

4.5 Other Applications of Green Nanomaterials

Green nanomaterials have several applications including antibacterial and antifungal activities [68], Pişkin et al. 2013; [35]. TiO_2 nanoparticles are viable for such applications due to the inherent antibacterial and antifungal characteristics of TiO_2. In addition, due to the nature of Ag, it has been reported that green nanoparticles based on Ag have been used for antimicrobial activity control [46]. Furthermore, TiO_2-based nanoparticles have been used as a photocatalytic degradation of tannery wastewater [33, 64, 107]. Green titanium dioxide (TiO_2) nanoparticles (NPs) evaluate its performance for the photocatalytic treatment of tannery waste after the secondary (biological) treatment process. Tannery wastes are full of chromium-based compounds, and this toxic metal can be removed using TiO_2 NPs. In general, there are several applications of green nanomaterials and hence green nanomaterials are featured to be the next-generation technologies, which can save the world from toxic materials.

5 Conclusions and Future Outlook

Synthesis of green nanomaterials is becoming the centre point in the era of nanotechnology due to sustainability of manufacturing, eco-friendly and environmentally friendly nature. In this chapter, we addressed the various processing mechanisms of green nanomaterials, characterization techniques and their applications in different fields. One of our contributions is to show how green nanomaterials are becoming the centre point in every aspect of life and their combination with advanced technologies. We showed also why green nanotechnologies are environmentally friendly processes and their importance in the sustainable manufacturing of materials in the industry.

A discussion on different applications of green nanomaterials has been provided. In particular, we detailed the applications of green nanomaterials in packaging, health, coatings and other few applications areas. Finally, we believe that, as far as sustainability and eco-friendly manufacturing are concerned, green nanomaterials will be the future directions for several decades ahead.

Acknowledgements The authors gratefully acknowledge the financial support from Alexander von Humboldt Foundation for the Georg Forster Research Fellowship, which is internally referred to as SmartFib project at Albstadt-Sigmaringen University.

Conflicts of Interest The authors declare no conflict of interest.

References

1. AboDalam H, Devra V, Ahmed FK, Li B, Abd-Elsalam KA (2022) Rice wastes for green production and sustainable nanomaterials: an overview. Agri-Waste and Microbes for Prod Sustain Nanomater 707–728
2. Ahmad N, Sharma S, Alam MK, Singh V, Shamsi S, Mehta B, Fatma A (2010) Rapid synthesis of silver nanoparticles using dried medicinal plant of basil. Colloids Surf B Biointerfaces 81(1):81–86
3. Ali M, Kim B, Belfield KD, Norman D, Brennan M, Ali GS (2016) Green synthesis and characterization of silver nanoparticles using Artemisia absinthium aqueous extract—a comprehensive study. Mater Sci Eng C 58:359–365
4. Anukiruthika T, Priyanka S, Moses J, Anandharamakrishnan C (2020) Characterisation of green nanomaterials. In: Green nanomaterials. Springer, pp 43–79
5. Arunachalam KD, Annamalai SK, Hari S (2013) One-step green synthesis and characterization of leaf extract-mediated biocompatible silver and gold nanoparticles from Memecylon umbellatum. Int J Nanomed 1307–1315
6. Azizi S, Ahmad MB, Namvar F, Mohamad R (2014) Green biosynthesis and characterization of zinc oxide nanoparticles using brown marine macroalga Sargassum muticum aqueous extract. Mater Lett 116:275–277
7. Balamurugan M, Saravanan S, Soga T (2017) Coating of green-synthesized silver nanoparticles on cotton fabric. J Coat Technol Res 14(3):735–745
8. Bartolucci C, Antonacci A, Arduini F, Moscone D, Fraceto L, Campos E, Attaallah R, Amine A, Zanardi C, Cubillana-Aguilera LM (2020) Green nanomaterials fostering agrifood sustainability. TrAC, Trends Anal Chem 125:115840
9. Bolade OP, Williams AB, Benson NU (2020) Green synthesis of iron-based nanomaterials for environmental remediation: a review. Environ Nanotechnol, Monit Manage 13:100279
10. Bumbudsanpharoke N, Choi J, Ko S (2015) Applications of nanomaterials in food packaging. J Nanosci Nanotechnol 15(9):6357–6372
11. Bykkam S, Ahmadipour M, Narisngam S, Kalagadda VR, Chidurala SC (2015) Extensive studies on X-ray diffraction of green synthesized silver nanoparticles. Adv Nanopart 4(1):1–10

12. Channa IA, Ashfaq J, Gilani SJ, Shah AA, Chandio AD, Jumah MNb (2022) UV blocking and oxygen barrier coatings based on polyvinyl alcohol and zinc oxide nanoparticles for packaging applications. Coatings 12(7):897
13. Chausali N, Saxena J, Prasad R (2022) Recent trends in nanotechnology applications of bio-based packaging. J Agri Food Res 7:100257
14. Chen Q, Brett CJ, Chumakov A, Gensch M, Schwartzkopf M, Körstgens V, Söderberg LD, Plech A, Zhang P, Müller-Buschbaum P (2021) Layer-by-layer spray-coating of cellulose nanofibrils and silver nanoparticles for hydrophilic interfaces. ACS Appl Nano Mater 4(1):503–513
15. Cheng HN, Klasson K, Asakura T, Wu Q (2016) Nanotechnology in agriculture. In: Nanotechnology: delivering on the promise, vol 2. ACS Publications, pp 233–242
16. Cruz DM, Mostafavi E, Vernet-Crua A, Barabadi H, Shah V, Cholula-Díaz JL, Guisbiers G, Webster TJ (2020) Green nanotechnology-based zinc oxide (ZnO) nanomaterials for biomedical applications: a review. J Phys: Mater 3(3):034005
17. Cumkur EA, Baouz T, Yilmazer U (2015) Poly (lactic acid)–layered silicate nanocomposites: the effects of modifier and compatibilizer on the morphology and mechanical properties. J Appl Polym Sci 132(38)
18. Debiagi F, Faria-Tischer P, Mali S (2020) Nanofibrillated cellulose obtained from soybean hull using simple and eco-friendly processes based on reactive extrusion. Cellulose 27(4):1975–1988
19. Dias OAT, Konar S, Leão AL, Yang W, Tjong J, Sain M (2020) Current state of applications of nanocellulose in flexible energy and electronic devices. Front Chem 8:420
20. Divya K, Jisha M (2018) Chitosan nanoparticles preparation and applications. Environ Chem Lett 16(1):101–112
21. Dufresne A (2018) Cellulose nanomaterials as green nanoreinforcements for polymer nanocomposites. Philos Trans Royal Soc A: Math, Phys Eng Sci 376(2112):20170040
22. Dufresne A (2019) Nanocellulose processing properties and potential applications. Curr Forestry Rep 5(2):76–89
23. Dutta S, Jaiswal KK, Verma R, Basavaraju DM, Ramaswamy AP (2019) Green synthesis of zinc oxide catalyst under microwave irradiation using banana (Musa spp.) corm (rhizome) extract for biodiesel synthesis from fish waste lipid. Biocatal Agri Biotechnol 22:101390
24. Eichhorn SJ, Etale A, Wang J, Berglund LA, Li Y, Cai Y, Chen C, Cranston ED, Johns MA, Fang Z (2022) Current international research into cellulose as a functional nanomaterial for advanced applications. J Mater Sci 1–71
25. El-Megharbel SM, Alsawat M, Al-Salmi FA, Hamza RZ (2021) Utilizing of (zinc oxide nano-spray) for disinfection against "SARS-CoV-2" and testing its biological effectiveness on some biochemical parameters during (COVID-19 pandemic)—" ZnO nanoparticles have antiviral activity against (SARS-CoV-2)". Coatings 11(4):388
26. Ferroni C, Varchi G (2021) Keratin-based nanoparticles as drug delivery carriers. Appl Sci 11(20):9417
27. Gautam YK, Sharma K, Tyagi S, Kumar A, Singh BP (2022) Applications of green nanomaterials in coatings. In: Green nanomaterials for industrial applications. Elsevier, pp 107–152
28. Gebeyehu EK, Sui X, Adamu BF, Beyene KA, Tadesse MG (2022) Cellulosic-based conductive hydrogels for electro-active tissues: a review summary. Gels 8(3):140
29. Geisse NA (2009) AFM and combined optical techniques. Mater Today 12(7–8):40–45
30. Goldstein JI, Newbury DE, Michael JR, Ritchie NW, Scott JHJ, Joy DC (2017) Scanning electron microscopy and X-ray microanalysis. Springer
31. Gómez-López P, Puente-Santiago A, Castro-Beltrán A, do Nascimento LAS, Balu AM, Luque R, Alvarado-Beltrán CG (2020) Nanomaterials and catalysis for green chemistry. Curr Opin Green Sustain Chem 24:48–55
32. Gordobil O, Egüés I, Llano-Ponte R, Labidi J (2014) Physicochemical properties of PLA lignin blends. Polym Degradation Stab 108:330–338

33. Goutam SP, Saxena G, Singh V, Yadav AK, Bharagava RN, Thapa KB (2018) Green synthesis of TiO2 nanoparticles using leaf extract of Jatropha curcas L. for photocatalytic degradation of tannery wastewater. Chem Eng J 336:386–396
34. Guo Y, Zhang Y, Zheng D, Li M, Yue J (2020) Isolation and characterization of nanocellulose crystals via acid hydrolysis from agricultural waste-tea stalk. Int J Biol Macromol 163:927–933
35. Haghighi F, Roudbar Mohammadi S, Mohammadi P, Hosseinkhani S, Shipour R (2013) Antifungal activity of TiO2 nanoparticles and EDTA on Candida albicans biofilms. Infect, Epidemiol Microbiol 1(1):33–38
36. Haque MM-U, Puglia D, Fortunati E, Pracella M (2017) Effect of reactive functionalization on properties and degradability of poly (lactic acid)/poly (vinyl acetate) nanocomposites with cellulose nanocrystals. React Funct Polym 110:1–9
37. Hashmi A, Singh AK, Jain B, Singh A (2020) Muffle atmosphere promoted fabrication of graphene oxide nanoparticle by agricultural waste. Fullerenes, Nanotubes, Carbon Nanostruct 28(8):627–636
38. Hosseini F, Sadighian S, Hosseini-Monfared H, Mahmoodi NM (2016) Dye removal and kinetics of adsorption by magnetic chitosan nanoparticles. Desalination and Water Treatment 57(51):24378–24386
39. Husen A, Iqbal M (2019) Nanomaterials and plant potential: an overview. Springer
40. Jaafari J, Barzanouni H, Mazloomi S, Farahani NAA, Sharafi K, Soleimani P, Haghighat GA (2020) Effective adsorptive removal of reactive dyes by magnetic chitosan nanoparticles: kinetic, isothermal studies and response surface methodology. Int J Biol Macromol 164:344–355
41. Kaliva M, Vamvakaki M (2020) Nanomaterials characterization. In: Polymer science and nanotechnology. Elsevier, pp 401–433
42. Kasaw E, Haile A, Getnet M (2020) Conductive coatings of cotton fabric consisting of carbonized charcoal for E-Textile. Coatings 10(6):579
43. Katiyar V, Gerds N, Koch CB, Risbo J, Hansen HCB, Plackett D (2011) Melt processing of poly (L-lactic acid) in the presence of organomodified anionic or cationic clays. J Appl Polym Sci 122(1):112–125
44. Kaur PJ, Kaushik G, Siddiqui R, Goyal PK (2022) Biomedical applications of some green synthesized metal nanomaterials. In: Green nanomaterials for industrial applications. Elsevier, pp 71–91
45. Khan A, Wen Y, Huq T, Ni Y (2018) Cellulosic nanomaterials in food and nutraceutical applications: a review. J Agric Food Chem 66(1):8–19
46. Kim JS, Kuk E, Yu KN, Kim J-H, Park SJ, Lee HJ, Kim SH, Park YK, Park YH, Hwang C-Y (2007) Antimicrobial effects of silver nanoparticles. Nanomed Nanotechnol Biol Med 3(1):95–101
47. Kumari A, Kumar V, Yadav SK (2012) Plant extract synthesized PLA nanoparticles for controlled and sustained release of quercetin: a green approach. PLoS ONE 7(7):e41230
48. Li X, He J (2013) Synthesis of raspberry-like SiO2–TiO2 nanoparticles toward antireflective and self-cleaning coatings. ACS Appl Mater Interfaces 5(11):5282–5290
49. Liu C, Li B, Du H, Lv D, Zhang Y, Yu G, Mu X, Peng H (2016) Properties of nanocellulose isolated from corncob residue using sulfuric acid, formic acid, oxidative and mechanical methods. Carbohyd Polym 151:716–724
50. Liu K, Du H, Zheng T, Liu W, Zhang M, Liu H, Zhang X, Si C (2021) Lignin-containing cellulose nanomaterials: preparation and applications. Green Chem 23(24):9723–9746
51. Liu S, Qamar SA, Qamar M, Basharat K, Bilal M (2021) Engineered nanocellulose-based hydrogels for smart drug delivery applications. Int J Biol Macromol 181:275–290
52. Lu Y, Ozcan S (2015) Green nanomaterials: on track for a sustainable future. Nano Today 10(4):417–420
53. Lyu J, Wu B, Wu N, Peng C, Yang J, Meng Y, Xing S (2021) Green preparation of transparent superhydrophobic coatings with persistent dynamic impact resistance for outdoor applications. Chem Eng J 404:126456

54. Malladi R, Nagalakshmaiah M, Robert M, Elkoun S (2018) Importance of agriculture and industrial waste in the field of nano cellulose and its recent industrial developments: a review. ACS Sustain Chem Eng 6:2807–2828
55. Mandal A, Chakrabarty D (2011) Isolation of nanocellulose from waste sugarcane bagasse (SCB) and its characterization. Carbohyd Polym 86(3):1291–1299
56. Mariño MA, Rezende CA, Tasic L (2018) A multistep mild process for preparation of nanocellulose from orange bagasse. Cellu 25(10):5739–5750
57. Marius M, Bonnaud L, Fontaine G, Bourbigot S, Dubois P (2010) New trends in polylactide (PLA)-based materials:'Green'PLAeCalcium sulfate (nano) composites tailored with flame retardant properties. Polym Degradation Stab 95(3)
58. Melikoğlu AY, Bilek SE, Cesur S (2019) Optimum alkaline treatment parameters for the extraction of cellulose and production of cellulose nanocrystals from apple pomace. Carbohyd Polym 215:330–337
59. Mohd Yusof H, Abdul Rahman NA, Mohamad R, Zaidan UH (2020) Microbial mediated synthesis of silver nanoparticles by Lactobacillus Plantarum TA4 and its antibacterial and antioxidant activity. Appl Sci 10(19):6973
60. Mokhena TC, John MJ (2020) Cellulose nanomaterials: new generation materials for solving global issues. Cellulose 27(3):1149–1194
61. Mokhena TC, Sefadi JS, Sadiku ER, John MJ, Mochane MJ, Mtibe A (2018) Thermoplastic processing of PLA/cellulose nanomaterials composites. Polymers 10(12):1363
62. Moustafa H, Youssef AM, Darwish NA, Abou-Kandil AI (2019) Eco-friendly polymer composites for green packaging: future vision and challenges. Compos Part B: Eng 172:16–25
63. Murariu M, Galluzzi A, Paint Y, Murariu O, Raquez J-M, Polichetti M, Dubois P (2021) Pathways to green perspectives: production and characterization of polylactide (PLA) nanocomposites filled with superparamagnetic magnetite nanoparticles. Materials 14(18):5154
64. Mustapha S, Tijani J, Egbosiuba T, Sumaila A, Amigun T, Salihu A, Ibrahim Y, Ndamitso M, Abdulkareem S (2022) Sol–gel synthesis of kaolin/TiO2 nanocomposites for photocatalytic degradation of tannery wastewater. In: Industrial wastewater treatment: emerging technologies for sustainability. Springer, pp 323–343
65. Nakayama N, Hayashi T (2007) Preparation and characterization of poly (l-lactic acid)/TiO2 nanoparticle nanocomposite films with high transparency and efficient photodegradability. Polym Degrad Stab 92(7):1255–1264
66. Nehra P, Chauhan RP (2022) Facile synthesis of nanocellulose from wheat straw as an agricultural waste. Iran Polym J 31(6):771–778
67. Olivera S, Muralidhara HB, Venkatesh K, Guna VK, Gopalakrishna K, Kumar Y (2016) Potential applications of cellulose and chitosan nanoparticles/composites in wastewater treatment: a review. Carbohyd Polym 153:600–618
68. Othman SH, Abd Salam NR, Zainal N, Kadir Basha R, Talib RA (2014) Antimicrobial activity of TiO2 nanoparticle-coated film for potential food packaging applications. Int J Photoenergy 2014
69. Palani G, Arputhalatha A, Kannan K, Lakkaboyana SK, Hanafiah MM, Kumar V, Marella RK (2021) Current trends in the application of nanomaterials for the removal of pollutants from industrial wastewater treatment—a review. Molecules 26(9):2799
70. Patra JK, Baek K-H (2014) Green nanobiotechnology: factors affecting synthesis and characterization techniques. J Nanomater 2014
71. Pişkin S, Palantöken A, Yılmaz MS Antimicrobial activity of synthesized TiO2 nanoparticles. In: International conference on emerging trends in engineering and technology (ICETET'2013), 2013. pp 7–8
72. Pourhassan F, Khalifeh R, Eshghi H (2021) Well dispersed gold nanoparticles into the multi amine functionalized SBA-15 for green chemical fixation of carbon dioxide to cyclic carbonates under solvent free conditions. Fuel 287:119567
73. Prasher P, Sharma M, Mudila H (2022) Green nanomaterials produced by agro-waste and microbes: mechanisms and risk assessment. In: Agri-Waste and microbes for production of sustainable nanomaterials. Elsevier, pp 535–561

74. Rajinipriya M, Nagalakshmaiah M, Robert M, Elkoun S (2018) Importance of agricultural and industrial waste in the field of nanocellulose and recent industrial developments of wood based nanocellulose: a review. ACS Sustain Chem Eng 6(3):2807–2828
75. Rani M, Shanker U (2022) Green nanomaterials: an overview. Green Functionalized Nanomaterials for Environmental Applications, pp 43–80
76. Reig CS, Lopez AD, Ramos MH, Ballester VAC (2014) Nanomaterials: a map for their selection in food packaging applications. Packag Technol Sci 27(11):839–866
77. Rezanezhad S, Nazanezhad N, Asadpur G (2015) Isolation of nanocellulose from rice waste via ultrasonication. لیگنوسلولز
78. Romanovski V (2021) Agricultural waste based-nanomaterials: green technology for water purification. In: Aquananotechnology. Elsevier, pp 577–595
79. Rónavári A, Igaz N, Adamecz DI, Szerencsés B, Molnar C, Kónya Z, Pfeiffer I, Kiricsi M (2021) Green silver and gold nanoparticles: biological synthesis approaches and potentials for biomedical applications. Molecules 26(4):844
80. Salam HA, Sivaraj R, Venckatesh R (2014) Green synthesis and characterization of zinc oxide nanoparticles from Ocimum basilicum L. var. purpurascens Benth.-Lamiaceae leaf extract. Mater Lett 131:16–18
81. Saratale RG, Karuppusamy I, Saratale GD, Pugazhendhi A, Kumar G, Park Y, Ghodake GS, Bharagava RN, Banu JR, Shin HS (2018) A comprehensive review on green nanomaterials using biological systems: recent perception and their future applications. Colloids Surf, B 170:20–35
82. Shahabi-Ghahafarrokhi I, Khodaiyan F, Mousavi M, Yousefi H (2015) Preparation and characterization of nanocellulose from beer industrial residues using acid hydrolysis/ultrasound. Fibers Polym 16(3):529–536
83. Shak KPY, Pang YL, Mah SK (2018) Nanocellulose: recent advances and its prospects in environmental remediation. Beilstein J Nanotechnol 9(1):2479–2498
84. Simegnaw AA, Malengier B, Rotich G, Tadesse MG, Van Langenhove L (2021) Review on the integration of microelectronics for E-Textile. Materials 14(17):5113
85. Singh P, Yadav SK, Kuddus M (2020) Green nanomaterials for wastewater treatment. In: Green nanomaterials. Springer, pp 227–242
86. Smirnov MA, Sokolova MP, Tolmachev DA, Vorobiov VK, Kasatkin IA, Smirnov NN, Klaving AV, Bobrova NV, Lukasheva NV, Yakimansky AV (2020) Green method for preparation of cellulose nanocrystals using deep eutectic solvent. Cellulose 27(8):4305–4317
87. Su D (2017) Advanced electron microscopy characterization of nanomaterials for catalysis. Green Energy Environ 2(2):70–83
88. Sun Y-P, Li X-q, Cao J, Zhang W-x, Wang HP (2006) Characterization of zero-valent iron nanoparticles. Adv Coll Interface Sci 120(1–3):47–56
89. Tadesse MG, Dumitrescu D, Loghin C, Chen Y, Wang L, Nierstrasz V (2018) 3D printing of NinjaFlex filament onto PEDOT: PSS-coated textile fabrics for electroluminescence applications. J Electron Mater 47:2082–2092
90. Tadesse MG, Kasaw E, Fentahun B, Loghin E, Lübben JF (2022) Banana peel and conductive polymers-based flexible supercapacitors for energy harvesting and storage. Energies 15(7):2471
91. Tadesse MG, Kasaw E, Lübben JF (2023) Valorization of banana peel using carbonization: potential use in the sustainable manufacturing of flexible supercapacitors. Micromachines 14(2):330
92. Tadesse MG, Loghin C, Chen Y, Wang L, Catalin D, Nierstrasz V (2017) Effect of liquid immersion of PEDOT: PSS-coated polyester fabric on surface resistance and wettability. Smart Mater Struct 26(6):065016
93. Tadesse MG, Lübben JF (2023) Review on hydrogel-based flexible supercapacitors for wearable applications. Gels 9(2):106
94. Tadesse MG, Mengistie DA, Chen Y, Wang L, Loghin C, Nierstrasz V (2019) Electrically conductive highly elastic polyamide/lycra fabric treated with PEDOT: PSS and polyurethane. JMatS 54(13):9591–9602

95. Tafete GA, Abera MK, Thothadri G (2022) Review on nanocellulose-based materials for supercapacitors applications. J Energy Storage 48:103938
96. Tang F, Wang X, Xu X, Li L (2010) Phytic acid doped nanoparticles for green anticorrosion coatings. Colloids Surf, A 369(1–3):101–105
97. Tiwari K, Singh R, Negi P, Dani R, Rawat A (2021) Application of nanomaterials in food packaging industry: a review. Mater Today: Proc 46:10652–10655
98. Trache D, Tarchoun AF, Derradji M, Hamidon TS, Masruchin N, Brosse N, Hussin MH (2020) Nanocellulose: from fundamentals to advanced applications. Front Chem 8:392
99. Verma R, Pathak S, Srivastava AK, Prawer S, Tomljenovic-Hanic S (2021) ZnO nanomaterials: green synthesis, toxicity evaluation and new insights in biomedical applications. J Alloys Compounds 876:160175
100. Wu J-H, Hu T-G, Wang H, Zong M-H, Wu H, Wen P (2022) Electrospinning of PLA nanofibers: recent advances and its potential application for food packaging. J Agric Food Chem 70(27):8207–8221
101. Yan S, Yin J, Yang Y, Dai Z, Ma J, Chen X (2007) Surface-grafted silica linked with L-lactic acid oligomer: a novel nanofiller to improve the performance of biodegradable poly (L-lactide). Polymer 48(6):1688–1694
102. Yanat M, Schroën K (2021) Preparation methods and applications of chitosan nanoparticles; with an outlook toward reinforcement of biodegradable packaging. React Funct Polym 161:104849
103. Yaragalla S, Rajendran R, Jose J, AlMaadeed MA, Kalarikkal N, Thomas S (2016) Preparation and characterization of green graphene using grape seed extract for bioapplications. Mater Sci Eng, C 65:345–353
104. Zamani A, Marjani AP, Mousavi Z (2019) Agricultural waste biomass-assisted nanostructures: synthesis and application. Green Process Synthesis 8(1):421–429
105. Zhang L, Jia X, Ai Y, Huang R, Qi W, He Z, Klemeš JJ, Su R (2022) Greener production of cellulose nanocrystals: an optimised design and life cycle assessment. J Clean Prod 345:131073
106. Zhang Y, Xu W, Wang X, Ni S, Rosqvist E, Smått J-H, Peltonen J, Hou Q, Qin M, Willför S (2019) From biomass to nanomaterials: a green procedure for preparation of holistic bamboo multifunctional nanocomposites based on formic acid rapid fractionation. ACS Sustain Chem Eng 7(7):6592–6600
107. Zhao Y, Huang Z, Chang W, Wei C, Feng X, Ma L, Qi X, Li Z (2017) Microwave-assisted solvothermal synthesis of hierarchical TiO2 microspheres for efficient electro-field-assisted-photocatalytic removal of tributyltin in tannery wastewater. Chemosphere 179:75–83
108. Zhou L, Jin J, Liu Z, Liang X, Shang C (2011) Adsorption of acid dyes from aqueous solutions by the ethylenediamine-modified magnetic chitosan nanoparticles. J Hazard Mater 185(2–3):1045–1052

Nanocellulose: Extraction, Mechanical Properties, and Applications

S. Aboul Hrouz, O. Amadine, S. Sair, K. Dânoun, Y. Essemlali, and M. Zahouily

Abstract Cellulose nanocrystals are considered as one of the most attractive renewable reinforcements due to their high crystallinity, high Young's modulus, and large surface area, which provides them excellent mechanical and physicochemical properties. Herein, we focus on the different techniques used for the extraction of nanocelluloses from lignocellulosic biomass. Furthermore, their application in different fields as well as their effect on the mechanical properties of nanocomposites have been highlighted.

Keywords Cellulose nanocrystals · Nanocellulose extraction · Biomass · Nanocellulose application · Mechanical properties of nanocomposites

1 Introduction

Advances in nanotechnology research are driving scientists to develop renewable bio-based nanomaterials with significantly reduced environmental impact to replace synthetic petroleum-based products. These bio-based materials are promoted due to their low cost, availability, eco-friendly, renewable, and non-toxic properties. Lignocellulosic biomass is a complex biomaterial that includes cellulose, hemicellulose, and lignin.

Cellulose is one of the most abundant bio-based polymers on the planet. Cellulose fibers are known to possess outstanding physical and mechanical properties due to the presence of various hydroxyl groups and the strong hydrogen bonding network.

S. Aboul Hrouz (✉) · O. Amadine · S. Sair · K. Dânoun · Y. Essemlali · M. Zahouily
MASCIR Foundation-Mohammed VI Polytechnic University, Lot 660-Hay Moulay Rachid, 43150 Ben Guerir, Morocco
e-mail: s.aboulhrouz@mascir.ma

M. Zahouily
e-mail: m.zahouily@mascir.ma

M. Zahouily
Laboratory of Materials, Catalysis and Valorization of Natural Resources, University of Hassan II Casablanca, 20000 Casablanca, Morocco

© The Author(s), under exclusive license to Springer Nature Singapore Pte Ltd. 2023
H. Abdellaoui et al. (eds.), *Mechanics of Nanomaterials and Polymer Nanocomposites*, Smart Nanomaterials Technology,
https://doi.org/10.1007/978-981-99-2352-6_6

They can play a crucial role in nanotechnology research due to their low density, high tensile strength, high modulus of elasticity, biocompatibility, high crystallinity, and surface rich in hydroxyl groups [1, 2]. Cellulosic nanomaterials present a crystalline structure with rigid rods of several hundred nanometers in length and less than 100 nm in diameter, which can be extracted using different techniques from natural cellulose [3, 4]. The most frequent nomination of nanocelluloses is CNC, although their distribution varies according to the different types of nanocrystals studied, such as nanofibers, nanowhiskers, nanoparticles, and bacterial nanocellulose. Different fields are using these nanomaterials, including the biomedical sector, nanocomposites, energy, and the environment [5]. These nanocelluloses were first introduced in the end of the 1970s to describe a gel produced by homogenizing wood pulp at high temperature through a Gaulin-type milk homogenizer [6]. It has been reported that CNCs present remarkable properties compared to CNFs, owing to their size which gives them a high specific surface, suggesting more important interactions at the particle/matrix interface. Moreover, the high crystallinity of CNCs promotes their rigidity and their permeability to water and gases. These nanoparticles can also be applied as multipurpose reinforcement in automotive, packaging, or construction industries [7].

In this chapter, we discuss the potential application of nanocelluloses and their effect on the mechanical properties of materials. First, a description of the isolation of nanocelluloses from different raw materials and using different methods will be discussed. Subsequently, we discuss the addition of nanocelluloses as a strategy to produce nanocomposites with improved performance by citing different applications. Finally, report the effect of these nanocelluloses as reinforcement on the mechanical properties of the materials.

2 Extraction Method

Cellulose comprises both amorphous and crystalline phases. Extraction methods are designed for the separation of the amorphous part from the crystalline fibers of cellulose. It is reported that the breaking of this amorphous phase is relatively easy via different methods compared to the crystalline phase due to the number of hydrogen bonds present in the backbone [5]. Many techniques have been employed to extract highly crystalline nanocelluloses from cellulose, including chemical treatments, mechanical methods, enzymatic methods, and method combining these techniques (Fig. 1). The final morphology and properties of nanocelluloses depend on the raw materials used and the process for production (Table 1).

Excluding cotton, which has the most purified cellulosic fibers, obtaining pure cellulose from other plant fibers requires several pretreatments to eliminate non-cellulosic substances (extractable, pectin, wax…) in combination with lignin and hemicelluloses, which significantly impact the properties of the fibers, and to obtain highly purified fibers. These pretreatments improve the reactivity of the fibers by promoting the accessibility of hydroxyl groups, increasing the crystallinity and

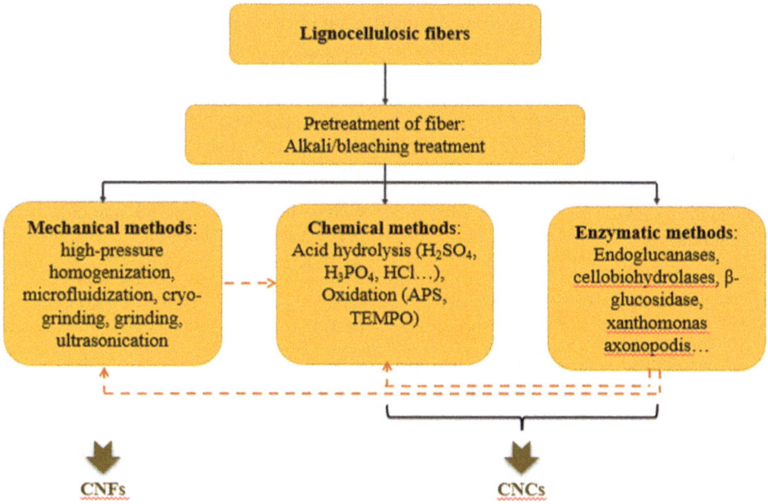

Fig. 1 Methods of extraction of nanocelluloses from lignocellulosic fibers

internal surface area, and breaking the hydrogen bonds of cellulose fibers [4, 8]. Various protocols have been studied in the literature with variable efficiencies. Among them, solvent treatments (toluene/ethanol mixture) have been examined for the removal of the extractables which are generally present on the surface of the fibers. It was mentioned that the hydrogen peroxide provides the lignin oxidation and solubilization in water [9]. For this purpose, Malladi Rajinipriya et al. 2017 reported the purification of flax, hemp and milkweed fibers in a single step using 7.5% hydrogen peroxide at 65 °C without stirring for 4 h. The results obtained showed efficient removal of lignin, significant improvement in thermal stability and morphology, and the crystallinity of these fibers was not affected by the treatment via hydrogen peroxide [9]. Despite of these pre-treatments, many impurities are still present in the fibers. For this reason, most of the researchers have used a series of treatments in order to purify the cellulosic fibers and allow a specific extraction of the compounds according to their accessibility and reactivity. Previous studies have shown that alkali and bleaching are effective for the removal of lignin and hemicelluloses. In addition to the removal of non-cellulosic compounds, these processes promote the formation of cellulosic fibers with a more uniform width, less rough surface, and uniform microfibrils structure [10]. Jessica de Aguiar et al. 2020 subjected sugarcane bagasse to an alkaline treatment using 10% of NaOH, followed by bleaching via a mixture of sodium chlorite and acetate buffer solution. The results showed 89% and 65% removal of hemicelluloses and lignin, respectively. Moreover, the treatment resulted in cellulosic microfibrils with even smoother surfaces, and high α-cellulose content (Fig. 2). Previous studies have shown that the combination of alkaline pretreatment and sodium chlorite bleaching in the presence of acid is able to remove lignin and hemicelluloses [11]. Jiyoo Baek et al. 2018 treated hardwood kraft pulp with 6% NaOH followed by a mixture of 1.25% $NaClO_2$ and glacial acetic acid [12].

Table 1 The effect of source and extraction method on the dimensions of nanocelluloses

Source	Extraction method	Dimensions of nanocelluloses	Refs.
Raw cotton linter	Acid hydrolysis (H_2SO_4 60%)	CNCs: length of 177 nm, width of 12 nm	Morais et al. [69]
Dunchi fiber *Sesbania bispinosa* stalks	Acid hydrolysis (H_2SO_4 64%)	CNCs: length of 202.87 nm and width of 20.67 nm	Khan et al. [70]
Residues of peach palm	Colloidal grinder	CNFs: average widths of 100 nm	Franco et al. [71]
Tomato plant residue	Acid hydrolysis (citric acid (3 M) and HCl (6 M))	CNCs: diameter of 4.7 nm and length of 514 nm	Kassab et al. [45]
Bamboo pulp	Acid hydrolysis (H_2SO_4 64%)	CNCs: from 500 to 600 nm in length and 70 to 98 nm in diameter	Borkotoky et al. [72]
Rice straw	Oxidation (TEMPO-mediated oxidation)	CNFs: 1.7 nm of wide	Jiang and Lo [73]
Rice straw	High-speed blending	CNFs: between 2.7 and 8.5 mm wide and 100 to a few micrometers long	Jiang and Lo [73]
Rice straw	Acid hydrolysis (H_2SO_4 64%)	CNCs: 4.7 nm wide and 143 nm long	Jiang and Lo [73]
Balsa fibers	Oxidation (ammonium persulfate)	CNCs: diameter ranging 1.77–10.02 nm and long up to 100 nm	Marwanto et al. [74]
Kapok fibers	Oxidation (ammonium persulfate)	CNCs: diameter ranging 1.25–11.87 and long up to 100 nm	Marwanto et al. [74]
Miscanthus fibers	Acid hydrolysis (H_2SO_4 64%)	CNCs: diameter of 18 nm and length of 680 nm	El Achaby et al. [75]

In another study, coconut husk fiber was delignified with an organosolv process, followed by alkaline bleaching using 5% hydrogen peroxide and 4% NaOH at 50 °C for 90 min [13].

Alternatively, steam explosion is commonly used as environmentally friendly pretreatments for cellulosic fiber extraction [14, 15]. Francisco Pereira Marques et al. 2020 reported a pretreatment of oil palm mesocarp fiber using steam explosion. The results obtained showed high quality of the cellulose pulp, with lower hemicellulose content, higher α-cellulose content, higher crystallinity, and good

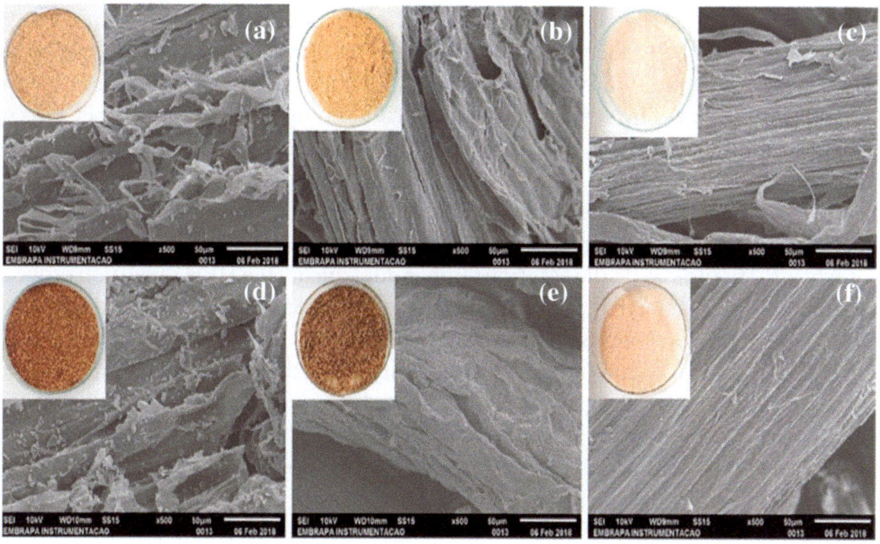

Fig. 2 Scanning electron microscopy micrographs and visual aspect images of **a** raw sugarcane bagasse, **b** alkali treated sugarcane bagasse, **c** bleached sugarcane bagasse, **d** raw sugarcane straw, **e** alkali treated sugarcane straw, and **f** bleached sugarcane straw. "Reprinted with permission from [10] Copyright 2020 American Chemical Society"

thermal stability [14]. The combination of these pretreatments resulted in efficient fiber refining, removing most of hemicelluloses and destroying the lignin structure [16]. It is reported that cellulose nanocrystals were initially extracted from natural fibers by Mukherjee and Woods in 1953 [8]. Nanocelluloses have been isolated from various sources including wood pulp, tree leaves [17], fruit peels [18], plant fibers (hemp, sisal, flax, ...) [19], marine species [20], fruit shells [21], or even agricultural products [22].

2.1 Mechanical Method

The mechanical treatments used to obtain CNCs or CNFs are vigorous fibrillation processes that allow the destruction of the fiber network in order to isolate them by disintegrating the cellulose along their longitudinal axis. The resulting CNFs after treatment have usually a final diameter ranging from 10 nm to hundreds of nanometers and a length in the micrometer scale [23]. The most commonly used extraction methods to isolate nanocellulose comprise high-pressure homogenizers [24, 25], microfluidization [26, 27], cryo-grinding [28], as well as ultrasonic processing [29, 30]. These methods provide high efficiency, howbeit, their implementation usually requires very high energy. Among these many treatments, high-pressure homogenization can be considered the most exploited mechanical method for the production of cellulose nanofibers, owing to its high efficiency and good product homogeneity [31]. It was firstly introduced in 1983 [32]. It is an efficient biomass refining technique, which generates forces of turbulence, cavitation, shear, and impact by the rapid change of pressure leading to microfibrillation and reduction of fiber size to nanoscale [33, 34]. Nae-Man Park et al. 2019 evaluated the properties of nanocellulose fibrils from cotton using high-pressure homogenization (HPH) by controlling the process temperature. The obtained CNCs have rod-like shapes with a size distribution of 4–14 nm for width and 60–320 nm for length. The crystallinity increases linearly up to twofold with the solution temperature during the HPH repetition process compared to the starting material [35]. Huiqiang Wang et al. 2019 extracted cellulose nanofibers from wood and bamboo waste pretreated by high-pressure homogenization (HPH) in homogeneous medium. The results showed that the crystallinity indices increased with good suspension stability. Nevertheless, the thermal stability of the nanocellulose was lower than that of the raw fibers [24]. It has been reported that the morphology and dimensions of CNFs mainly depend on the number of passes and pressure used during the fibrillation process. In the study conducted by Xianxing Luo and Xiwen Wang 2017, CNF fibers obtained from bleached OPT pulp and pretreated with NaOH/urea underwent high-pressure homogenization at 1000 bar for 1 h. Evaluation of the final diameter of these CNFs showed a size between 10 and 100 nm with a few hundred nanometers to several micrometers in length. However, the crystallinity index of the CNFs was quite low (59.35%) due to the high shear force exerted during the homogenization process which destroys the crystalline regions of the cellulose [25].

Ultrasonication is another technique used for the preparation of CNFs by means of hydrodynamic forces [36]. This technique is usually used after pretreatment of the fibers by TEMPO or other oxidations [4]. In order to produce nanofibers from oil palm mesocarp, TAT Yasim-Anuar et al. 2017 treated the fibers with $NaClO_2$ and KOH to remove non-cellulosic components, after the treated cellulose fibers were subjected to ultrasonication in an ice bath for 3, 6, 9, and 12 h. The results obtained showed that the diameter of the fibers ranged from 40 to 200 nm by increasing the sonication time from 3 to 9 h. This proves that sonication can break the relatively weak interfaces between cellulose fibers, which are bound to each other by hydrogen bonds. Thus, it can disintegrate cellulose microfibers into nanocellulose. Furthermore, the results obtained show that the applied acoustic cavitation reduces the crystallinity index and thermal degradation temperature [29]. Some other investigators have combined this technique with sulfuric acid hydrolysis for efficient production of CNFs. Wei Li et al. 2011 prepared cellulose nanofibers from bleached kraft pulp (BSKP) using hydrolysis with 64% sulfuric acid for 1 h followed by ultrasonication at 45 °C for 1.5 h. SEM and TEM analysis revealed rod-shaped CNF particles with a diameter of 10–20 nm and length of 90 nm. Thermal stability evaluation showed that the nanofibers are less stable than BSKP fibers due to the smaller size and larger number of free chain ends. In addition, the resulting CNFs possess higher crystallinity (82%) of the BSKP fibers (74%). Thus, the researchers deduced that ultrasonication-induced cellulose folding, surface erosion, and external fibrillation of BSKP are beneficial for acid penetration and provide more reactive sites [30]. In contrast, the microfluidizer is a technique

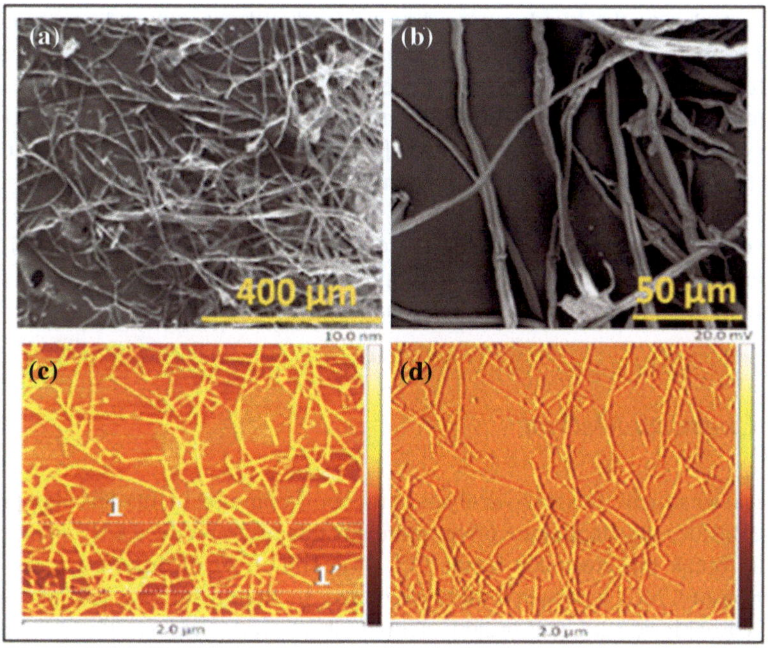

Fig. 3 SEM images of (**a**, **b**) CNFs extracted using TEMPO-mediated oxidation CMF samples, and AFM images at two magnifications (**c**, **d**) of extracted CNFs [26]

similar to HPH that can be used to produce CNFs. This technique applies high shear rates and impacts forces against the colliding flows and channel walls, which defibrates the fibers to nanoscale dimensions [23, 37]. The resulting nanofibers have homogeneous diameters below 100 nm. In this regard, Zineb Kassab et al. 2020 used microfluidization for the production of *Juncs* nanofibers with a diameter of 2.5 nm. The fibers were treated with oxidation-TEMPO to facilitate fiber separation and mechanical disintegration of CNFs by driving negatively charged carboxyl groups on the surface of microfibrils, which promotes their fibrillation through the repulsion mechanism. Then, the suspension of T-CMF microfibrils was passed several times through a high-pressure microfluidizer. The results obtained showed that the small diameter and negatively charged surfaces of CNFs help them to disperse in a variety of media (Fig. 3) [26].

Moreover, it is possible to combine grinding and the microfluidization process to produce nanofibers. Kenly Araya-Chavarría et al. 2021 produced nanofibers from TEMPO oxidation-treated pineapple stubble by combining grinding and microfluidization. Grinding is another common approach for the production of nanocellulose. The concept of this technique is to destroy plant cell wall structures by generating high shear forces to produce individual nanofibrils [23]. The resulting fibers have fibril lengths and widths between 481 and 746 nm and between 16 and 48 nm, respectively [27].

Indeed, due to the highly ordered hydrogen bond network, a huge amount of energy is required to extract cellulose nanofibers by mechanical treatments [34]. Therefore, many pretreatment techniques have been proposed to reduce energy consumption, namely: enzymatic hydrolysis [38], TEMPO-mediated oxidation [39], carboxymethylation [40], quaternization [41]. The negative or positive charge introduced on the surface of cellulose fibers after pretreatment could reduce the energy requirement for fibrillation and improve the colloidal stability of final CNFs [42]. However, the absence of the strong hydrolysis treatment may negatively affect the mechanical, thermal, and structural properties of the nanofibers, as there can be residual lignin and hemicellulose in the nanocellulose structure.

2.2 *Chemical Method*

Chemical treatment by acid hydrolysis is the most common method used to produce CNCs. Besides, this technique offers more advantages to produce CNCs since it can partially break the glucoside bonds [43, 44]. The effect of different acids on the extraction of CNCs has been widely evaluated, including sulfuric acid, phosphoric acid, hydrochloric acid, and organic acids such as acetic acid, oxalic acid, etc. [12, 36, 45–47]. Indeed, the hydrolysis by the acid causes the rupture and the destruction of the most accessible glycosidic bonds in the amorphous regions of the microfibrils which leads to the isolation of the fragments formed by the crystalline parts with a reduced degree of polymerization [48, 49]. This results in a stable colloidal dispersion of cellulose nanocrystals with high crystallinity due to the repulsive forces induced by the electrical double layers [37]. Usually, a neutralization with deionized water is performed to prevent the degradation of the nanocrystals by the residual acid [50]. It has been noted that the resulting CNCs after acid hydrolysis are more reactive due to the abundance of hydroxyl groups [37]. The type of acid, its concentration, the time of reaction, and the temperature are parameters that greatly affect the properties of the obtained nanocrystals [50].

Hydrolysis with sulfuric acid has been widely used by researchers. The first isolation of CNCs using sulfuric acid was reported by Ranby et al. 1949 [4]. H_2SO_4 can react with the hydroxyl groups of cellulose via an esterification process, allowing the grafting of anionic sulfate ester groups [51]. It has to be mentioned that CNCs produced by hydrolysis via H_2SO_4 have less tendency to aggregate compared to nanocellulose hydrolyzed with hydrochloric acid with a neutral surface [49]. The most commonly adopted hydrolysis protocol with a concentration of sulfuric acid ranging from 60 to 65% by weight depending on the source of the raw materials and the pretreatment conditions. The treatment temperature varies from 45 to 60 °C and the duration from 1 to 4 h [8, 49]. CNCs obtained in this way are generally characterized by a relative width of 3–50 nm, a length varying between 100 nm and several micrometers, a high crystallinity varying from 54 to 88%, and a low thermal stability [52]. It is reported that at low acid concentration, H^+ ions are insufficient to effectively penetrate the internal structure of cellulose fibers, and therefore no hydrolysis reaction occurs. When the acid concentration is optimal, the H^+ ions dissolve the amorphous parts of the cellulose to release the individual nanocrystals [23]. Therefore, a high concentration of acid can severely degrade the cellulose by destroying the amorphous and crystalline regions of the cellulose to form water-soluble oligomers [8, 23, 49]. Malladi Rajinipriya et al. 2018 extracted CNCs from carrot pulp using a hydrolysis process with 64 wt% sulfuric acid solutions for 1.30 h at 50 °C. First, the raw carrot fibers were pretreated with hot water and bleached with hydrogen peroxide. The resulting CNCs, in the form of rods, had a fiber length between 84 and 590 nm and a diameter ranging from 10 to 36 nm, a crystallinity of 78%. Furthermore, the degradation of CNCs started between 160 and 200 °C due to the sulfate groups present in the CNCs prepared by sulfuric acid hydrolysis [53]. Unlike the CNCs prepared from bamboo by Masrat Rasheed et al. 2020 showed higher thermal stability (between 272 and 378 °C). These authors also found CNCs with higher crystallinity (86.96%) and rod-like morphology with length ranging from 200 to 500 nm, and the crystal has a diameter less than 10 nm [36]. However, the acid hydrolysis method has some drawbacks such as the time required for hydrolysis and the presence of sulfate groups that can lead to dehydration and decomposition reaction of CNCs, which reduces their thermal stability [23, 37]. To overcome these drawbacks, researchers have combined mechanical treatments such as ultrasound, microwave, and homogenization with acid hydrolysis [54, 55]. Successfully, Azrina et al. 2017 produced spherical CNCs from EFB paste by ultrasonically assisted sulfuric acid hydrolysis. The unique spherical morphology of the CNCs could be attributed to the ultrasonic treatment used during the acid hydrolysis process, with an average diameter of 30–40 nm and a crystallinity of 80%. In addition, the obtained CNCs exhibited a higher thermal degradation temperature of about 362.17 °C. The authors showed that the use of ultrasound in the hydrolysis reaction promotes the penetration of acid molecules into the cellulose structure, thus increasing the rate of the hydrolysis reaction. The cavitation generated by ultrasound produces high mechanical forces during hydrolysis to generate cellulose nanocrystals with spherical shapes [54]. Other acids are also able to produce crystalline CNCs including hydrochloric acid, phosphoric acid, nitric acid, citric acid, or oxalic acid. Bondancia et al. 2020 have used 64% citric acid for 1.5–6 h at a temperature of 120 °C to produce nanocrystals and nanofibrils from Eucalyptus cellulose kraft pulp. The results indicated that hydrolysis was incomplete when a shorter reaction time was used (1.5 h). After 6 h of hydrolysis, the length of the CNCs was approximately 215 $\pm$ 89 nm, and the diameters of the structures obtained were between 9 and 10 nm. The crystallinity increased from 63% for the raw material to 83% for the CNCs. The authors found a stable colloidal suspension that avoids cellulose aggregation due to the carboxyl groups present on the nanocellulose surface. This stability may vary depending on the carboxylate content [56]. Cellulose nanocrystals with a diameter of 16 nm and a crystallinity of 88.6%, isolated from commercial microcrystalline cellulose, were obtained by hydrolysis with hydrochloric acid under hydrothermal conditions. Neutralization of the acid was carried out by ammonia, which improved the thermal stability and stability of aqueous CNC suspensions [57]. Furthermore, CNCs with dimensions of 7–8 nm and lengths of 100–200 nm were obtained from Whatman #1 filter paper via hydrolysis with 1.5–4 M boric acid followed by ultrasonication [55]. In another research work,

Jiyoo Baek et al. 2018 extracted CNCs from hardwood Kraft pulp using phosphoric acid. The hydrolysis reaction was performed with 75% H_3PO_4 at 70 °C for 5 h. The average length of the CNCs obtained was about 200 nm. However, the authors deduce that the stability of colloidal suspension is less compared to the CNC extracted by H_2SO_4 [12].

Other than acids, oxidation with ammonium persulfate (APS) can be considered as promising approach for the isolation of CNCs. For this purpose, Kar Yin Goh et al., 2015 evaluated the properties of CNCs extracted from the oxidation of oil palm empty fruit bunches by APS and by H_2SO_4 hydrolysis. They found that the diameter of the CNFs hydrolyzed with sulfuric acid ranged from 8 to 26 nm while the APS-derived nanofibers had a diameter of 23 nm. Even though, the yield and crystallinity of the CNFs derived from APS oxidation in order of 40% and 76%, respectively, were higher than those obtained by acid hydrolysis (25% yield and 72% crystallinity) [58].

Oxidation by the 2,2,6,6-tetramethylpiperidine-1-oxyl radical (TEMPO) is a chemical modification used to provide the surface of CNCs with negative charges (sodium carboxylate groups), which create interfibrillar repulsive forces between the fibrils, resulting in simple and rapid individualization [49, 59]. This leads to the improvement of CNC stability in aqueous solutions. In addition to the treatment by acid hydrolysis, this technique is usually used for the extraction of nanocelluloses followed by mechanical defibrillation. It has been claimed that only the C6 hydroxyl groups present on the surface of cellulose microfibrils are converted to sodium carboxylate groups [8]. The research study conducted by Kumar et al. 2020 revealed that oxidation of cellulose nanofibril from bleached bisulfite pulp induced modification of the C6 carbon, which resulted in the presence of 24% carboxylate groups in the obtained CNFs, while 74% of the C6 carbons remained intact and did not undergo any modification [60]. In another research work, Xiuxuan Sun et al. 2015 compared the morphological and structural properties of TOCNs extracted by acid hydrolysis and TEMPO followed by HPH. The results showed that TOCNs extracted by TEMPO had a broader distribution of crystal length, while those obtained by H_2SO_4 hydrolysis had a much larger distribution of crystal diameter. The authors deduced that these differences between the length and morphology of the two types of NOCs are mainly due to different reaction mechanisms. It is stated that during the TEMPO oxidation process, the glucosyl units are oxidized when TEMPO molecules penetrate the cellulose microfibrils. Whereas sulfuric acid acts on the cellulose molecules by decreasing the degree of polymerization (DP) [39].

2.3 *Enzymatic Methods*

Other than chemical and mechanical treatment, the extraction of nanocelluloses by enzymatic hydrolysis is considered the most environmentally friendly with reduced energy consumption due to the availability of nanocelluloses under mild hydrolysis conditions [61–63]. Nanocelluloses extracted by biological treatment are characterized by a ribbon shape with length less than 100 nm and width of 2–4 nm, as well as high purity, high flexibility, high mechanical property, and high-water absorption capacity [5, 50]. Moreover, it enhances the biocompatibility of CNCs and their scope in biomedical applications and pharmaceuticals. The most used enzyme in enzymatic hydrolysis is Cellulase, which contains three active compounds: *endoglucanases* (EG), *cellobiohydrolases* (CBH), and *β-glucosidase* (GB) [23, 37, 44]. *Endoglucanases* affect the amorphous part of cellulose by cleaving the internal β-1,4-glycosidic bond to transform oligosaccharides of different lengths into smaller chains [48]. The *cellobiohydrolases* dissociate mainly the terminal glycosidic bond on the reducing or non-reducing ends of the cellulose, while *β-glucosidase* is used to hydrolyze cellulose into nanocellulose or small molecular units of glucose [64]. Typically, enzymatic processing is performed by a combination of enzymes that have highly synergistic effects to break down cellulose. Enzymatically produced CNFs from cellulosic wood fibers have been shown to have better structure, higher aspect ratio, and higher average molecular weight than those obtained by acid hydrolysis [65]. Other researchers have used *Xanthomonas axonopodis* enzymatic lysate from orange pulp to produce 60% crystalline CNFs [66]. In addition, the study conducted by Chen et al. 2019 showed the successful production of nanocelluloses with diameter ranging from 30 to 45 nm and length from 250 to 900 using the enzyme cellulase (Ningxia Xiasheng) from cotton pulp [67]. Priyanka Kumari et al. 2019 have explored a mild operating condition of enzymatic to prepare CNFs from lemongrass. The fibers were first exposed to steam explosion with 2% NaOH (fiber/solution ratio 1:10 g/ml) in an autoclave for 3 h. Then, the cellulose residues were delignified with 1% acidified sodium chlorite solution (15 ml/g fiber) at 70 °C for 3 h to remove non-cellulosic parts. Then, the enzyme *Viscozyme® L* was used to prepare the CNFs. The achieved results showed that the enzymatic approach led to a higher yield of 57% of CNFs, with moderate stability. In addition, the obtained CNFs had a web-like structure with long entangled cellulosic segments, as well as a low crystallinity in the order of 48.9%, compared to cellulose (66.51%). This was explained by the fact that the surface of the fibers detaches after the sonication process, which leads to a loss of its crystalline behavior [64]. In the research performed by Rossi et al. 2020, three different recombinant enzymes, an endoglucanase, a xylanase, and a lytic polysaccharide monooxygenase, were combined to improve cellulose fibrillation and produce CNFs from sugarcane bagasse (SCB). The results of the evaluation of the properties of these CNFs were compared to CNFs obtained by TEMPO-mediated oxidation. The authors have found that the CNFs obtained by enzymatic treatment are longer and more thermally stable. Furthermore, they showed that enzyme treatment and TEMPO-mediated oxidation can be used as a fine treatment before the sonication step to produce cellulosic CNFs with advanced properties [68].

3 Applications

Due to the nanometric size as well as the good mechanical properties (tensile strength 2–3 GPa, modulus of elasticity 110–220 GPa), optical activity, non-toxicity, better thermal conductivity, higher thermal stability, high specific surface area (150–300 m^2g^{-1}) [49], high crystallinity (>85%), biodegradability, CNCs provides significant potential in the development of multifunctional materials for a myriad of applications. Several research studies have shown that these nanocelluloses can be applied in different sectors (Fig. 4), including the plastics industry [76], aeronautics [77], food [12], paper, biomedical [78, 79], wastewater treatment [80], automotive [81], cosmetics as well as construction. Among these various fields of application, we have chosen to highlight some examples of nanocelluloses applied as nanocomposites for food packaging, cosmetics, water treatment, and biomedical applications.

The use of CNCs as a reinforcement in various polymer matrices is a fast-expanding field. This growth stems from the possibility of preparing nanocomposites with improved properties while maintaining the biocompatibility and biodegradability of the matrix. In this case, Yasmein Hussein et al. 2021 developed a curcumin-loaded polyvinyl alcohol/cellulose nanocrystal (PVA/CNCs) membrane as a localized delivery system for breast/liver cancer. The membranes were prepared by the solution casting method using citric acid as a crosslinker. The results revealed enhanced anticancer properties of the membrane selectively, due to its cytotoxic effect on MCF-7 and Huh-7 cells, without affecting the percentage of viability of HFB-4 cells which was above 80%. Thus, these membranes could be promising anti-infective biomaterials for breast and liver cancer wound healing [82]. For instance, Malladi Rajinipriya et al. 2018 made a homogeneous and transparent nanocellulosic film of NFC and NCCs were also prepared by simple casting evaporation. These films

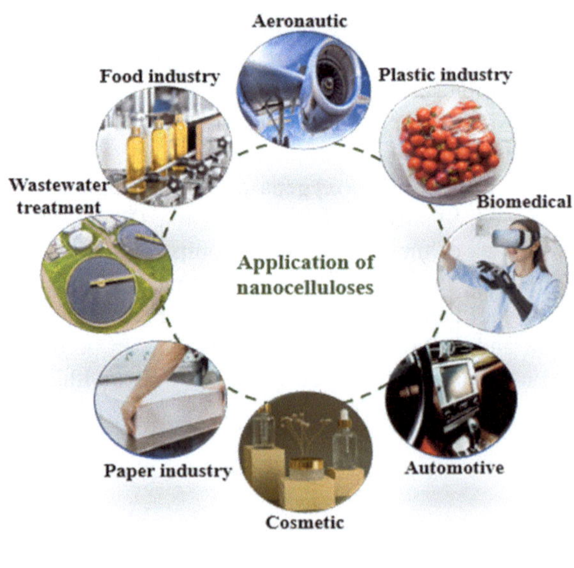

Fig. 4 Application of nanocelluloses in various fields

showed significant improvement in transparency and homogeneity with increasing milling time. However, the mechanical properties of the carrot CNFs films were low compared to the other investigations. This could be explained by the different preparation methods of the nanocellulosic films [53].

Nanocelluloses, including nanofibers and cellulose nanocrystals isolated from natural resources, have been widely used for the last decade in the biomedical field, as tissue engineering, implants, drug delivery systems, cardiovascular devices, and wound healing [83, 84]. In the research study conducted by Fatima A. Husseinet al. 2021, the authors succeeded in synthesizing a new organic material by the reaction of CNC with the antibiotic "cefixime". This nanocomposite showed a good biological effect against three types of bacteria (*Escherichia coli, Staphylococcus epidermidis, and Klebsilla sp*) [85]. In addition, nanocelluloses are used in the treatment of chronic wounds, the preparation of hydrogels and synthetic active extracellular matrices and modern wound dressings. These nanomaterials promoting tissue regeneration are developed to guide cell growth and differentiation. In a research work investigating conducted by Ruut Kummala et al., 2020, the effect of surface properties of four types of nanocelluloses (CMF, CNC, oxidized CNF-TEMPO) on the behavior of human dermal fibroblasts (HDF) was performed. The authors deduced that the highest viability and the most stable adhesion were observed on the anionic CNFs coating with a surface charge of 1.14 mmol/g. In addition, the Young's modulus and morphology of the films were significantly closer to the characteristics of soft tissue, endothelial, muscle, and cartilage, than traditional plastic plates, creating a more natural mechanical environment for soft tissue [86].

Wastewater purification and desalination of seawater to produce freshwater is one of the most important topics in the fields of engineering and environmental sciences. For this reason, many research studies have been focused on the development of new materials based on nanocelluloses, which are able to treat water efficiently [80, 86, 87]. For example, Langming Bai et al. 2018 incorporated cellulose nanocrystals (CNCs) into a polyamide layer to prepare novel thin film composite nanofiltration membranes (CNC-TFC-Ms). The CNC-TFC-Ms exhibited successful removal of anionic and cationic dyes due to their reduced negative charge and smaller pore size than the control membrane (without CNCs). Furthermore, CNC-TFC-Ms showed high rejection performance for divalent salts with rejection rates of more than 98.0% and 97.5% for Na_2SO_4 and $MgSO_4$, respectively. Therefore, the authors inferred that the permeate flux and NaCl rejection were simultaneously improved when the CNCs content increased, demonstrating that CNCs-TFC-Ms can overcome the trade-off limitation [88].

Many researchers have studied the replacement of packaging and food additives based on synthetic materials by biodegradable matrices reinforced with nanocelluloses to improve their physicochemical and mechanical properties [88–90]. V. A. Barbash et al. 2022 food packaging paper reinforced with nanocellulose extracted from hemp. The results showed that the addition of 2% hemp nanocellulose improved the mechanical properties of the paper by exceeding 40% of the breaking strength requirements of the higher-grade paper and increasing the breaking length by 42% compared to paper without chemical additives [91]. In other work, nanocellulose extracted from cotton linter was used as a reinforcement to improve the functionalities of starch-based composite films. The authors used a response surface methodology to investigate and optimize the effect of film composition on the tensile strength (TS), elongation at break (EB), burst strength (BS), and water vapor permeability (WVP) of the films. The results obtained revealed a significant improvement in the mechanical and barrier properties of the nanocellulose-reinforced films. In addition, the films also exhibited high efficiency in preserving the quality of the edible oil for more than 3 months under ambient storage conditions. These results prove the promising potential of these films as an alternative source for edible oil storage [92].

On the other hand, in cosmetic domains, the use of biopolymers has attracted increasing attention, and it will become more and more relevant due to their biocompatibility. As emulsifiers used for emulsions, cellulose, and its derivatives have been favored by researchers. In the study by Jiyoo Baek et al. 2019, phosphorylated-CNC/modified-chitosan nanocomplexes for the stabilization of Pickering emulsions were prepared. The CNCs were extracted by hydrolysis with phosphoric acid because it is not harmful and can be used in food processing. The results showed that the Pickering emulsion containing phosphorylated CNCs is more stable than those containing only tripolyphosphate (TPP) crosslinked chitosan. Therefore, these new nanostructures provide a new approach to stabilize emulsions for functional food formulation [12]. Similar studies have focused on the ability of nanocelluloses to stabilize Pickering emulsions. The authors have found that the stability of the emulsions can be controlled by the surface charge density of the CNCs since the repulsive forces between the particles prevent droplet coalescence [92–94].

4 Mechanical Properties

It is widely recognized that CNCs act as an effective reinforcement due to their outstanding mechanical properties, which are significantly improved over those of native fibers due to their more uniform morphological structure [5]. Various parameters impact the mechanical properties of nanocelluloses including isolation method, source, chemical composition, morphology, surface modification, and so on [77, 95]. Another factor that affects the CNC mechanical properties is agglomeration. This phenomenon is proven to influence the dissipation of external stresses by the particle-polymer interaction, also induces stress concentration points in the matrix, causing a decrease in the value of the elongation at the break of the matrices [77]. It is reported that the average Young's modulus of CNCs is around 150 GPa, and the tensile strength is between 7.5 and 15 GPa for higher aspect ratios. While CNFs showed tensile modulus values below 1 GPa [50, 96]. The relatively lower modulus of CNFs is due to the presence of amorphous parts in the structure [50]. The mechanical properties of nanocelluloses as reinforcement in nanocomposites are widely evaluated by the scientific community (Table 2). For instance, Mohammad Asad et al. 2018 demonstrated the reinforcing properties of CNCs by preparing nanocomposite films with polyvinyl alcohol using the casting method. The improvement in mechanical properties was dependent on the content of CNCs. It was observed that the tensile strength and Young's modulus were improved by 122% and 291%, respectively, when the CNCs content increased to 4% (w/w). This improvement in mechanical properties is likely due to the formation of a network structure above the percolation threshold resulting from the increased hydrogen bonds' formation between these materials owing to the large surface area of CNCs. However, the elongation at break decreased continuously with the addition of CNCs, reaching a lower value at 4% (about 3.5%) [97]. In the same tendency, Mohamed S. Hasanin and Ahmed M. Youssef 2022 prepared bioactive and biodegradable packaging films based on poly (butylene adipate-co-terephthalate (PBAT) containing copper oxide nanoparticles (CuO-NPs) and loaded with enzymatically prepared nanocellulose (ENC) to improve the mechanical properties of the obtained films. The fabricated bioactive films exhibited good mechanical properties compared to pure PABT, and the tensile strength was increased by increasing the addition of ENC to 3%. Nevertheless, the elongation and Young's modulus were going in the opposite trend [98]. In another study constructed by Milena Martelli-Tosi et al. 2018, they compared the effect of CNCs, and CNFs extracted from mercerized soybean straw on the properties of soy protein isolate (SPI) films. Incorporation of 5% (g/100 g) of CNFs or CNCs improved the mechanical properties of the resulting film. The tensile strength (TS) increased from 6.1 $\pm$ 0.6 MPa to 9.0 for CNFs and 8.4 MPa for CNCs. The Young's modulus followed the same trend with an increase from 459 MPa to 575 and 537 MPa for CNFs and CNCs, respectively. However, the flexibility of the films decreased with the incorporation of the fillers. The authors recorded a significant loss in elongation at break upon the addition of CNCs while the addition of CNFs caused a lesser effect on the flexibility of the nanocomposites. This was explained by the fact that CNFs are longer in size and display an entangled structure which has not affected the elongation at break in the same manner as CNCs [66]. On the other hand, Amir Asadi et al. 2016 evaluated the mechanical properties of short glass fiber/epoxy composites containing cellulose nanocrystals (CNCs). These nanocomposites were fabricated using the sheet molding compound (SMC) fabrication method. They deduced that incorporation of 0.9 wt% CNCs in the SMC composite increased the tensile modulus by 25%, tensile strength by 30%, strain at break by 22%, work to break by 49%, flexural modulus by 44%, and flexural strength by 33% [2]. Nanocellulose-based aerogels are widely used in diverse applications owing to their lightweight and porous characteristics [98, 99]. In this regard, a study by Meiling Zhang et al. 2022 focuses on the preparation of CNFs/chitosan-based aerogel with a conceivable pore structure by ice-induced self-assembly. Following the results, the authors provided a theoretical basis for the preparation of nanocellulose-based aerogel with satisfactory mechanical properties and enabling its application in the field of thermal management [100].

Howbeit, the hydrophilic nature of nanocelluloses due to numerous hydroxyl groups limits their effectiveness as reinforcement in hydrophobic matrices. For this purpose, many researchers have modified the surface of nanocelluloses to produce nanocomposites with enhanced performance [100, 101]. There are two types of modifications: covalent that is responsible for the formation of irreversible bonds, and non-covalent linked to reversible secondary interactions such as van der Waals forces or hydrogen bonding [102, 103]. In this context, a study conducted by Caihong

Table 2 Effect of CNCs and extraction method on the mechanical properties of nanocomposites

Matrix	Type of nanocellulose (extraction method)	Content %	Tensile strength	Young's modulus	Elongation at break (%)	Refs.
Starch	CNCs (acid hydrolysis)	8	19.5 ± 1.3 MPa	720.9 ± 26.1 MPa	–	El Achaby et al. [75]
Poly (butylene adipate-co-terephthalate)	Modified CNCs (acid hydrolysis)	7	12.4 ± 0.7 MPa	79.2 ± 2.0 MPa	567.4 ± 38.7	Pinheiro et al. [101]
Poly (vinyl alcohol)	CNFs (high-intensity ultrasonication)	4	~34.16 MPa	~96.66 MPa	–	Li et al. [107]
Hydroxypropyl cellulose	CNFs (TEMPO-oxidation)	5	~38	~1.250 GPa	~425	Okahashi et al. [108]
Hydroxyethyl cellulose	CNFs (TEMPO-oxidation)	5	~55 MPa	~1 GPa	~67	Okahashi et al. [108]
Methyl cellulose	CNFs (TEMPO-oxidation)	5	~100 MPa	~1.9 GPa	~100	Okahashi et al. [108]
Carboxymethyl cellulose	CNFs (TEMPO-oxidation)	5	~50 MPa	~2.5 GPa	~20	Okahashi et al. [108]
Starch	CNCs (TEMPO-oxidation/homogenized by a high-pressure homogenizer	4	12.4 ± 2.70 MPa	–	30.48 ± 1.06	Chen et al. [89]
Polylactic acid	CNCs (acid hydrolysis)	3	78 ± 3 MPa	4.1 ± 1.2 GPa	2.6 ± 0.2	De Souza et al. [90]
Polylactic acid/poly(3-hydroxybutyrate)	CNCs (acid hydrolysis)	1	15 ± 2.5	2791 ± 308 MPa	10 ± 6	Frone et al. [109]
Polylactic acid/poly(3-hydroxybutyrate)	CNFs (ultrasonication)	1	11 ± 1.6 MPa	1657 ± 243 MPa	75 ± 14.6	Frone et al. [109]

Chen et al. 2022 focused on the preparation of cellulose nanofibril (CNF) aerogels that were prepared by freeze-drying and then modified with polysiloxane functions to reduce the hydroxyl groups on the surface and improve their hydrophobic properties. The results revealed that the compressive modulus was improved by incorporating 4.5% of the CNFs. Moreover, compared to the pure CNF aerogels, the modified CNF aerogels showed an improved compressive modulus due to the cross-linked Si–O–Si bonds in the polysiloxanes [104]. Aihua Pei et al. 2010 were prepared CNCs by acid hydrolysis of cotton and functionalized by partial silylation by reacting with *n*-dodecyldimethylchlorosilane in toluene and incorporated into PLA-based matrices. The results revealed that the unmodified CNCs formed aggregates in the composite films, while the modified CNCs were well dispersed and individualized in the matrix. As a result, the tensile modulus and tensile strength of the nanocomposite films were improved by more than 20% due to the good dispersion of the modified CNCs in the matrix and their crystallinity [105]. The effect of modified CNCs on the mechanical properties of a similar matrix was studied by Fortunati et al. 2012. In this study, the authors modified CNCs using a surfactant (ethoxylated nonylphenol acid phosphate ester) to increase the dispersion in the polymer matrix and improve the final performance of the nanocomposites. The films showed 83% higher Young's modulus than pure PLA after the addition of 5% modified CNCs. These results demonstrate the reinforcing effect exerted by CNCs, especially modified CNCs, and the effectiveness of the surfactant on the dispersion of cellulose nanostructures. However, a decrease in elongation at break was observed, which is a common trend observed for thermoplastic nanocomposites [106].

5 Conclusion

Nanocellulose is a promising natural material that has attracted considerable interest from the scientific community due to its powerful properties: biodegradability, biocompatibility, crystallinity, specific surface, mechanical properties, non-toxicity, and so on. These nanocelluloses can be extracted from different lignocellulosic materials. Researchers have conceived various pretreatments of lignocellulosic fibers in order to purify the cellulose and remove non-cellulosic parts affecting the final properties of nanocellulose. The various extraction methods have been mentioned in this chapter, including mechanical, chemical, and enzymatic methods. Nevertheless, it has been found that the use of these methods only is associated with problems of environmental pollution, high energy consumption, or long extraction time. Therefore, some researchers have combined these extraction techniques in order to reduce their negative impact. Thus, nanocelluloses have attracted a great interest, mainly in the food packaging, biomedical, cosmetic, automotive, aeronautical, water treatment fields thanks to their performing properties. Research studies have shown that nanocelluloses can improve the physicochemical properties of composites. Similarly, the evaluation of the effect of nanocellulose as filler on the mechanical properties of nanocomposites has shown an improvement in the performance of nanocomposites produced.

References

1. Laghaei R, Hejazi SM, Fashandi H et al (2023) Reinforcement contribution of cellulose nanocrystals (CNCs) to tensile properties and fracture behavior of triaxial E-glass fabric/epoxy composites. Compos Part A Appl Sci Manuf 164:107258. https://doi.org/10.1016/j.compositesa.2022.107258
2. Asadi A, Miller M, Sultana S et al (2016) Introducing cellulose nanocrystals in sheet molding compounds (SMC). Compos Part A Appl Sci Manuf 88:206–215. https://doi.org/10.1016/j.compositesa.2016.05.033
3. Liu S, Qamar SA, Qamar M et al (2021) Engineered nanocellulose-based hydrogels for smart drug delivery applications. Int J Biol Macromol 181:275–290. https://doi.org/10.1016/j.ijbiomac.2021.03.147
4. Farooq A, Patoary MK, Zhang M et al (2020) Cellulose from sources to nanocellulose and an overview of synthesis and properties of nanocellulose/zinc oxide nanocomposite materials. Int J Biol Macromol 154:1050–1073. https://doi.org/10.1016/j.ijbiomac.2020.03.163
5. Mondal S (2017) Preparation, properties and applications of nanocellulosic materials. Carbohydr Polym 163:301–316. https://doi.org/10.1016/j.carbpol.2016.12.050
6. Mu R, Hong X, Ni Y et al (2019) Recent trends and applications of cellulose nanocrystals in food industry. Trends Food Sci Technol 93:136–144. https://doi.org/10.1016/j.tifs.2019.09.013
7. Habibi Y, Lucia LA, Rojas OJ (2010) CNC—Johnsy - habibi2010 [2].pdf. Chem Rev 110:3479–3500
8. Liu D, Dong Y, Bhattacharyya D, Sui G (2017) Novel sandwiched structures in starch/cellulose nanowhiskers (CNWs) composite films. Compos Commun 4:5–9. https://doi.org/10.1016/j.coco.2017.03.001
9. Rajinipriya M, Nagalakshmaiah M, Astruc J et al (2018) Single stage purification of flax, hemp, and milkweed stem and their physical and morphological properties. Int J Polym Anal Charact 23:78–88. https://doi.org/10.1080/1023666X.2017.1387688
10. De Aguiar J, Bondancia TJ, Claro PIC et al (2020) Enzymatic deconstruction of sugarcane bagasse and straw to obtain cellulose nanomaterials. ACS Sustain Chem Eng 8:2287–2299. https://doi.org/10.1021/acssuschemeng.9b06806
11. Wahib SA, Da'na DA, Al-Ghouti MA (2022) Insight into the extraction and characterization of cellulose nanocrystals from date pits. Arab J Chem 15:103650. https://doi.org/10.1016/j.arabjc.2021.103650
12. Baek J, Wahid-Pedro F, Kim K, et al (2019) Phosphorylated-CNC/modified-chitosan nanocomplexes for the stabilization of Pickering emulsions. Carbohydr Polym 206:520–527. https://doi.org/10.1016/j.carbpol.2018.11.006
13. Nascimento DM, Almeida JS, Dias AF et al (2014) A novel green approach for the preparation of cellulose nanowhiskers from white coir. Carbohydr Polym 110:456–463. https://doi.org/10.1016/j.carbpol.2014.04.053
14. Marques FP, Silva LMA, Lomonaco D et al (2020) Steam explosion pretreatment to obtain eco-friendly building blocks from oil palm mesocarp fiber. Ind Crops Prod 143:111907. https://doi.org/10.1016/j.indcrop.2019.111907
15. Boonterm M, Sunyadeth S, Dedpakdee S et al (2016) Characterization and comparison of cellulose fiber extraction from rice straw by chemical treatment and thermal steam explosion. J Clean Prod 134:592–599. https://doi.org/10.1016/j.jclepro.2015.09.084
16. Shishehbor M, Son H, Nuruddin M et al (2021) Influence of alignment and microstructure features on the mechanical properties and failure mechanisms of cellulose nanocrystals

(CNC) films. J Mech Behav Biomed Mater 118:104399. https://doi.org/10.1016/j.jmbbm.2021.104399
17. Sheltami RM, Abdullah I, Ahmad I et al (2012) Extraction of cellulose nanocrystals from mengkuang leaves (Pandanus tectorius). Carbohydr Polym 88:772–779. https://doi.org/10.1016/j.carbpol.2012.01.062
18. Oliveira TÍS, Rosa MF, Ridout MJ et al (2017) Bionanocomposite films based on polysaccharides from banana peels. Int J Biol Macromol 101:1–8. https://doi.org/10.1016/j.ijbiomac.2017.03.068
19. Do Nascimento JHO, Luz RF, Galvão FMF et al (2015) Extraction and characterization of cellulosic nanowhisker obtained from discarded cotton fibers. Mater Today Proc 2:1–7. https://doi.org/10.1016/j.matpr.2015.04.001
20. Bettaieb F, Khiari R, Dufresne A et al (2015) Mechanical and thermal properties of Posidonia oceanica cellulose nanocrystal reinforced polymer. Carbohydr Polym 123:99–104. https://doi.org/10.1016/j.carbpol.2015.01.026
21. Hiasa S, Iwamoto S, Endo T, Edashige Y (2014) Isolation of cellulose nanofibrils from mandarin (Citrus unshiu) peel waste. Ind Crops Prod 62:280–285. https://doi.org/10.1016/j.indcrop.2014.08.007
22. Akatan K, Kabdrakhmanova S, Kuanyshbekov T et al (2022) Highly-efficient isolation of microcrystalline cellulose and nanocellulose from sunflower seed waste via environmentally benign method. Cellulose 29:3787–3802. https://doi.org/10.1007/s10570-022-04527-4
23. Lim HJ, Cheng WK, Tan KW, Yu LJ (2022) Oil palm-based nanocellulose for a sustainable future: where are we now? J Environ Chem Eng 10:107271. https://doi.org/10.1016/j.jece.2022.107271
24. Wang H, Zuo M, Ding N et al (2019) Preparation of nanocellulose with high-pressure homogenization from pretreated biomass with cooking with active oxygen and solid alkali. ACS Sustain Chem Eng 7:9378–9386. https://doi.org/10.1021/acssuschemeng.9b00582
25. Luo X, Wang X (2017) Preparation and characterization of nanocellulose fibers from NaOH/urea pretreatment of oil palm fibers. BioResources 12:5826–5837. https://doi.org/10.15376/biores.12.3.5826-5837
26. Kassab Z, Mansouri S, Tamraoui Y et al (2020) Identifying Juncus plant as viable source for the production of micro- and nano-cellulose fibers: application for PVA composite materials development. Ind Crops Prod 144:112035. https://doi.org/10.1016/j.indcrop.2019.112035
27. Araya-Chavarría K, Rojas R, Ramírez-Amador K, et al (2022) Cellulose nanofibers as functional biomaterial from pineapple stubbles via TEMPO oxidation and mechanical process. Waste and Biomass Valorization 13:1749–1758.https://doi.org/10.1007/s12649-021-01619-3
28. Mahrous F, Koshani R, Tavakolian M et al (2021) Formation of hairy cellulose nanocrystals by cryogrinding. Cellulose 28:8387–8403. https://doi.org/10.1007/s10570-021-04092-2
29. Yasim-Anuar TAT, Ariffin H, Hassan MA (2018) Characterization of cellulose nanofiber from oil palm mesocarp fiber produced by ultrasonication. IOP Conf Ser Mater Sci Eng 368. https://doi.org/10.1088/1757-899X/368/1/012033
30. Li W, Wang R, Liu S (2011) Nanocrystalline cellulose prepared from softwood kraft pulp via ultrasonic-assisted acid hydrolysis. BioResources 6:4271–4281
31. Du H, Liu C, Zhang Y et al (2016) Preparation and characterization of functional cellulose nanofibrils via formic acid hydrolysis pretreatment and the followed high-pressure homogenization. Ind Crops Prod 94:736–745. https://doi.org/10.1016/j.indcrop.2016.09.059
32. Chaka KT (2022) Extraction of cellulose nanocrystals from agricultural by-products: a review. Green Chem Lett Rev 15:582–597. https://doi.org/10.1080/17518253.2022.2121183
33. Li X, Li J, Kuang Y et al (2020) Stabilization of pickering emulsions with cellulose nanofibers derived from oil palm fruit bunch. Cellulose 27:839–851. https://doi.org/10.1007/s10570-019-02803-4
34. Li Z, Zhang Y, Anankanbil S, Guo Z (2021) Applications of nanocellulosic products in food: manufacturing processes, structural features and multifaceted functionalities. Trends Food Sci Technol 113:277–300. https://doi.org/10.1016/j.tifs.2021.03.027

35. Park NM, Choi S, Oh JE, Hwang DY (2019) Facile extraction of cellulose nanocrystals. Carbohydr Polym 223:115114. https://doi.org/10.1016/j.carbpol.2019.115114
36. Rasheed M, Jawaid M, Parveez B et al (2020) Morphological, chemical and thermal analysis of cellulose nanocrystals extracted from bamboo fibre. Int J Biol Macromol 160:183–191. https://doi.org/10.1016/j.ijbiomac.2020.05.170
37. Noremylia MB, Hassan MZ, Ismail Z (2022) Recent advancement in isolation, processing, characterization and applications of emerging nanocellulose: a review. Int J Biol Macromol 206:954–976. https://doi.org/10.1016/j.ijbiomac.2022.03.064
38. Chen Y, Fan D, Han Y et al (2017) Length-controlled cellulose nanofibrils produced using enzyme pretreatment and grinding. Cellulose 24:5431–5442. https://doi.org/10.1007/s10570-017-1499-z
39. Sun X, Wu Q, Ren S, Lei T (2015) Comparison of highly transparent all-cellulose nanopaper prepared using sulfuric acid and TEMPO-mediated oxidation methods. Cellulose 22:1123–1133. https://doi.org/10.1007/s10570-015-0574-6
40. Im W, Lee S, Rajabi Abhari A, et al (2018) Optimization of carboxymethylation reaction as a pretreatment for production of cellulose nanofibrils. Cellulose 25:3873–3883.https://doi.org/10.1007/s10570-018-1853-9
41. Saini S, Yücel Falco Ç, Belgacem MN, Bras J (2016) Surface cationized cellulose nanofibrils for the production of contact active antimicrobial surfaces. Carbohydr Polym 135:239–247.https://doi.org/10.1016/j.carbpol.2015.09.002
42. Du H, Liu W, Zhang M et al (2019) Cellulose nanocrystals and cellulose nanofibrils based hydrogels for biomedical applications. Carbohydr Polym 209:130–144. https://doi.org/10.1016/j.carbpol.2019.01.020
43. Löbmann K, Svagan AJ (2017) Cellulose nanofibers as excipient for the delivery of poorly soluble drugs. Int J Pharm 533:285–297. https://doi.org/10.1016/j.ijpharm.2017.09.064
44. Abitbol T, Kloser E, Gray DG (2013) Estimation of the surface sulfur content of cellulose nanocrystals prepared by sulfuric acid hydrolysis. Cellulose 20:785–794. https://doi.org/10.1007/s10570-013-9871-0
45. Kassab Z, Kassem I, Hannache H et al (2020) Tomato plant residue as new renewable source for cellulose production: extraction of cellulose nanocrystals with different surface functionalities. Cellulose 27:4287–4303. https://doi.org/10.1007/s10570-020-03097-7
46. Jia W, Liu Y (2019) Two characteristic cellulose nanocrystals (CNCs) obtained from oxalic acid and sulfuric acid processing. Cellulose 26:8351–8365. https://doi.org/10.1007/s10570-019-02690-9
47. Nagarajan KJ, Balaji AN, Kasi Rajan ST, Ramanujam NR (2020) Preparation of bio-eco based cellulose nanomaterials from used disposal paper cups through citric acid hydrolysis. CarbohydrPolym 235:115997. https://doi.org/10.1016/j.carbpol.2020.115997
48. Marakana PG, Dey A, Saini B (2021) Isolation of nanocellulose from lignocellulosic biomass: synthesis, characterization, modification, and potential applications. J Environ Chem Eng 9:106606. https://doi.org/10.1016/j.jece.2021.106606
49. Ghasemlou M, Daver F, Ivanova EP et al (2021) Surface modifications of nanocellulose: from synthesis to high-performance nanocomposites. Prog Polym Sci 119:101418. https://doi.org/10.1016/j.progpolymsci.2021.101418
50. Peter S, Lyczko N, Gopakumar D et al (2022) Nanocellulose and its derivative materials for energy and environmental applications. J Mater Sci 57:6835–6880. https://doi.org/10.1007/s10853-022-07070-6
51. Dai H, Ou S, Huang Y, Huang H (2018) Utilization of pineapple peel for production of nanocellulose and film application. Cellulose 25:1743–1756. https://doi.org/10.1007/s10570-018-1671-0
52. Tang J, Sisler J, Grishkewich N, Tam KC (2017) Functionalization of cellulose nanocrystals for advanced applications. J Colloid Interface Sci 494:397–409. https://doi.org/10.1016/j.jcis.2017.01.077
53. Rajinipriya M, Nagalakshmaiah M, Robert M, Elkoun S (2018) Homogenous and transparent nanocellulosic films from carrot. Ind Crops Prod 118:53–64. https://doi.org/10.1016/j.indcrop.2018.02.076

54. Zianor Azrina ZA, Beg MDH, Rosli MY, et al (2017) Spherical nanocrystalline cellulose (NCC) from oil palm empty fruit bunch pulp via ultrasound assisted hydrolysis. Carbohydr Polym 162:115–120.https://doi.org/10.1016/j.carbpol.2017.01.035
55. Sadeghifar H, Filpponen I, Clarke SP et al (2011) Production of cellulose nanocrystals using hydrobromic acid and click reactions on their surface. J Mater Sci 46:7344–7355. https://doi.org/10.1007/s10853-011-5696-0
56. Bondancia TJ, De Aguiar J, Batista G, et al (2020) Production of nanocellulose using citric acid in a biorefinery concept: effect of the hydrolysis reaction time and techno-economic analysis
57. Yu H, Qin Z, Liang B et al (2013) Facile extraction of thermally stable cellulose nanocrystals with a high yield of 93% through hydrochloric acid hydrolysis under hydrothermal conditions. J Mater Chem A 1:3938–3944. https://doi.org/10.1039/c3ta01150j
58. Goh KY, Ching YC, Chuah CH et al (2016) Individualization of microfibrillated celluloses from oil palm empty fruit bunch: comparative studies between acid hydrolysis and ammonium persulfate oxidation. Cellulose 23:379–390. https://doi.org/10.1007/s10570-015-0812-y
59. Pierre G, Punta C, Delattre C et al (2017) TEMPO-mediated oxidation of polysaccharides: an ongoing story. Carbohydr Polym 165:71–85. https://doi.org/10.1016/j.carbpol.2017.02.028
60. Kumar A, Durand H, Zeno E et al (2020) The surface chemistry of a nanocellulose drug carrier unravelled by MAS-DNP. Chem Sci 11:3868–3877. https://doi.org/10.1039/c9sc06312a
61. Thomas B, Raj MC, Athira BK et al (2018) Nanocellulose, a versatile green platform: from biosources to materials and their applications. Chem Rev 118:11575–11625. https://doi.org/10.1021/acs.chemrev.7b00627
62. Aditiawati P, Dungani R, Amelia C (2018) Enzymatic production of cellulose nanofibers from oil palm empty fruit bunch (EFB) with crude cellulase of Trichoderma sp. Mater Res Express 5:0–8.https://doi.org/10.1088/2053-1591/aab449
63. Reshmy R, Philip E, Paul SA et al (2020) Nanocellulose-based products for sustainable applications-recent trends and possibilities. Rev Environ Sci Biotechnol 19:779–806. https://doi.org/10.1007/s11157-020-09551-z
64. Kumari P, Pathak G, Gupta R et al (2019) Cellulose nanofibers from lignocellulosic biomass of lemongrass using enzymatic hydrolysis: characterization and cytotoxicity assessment. DARU, J Pharm Sci 27:683–693. https://doi.org/10.1007/s40199-019-00303-1
65. Henriksson M, Henriksson G, Berglund LA, Lindström T (2007) An environmentally friendly method for enzyme-assisted preparation of microfibrillated cellulose (MFC) nanofibers. Eur Polym J 43:3434–3441. https://doi.org/10.1016/j.eurpolymj.2007.05.038
66. Martelli-Tosi M, Masson MM, Silva NC, et al (2018) Soybean straw nanocellulose produced by enzymatic or acid treatment as a reinforcing filler in soy protein isolate films. Carbohydr Polym 198:61–68. https://doi.org/10.1016/j.carbpol.2018.06.053
67. Chen XQ, Pang GX, Shen WH et al (2019) Preparation and characterization of the ribbon-like cellulose nanocrystals by the cellulase enzymolysis of cotton pulp fibers. Carbohydr Polym 207:713–719. https://doi.org/10.1016/j.carbpol.2018.12.042
68. Rossi BR, Pellegrini VOA, Cortez AA et al (2021) Cellulose nanofibers production using a set of recombinant enzymes. Carbohydr Polym 256:117510. https://doi.org/10.1016/j.carbpol.2020.117510
69. Morais JPS, Rosa MDF, De Souza FMDSM et al (2013) Extraction and characterization of nanocellulose structures from raw cotton linter. Carbohydr Polym 91:229–235. https://doi.org/10.1016/j.carbpol.2012.08.010
70. Khan MN, Rehman N, Sharif A et al (2020) Environmentally benign extraction of cellulose from dunchi fiber for nanocellulose fabrication. Int J Biol Macromol 153:72–78. https://doi.org/10.1016/j.ijbiomac.2020.02.333
71. Franco TS, Potulski DC, Viana LC et al (2019) Nanocellulose obtained from residues of peach palm extraction (Bactris gasipaes). Carbohydr Polym 218:8–19. https://doi.org/10.1016/j.carbpol.2019.04.035
72. Borkotoky SS, Dhar P, Katiyar V (2018) Biodegradable poly (lactic acid)/Cellulose nanocrystals (CNCs) composite microcellular foam: effect of nanofillers on foam cellular morphology,

thermal and wettability behavior. Int J Biol Macromol 106:433–446. https://doi.org/10.1016/j.ijbiomac.2017.08.036
73. Jiang F, Lo HY (2013) Chemically and mechanically isolated nanocellulose and their self-assembled structures. Carbohydr Polym 95:32–40. https://doi.org/10.1016/j.carbpol.2013.02.022
74. Marwanto M, Maulana MI, Febrianto F et al (2021) Characteristics of nanocellulose crystals from balsa and kapok fibers at different ammonium persulfate concentrations. Wood Sci Technol 55:1319–1335. https://doi.org/10.1007/s00226-021-01319-0
75. El Achaby M, El Miri N, Hannache H et al (2018) Cellulose nanocrystals from Miscanthus fibers: insights into rheological, physico-chemical properties and polymer reinforcing ability. Cellulose 25:6603–6619. https://doi.org/10.1007/s10570-018-2047-1
76. Attaran SA, Hassan A, Wahit MU (2017) Materials for food packaging applications based on bio-based polymer nanocomposites. J Thermoplast Compos Mater 30:143–173. https://doi.org/10.1177/0892705715588801
77. Ferreira FV, Dufresne A, Pinheiro IF et al (2018) How do cellulose nanocrystals affect the overall properties of biodegradable polymer nanocomposites: a comprehensive review. Eur Polym J 108:274–285. https://doi.org/10.1016/j.eurpolymj.2018.08.045
78. Baek J, Ramasamy M, Willis NC et al (2021) Encapsulation and controlled release of vitamin C in modified cellulose nanocrystal/chitosan nanocapsules. Curr Res Food Sci 4:215–223. https://doi.org/10.1016/j.crfs.2021.03.010
79. Mehanny S, Abu-El Magd EE, Ibrahim M, et al (2021) Extraction and characterization of nanocellulose from three types of palm residues. J Mater Res Technol 10:526–537. https://doi.org/10.1016/j.jmrt.2020.12.027
80. Song Y, Jiang M, Zhou L et al (2022) Rapidly regenerated CNC/TiO2/MnO2 porous microspheres for high-efficient dye removal. Carbohydr Polym 292:1–8. https://doi.org/10.1016/j.carbpol.2022.119644
81. Naskar AK, Keum JK, Boeman RG (2016) Polymer matrix nanocomposites for automotive structural components. Nat Nanotechnol 11:1026–1030. https://doi.org/10.1038/nnano.2016.262
82. Hussein Y, Loutfy SA, Kamoun EA et al (2021) Enhanced anti-cancer activity by localized delivery of curcumin form PVA/CNCs hydrogel membranes: preparation and in vitro bioevaluation. Int J Biol Macromol 170:107–122. https://doi.org/10.1016/j.ijbiomac.2020.12.133
83. Sultan S, Siqueira G, Zimmermann T, Mathew AP (2017) 3D printing of nano-cellulosic biomaterials for medical applications. Curr Opin Biomed Eng 2:29–34. https://doi.org/10.1016/j.cobme.2017.06.002
84. Supramaniam J, Adnan R, Mohd Kaus NH, Bushra R (2018) Magnetic nanocellulose alginate hydrogel beads as potential drug delivery system. Int J Biol Macromol 118:640–648.https://doi.org/10.1016/j.ijbiomac.2018.06.043
85. Hussein FA, Jaafar SS, Salih WM (2021) Synthesis and characterization of CNC-cefexime nanohybrid as a potential antibacterial agent. Mater Today Proc 45:5564–5568. https://doi.org/10.1016/j.matpr.2021.02.308
86. Kummala R, Soto Véliz D, Fang Z, et al (2020) Human dermal fibroblast viability and adhesion on cellulose nanomaterial coatings: influence of surface characteristics. Biomacromolecules 21:1560–1567. https://doi.org/10.1021/acs.biomac.0c00107
87. Cruz-Tato P, Ortiz-Quiles EO, Vega-Figueroa K, et al (2017) Metalized nanocellulose composites as a feasible material for membrane supports: design and applications for water treatment. Environ Sci Technol 51:4585–4595.https://doi.org/10.1021/acs.est.6b05955
88. Bai L, Liu Y, Ding A et al (2019) Fabrication and characterization of thin-film composite (TFC) nanofiltration membranes incorporated with cellulose nanocrystals (CNCs) for enhanced desalination performance and dye removal. Chem Eng J 358:1519–1528. https://doi.org/10.1016/j.cej.2018.10.147
89. Chen Q, Liu Y, Chen G (2019) A comparative study on the starch-based biocomposite films reinforced by nanocellulose prepared from different non-wood fibers. Cellulose 26:2425–2435. https://doi.org/10.1007/s10570-019-02254-x

90. de Souza AG, Barbosa RFS, Rosa DS (2020) Nanocellulose from industrial and agricultural waste for further use in PLA composites. J Polym Environ 28:1851–1868. https://doi.org/10.1007/s10924-020-01731-w
91. Barbash VA, Yashchenko OV, Yakymenko OS et al (2022) Preparation of hemp nanocellulose and its use to improve the properties of paper for food packaging. Cellulose 29:8305–8317. https://doi.org/10.1007/s10570-022-04773-6
92. Patil S, Bharimalla AK, Nadanathangam V et al (2022) Nanocellulose reinforced corn starch-based biocomposite films: composite optimization, characterization and storage studies. Food Packag Shelf Life 33:100860. https://doi.org/10.1016/j.fpsl.2022.100860
93. Bai L, Lv S, Xiang W et al (2019) Oil-in-water Pickering emulsions via microfluidization with cellulose nanocrystals: 1. Formation and stability. Food Hydrocoll 96:699–708. https://doi.org/10.1016/j.foodhyd.2019.04.038
94. Kalashnikova I, Bizot H, Cathala B, Capron I (2012) Modulation of cellulose nanocrystals amphiphilic properties to stabilize oil/water interface. Biomacromol 13:267–275. https://doi.org/10.1021/bm201599j
95. Hossain KMZ, Deeming L, Edler KJ (2021) Recent progress in pickering emulsions stabilised by bioderived particles. RSC Adv 11:39027–39044. https://doi.org/10.1039/d1ra08086e
96. Trigueiro JPC, Silva GG, Pereira FV, Lavall RL (2014) Layer-by-layer assembled films of multi-walled carbon nanotubes with chitosan and cellulose nanocrystals. J Colloid Interface Sci 432:214–220. https://doi.org/10.1016/j.jcis.2014.07.001
97. Asad M, Saba N, Asiri AM et al (2018) Preparation and characterization of nanocomposite films from oil palm pulp nanocellulose/poly (Vinyl alcohol) by casting method. Carbohydr Polym 191:103–111. https://doi.org/10.1016/j.carbpol.2018.03.015
98. Hasanin MS, Youssef AM (2022) Ecofriendly bioactive film doped CuO nanoparticles based biopolymers and reinforced by enzymatically modified nanocellulose fibers for active packaging applications. Food Packag Shelf Life 34:100979. https://doi.org/10.1016/j.fpsl.2022.100979
99. Wang C, Cao H, Jia L, et al (2022) Characterization of antibacterial aerogel based on ε-poly-L-lysine/nanocellulose by using citric acid as crosslinker. CarbohydrPolym 291.https://doi.org/10.1016/j.carbpol.2022.119568
100. Zhang M, Li M, Xu Q et al (2022) Nanocellulose-based aerogels with devisable structure and tunable properties via ice-template induced self-assembly. Ind Crops Prod 179:114701. https://doi.org/10.1016/j.indcrop.2022.114701
101. Pinheiro IF, Ferreira FV, Souza DHS et al (2017) Mechanical, rheological and degradation properties of PBAT nanocomposites reinforced by functionalized cellulose nanocrystals. Eur Polym J 97:356–365. https://doi.org/10.1016/j.eurpolymj.2017.10.026
102. Dehkhoda S, Bagheri M, Heydari M, Rabieh S (2022) Extraction of carboxylated nanocellulose from oat husk: characterization, surface modification and in vitro evaluation of indomethacin drug release. Int J Biol Macromol 212:165–171. https://doi.org/10.1016/j.ijbiomac.2022.05.125
103. Feng X, Dai H, Fu Y, et al (2022) Regulation mechanism of nanocellulose with different morphologies on the properties of low-oil gelatin emulsions: interfacial adsorption or network formation? Food Hydrocoll 133. https://doi.org/10.1016/j.foodhyd.2022.107960
104. Chen C, Li C, Yu D, Wu M (2022) A facile method to prepare superhydrophobic nanocellulose-based aerogel with high thermal insulation performance via a two-step impregnation process. Cellulose 29:245–257. https://doi.org/10.1007/s10570-021-04275-x
105. Pei A, Zhou Q, Berglund LA (2010) Functionalized cellulose nanocrystals as biobased nucleation agents in poly(l-lactide) (PLLA)—crystallization and mechanical property effects. Compos Sci Technol 70:815–821. https://doi.org/10.1016/j.compscitech.2010.01.018
106. Fortunati E, Armentano I, Zhou Q et al (2012) Multifunctional bionanocomposite films of poly(lactic acid), cellulose nanocrystals and silver nanoparticles. Carbohydr Polym 87:1596–1605. https://doi.org/10.1016/j.carbpol.2011.09.066
107. Li W, Zhao X, Huang Z, Liu S (2013) Nanocellulose fibrils isolated from BHKP using ultrasonication and their reinforcing properties in transparent poly (vinyl alcohol) films. J Polym Res 20. https://doi.org/10.1007/s10965-013-0210-9

108. Okahashi K, Takeuchi M, Zhou Y et al (2021) Nanocellulose-containing cellulose ether composite films prepared from aqueous mixtures by casting and drying method. Cellulose 28:6373–6387. https://doi.org/10.1007/s10570-021-03897-5
109. Frone AN, Panaitescu DM, Chiulan I et al (2019) Thermal and mechanical behavior of biodegradable polyester films containing cellulose nanofibers. J Therm Anal Calorim 138:2387–2398. https://doi.org/10.1007/s10973-019-08218-4

Nanomaterials Based Polymer Composites: Mechanical Properties

Melkie Getnet Tadesse, Aravin Prince Periyasamy, and Jörn Felix Lübben

Abstract The advancements of nanomaterials and polymer nanocomposites have been an area of hot issues in most recent times. This is due to the unique mechanical features possessed by nanomaterials and polymer nanocomposites. The unique mechanical performance of those materials includes but is not limited to tensile strength, abrasion resistance, bending resistance (stiffness), toughness, superior flexibility, compactness and high resistance against harsh environments. Nanomaterials also show special possibilities to form a matrix with polymers to form polymer nanocomposites which further creates an advantage to get the desired properties for unique applications. This chapter aims to review the mechanics of nanomaterials and polymer nanocomposites and the basic results obtained in this aspect. First, the main concept of nanomaterials and polymer nanocomposites will be presented. Furthermore, the basics of nanomaterials and polymer nanocomposites will be discussed. Finally, detailed mechanics of both nanomaterials and polymer nanocomposites will be highlighted. The review result indicated that nanomaterials and polymer nanocomposites are emerged as promising materials to be applied in different applications fields.

Keywords Nanomaterials · Polymer nanocomposites · Mechanical properties · Nanofillers · Polymers · Structure · Applications

M. G. Tadesse (✉) · J. F. Lübben
Sustainable Engineering (STE), Albstadt-Sigmaringen University, 72458 Albstadt, Germany
e-mail: melkiegetnet23@gmail.com

M. G. Tadesse
Ethiopian Institute of Textile and Fashion Technology, Bahir Dar University, Bahir Dar 1037, Ethiopia

A. Prince Periyasamy
Department of Bioproducts and Biosystems, School of Chemical Engineering, Aalto University, 02150 Espoo, Finland

© The Author(s), under exclusive license to Springer Nature Singapore Pte Ltd. 2023

H. Abdellaoui et al. (eds.), *Mechanics of Nanomaterials and Polymer Nanocomposites*, Smart Nanomaterials Technology,
https://doi.org/10.1007/978-981-99-2352-6_7

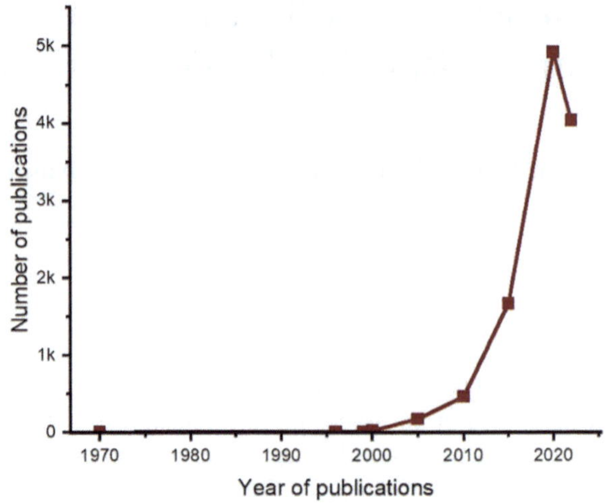

Fig. 1 Number of publications per year on the topic of nanomaterials, nanocomposites and their mechanical properties (as obtained from https://app.dimensions.ai on 21/09/2022)

1 Introduction

The use of wearable technologies are growing rapidly [1]. For these technologies, polymers play an important role [2–6]. However, the polymers used in such applications are not on the nanoscale and hence their applications are limited to the macroscale. Researchers have tried to solve such problems by introducing nanomaterials. Over the past five decades the number of publications related to nanomaterials and nanocomposites is dramatically increasing as depicted in Fig. 1. Based on the research data, the publications were realised in applied science, nanomaterials, composite and chemical related journals. This history can tell the scholars that there is a growing interest in the mechanics of nanomaterials and polymer composite materials. Nanomaterials are defined as a material that has at least a dimension in nanometre (1 nm, 10^{-9} m) scale [7]. The unique properties of nanomaterials are mainly dependent on the construction element that is dependent on the theory of nanoscience. While a polymer nanocomposite can be defined the same but it should comprise the polymer matrix in it [8]. Polymer nanocomposites are also one of the hot research areas in the nanotechnology area.

2 Nanomaterials

A solid material in the multiphase form where one among the phases has dimensions less than 100 nm can be called nanocomposite. The dimensions might be one, two or three in contrast to the micro composites. The phases have distances in the nanoscale range between them [9]. The basic difference between the micro composites with

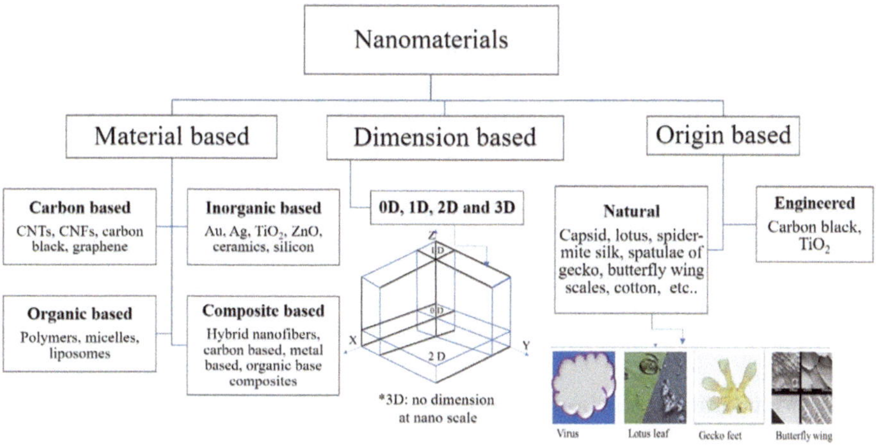

Fig. 2 Classifications of nanomaterials-based material, dimension and their origin

that of the nanocomposites is that nanocomposites can comprise properties such as mechanical, physical and chemical individually.

The use of nanomaterials is rapid for use in applications when unique electronic, optical, mechanical or thermo-physical properties are required. Nanomaterials can be categorised based on origin, source and based on their dimensions. Generally, nanomaterials can be categorised in various forms but the most common way of classification is indicated in Fig. 2 [10].

Nowadays, the commercial success of nanomaterials is increasing as the production and use of nanomaterials are environmentally safe as confirmed by both the users and the manufacturers [11]. This helps the rapid growth of the use of nanomaterials and the focus of research is going in this area. However, the mechanics of nanomaterials is still having unsolved problems and research has been performed in this area. The main focus of this sub-chapter is dealing with the mechanics of nanomaterials in detail.

2.1 Mechanics of Nanomaterials

Nanomaterials have been manipulated by nanotechnology at various scales to fabricate and create devices, systems or materials with completely different properties or functions. The mechanics of engineered nanomaterials is directly mimicked from those naturally occurring with direct resemblance properties. To the authors believe, every aspect of engineered products, materials or systems was mimicked by scientists from nature. While engineering is performed, some mechanics of materials in this aspect nanomaterials may not be directly imitated, which is why there is still research is the mechanics of nanomaterials.

The mechanics (physicochemical and structural) characteristics of nanomaterials help to put nanomaterials in various categories. Size and shape are considered to be the main characteristic features of nanomaterials which define the physical properties of it. The mechanical properties of nanomaterials are clearly discrete [12]. This distinction leads that very nanomaterials possess discrete mechanical properties. Therefore, it is very difficult to have comprehensive mechanics of nanomaterials rather the mechanics of nanomaterials should be investigated separately with respective nanomaterials. For instance, the mechanics of CNT and graphene should be dealt with separately. The following paragraph will discuss the mechanicals of the most important nanomaterials.

2.1.1 Mechanics of Carbon Nanotubes

Graphene is one of the most important nanomaterial due to the following reasons [12]:

- Have exceptional surface area to volume ratio- makes an excellent candidate for composite manufacturing;
- Possess high tensile strength-makes perfect candidate for various stress exposed applications;
- Possess high electrical conductivity- fit for electrical related purposes;
- Very high transparency-excellent candidate where transparency is required; and
- The thinnest 2D materials- excellent where thin sheets with excellent conductivity, tensile strength and transparency are required.

2.1.2 Mechanics of Graphene Sheets

Graphene is a 2D allotrope of carbon comprising of a single, carbon in a hexagonal and honeycomb lattice. The covalent bond formed between carbon atoms makes graphene 200 times stronger than steel with very lightweight [13]. In addition, graphene possesses the following mechanical properties [14–16]:

- Ultra-flexible and can be folded or bent several times;
- Has compact structure and tough materials;
- Can be stretched without inducing defects- however, during stretching its electrical and magnetic properties can be altered; and
- Have high in-plane stiffness for pristine graphene.

Various reports have shown that graphene is one of the strongest materials that existed. The mechanical properties of graphene sheets under tension have been investigated and the result confirmed that graphene has excellent tensile strength and found Young's modulus of 0.5TPa for the graphene sheet [17]. The strength of graphene has been also investigated against shear force [18] and reported 0.213, 0.228 and 0.233 TPa of shear modulus for zigzag, armchair and chiral graphene sheets, respectively. The mechanical properties of graphene sheets under shear deformation have been reported somewhere else [19] and the authors claimed that graphene performs well under shear deformation and can be a permissible candidate under such deformations.

The paper written somewhere else [15] addresses the various mechanical properties of graphene:

(1) *Elastic stiffness*: the extent to which a graphene material resists deformation with respect to an applied force called stiffness. The stiffness of graphene comes from bending and stretching actions.
(2) *Graphite 3D elastic modulus*: The elastic modulus in this regard has paramount importance for graphite materials.
(3) *In-plane graphene*: the in-plane Young's modulus properties have a great impact as one of the mechanical properties of graphene.
(4) *Out-of-plane*: in graphene, stiffness as out-of-plane properties are very important to fight back against challenges occurred.

2.1.3 Mechanics of Carbon Nanotube

Carbon nanotubes (CNTs) are long and hallow cylinder versions of graphene, but possess different properties in axial and radial directions. Therefore, there is no wonder that it is also one of the strongest materials that is found in nature. CNT possesses a Young's modulus of 270–950 GPa and a tensile strength of 11–63 GPa as claimed by Yu et al. [20]. This indicates that CNT has an outstanding mechanical property in terms of strength, stiffness and having very low density [21] which makes CNT a perfect candidate for a candidate where high performance applications are required. However, the direct measuring of the mechanical properties of CNT more specifically single walled carbon nanotubes (SWCNT) is very difficult due to their size [22]. Due to this reason, the indirect measurements such as mechanical simulation, using atomic force microscopy or any other theoretical modelling have been used to estimate the mechanical properties of CNT. Due to the fact that only estimation/simulation is possible on the mechanical characterization of CNT, uncertainty on the data and sample handling techniques are likely to occur [23]. Therefore, elastic modelling and micromechanical characterization of CNT needs careful attention and precise measurement and simulation. CNT is further divided into single walled carbon nanotubes (SWCNT) and multiwalled carbon nanotubes (MWCNT).

Mechanics of SWCNT: Formed by a single layer of graphite arranged in a cylindrical shape. SWCNTs are very important types of CNT which possess very special properties which might not be possible by other CNTs. Zhenhua Yao et al. [24] simulated the Young's modulus and tensile strength of armchair SWCNT using molecular dynamics (MD) and found that these mechanical properties are 1$\sim$2 order higher than those of regular metal materials. In another study [25], SWCNT possessed an average breaking strength and young's modulus of 30 GPa and 1002 GPa, respectively. In contrast to MD simulation, Cho [26] pointed out that finite element method was found to be very easy to compute shear and bending moduli of SWCNT with respect to simple use, efficiency and cost wise. On the other hand, Poisson's ratio, bulk modulus, and elastic modulus plays an important role in determining the mechanical properties of SWCNT [23]. Overall, the mechanical properties of SWCNT are dependent on the strength of atom–atom (zigzag strong CC bond) where the theoretical Young's modulus and the tensile strength of SWCNT are much higher while it is much lighter when compared with steel.

Mechanics of MWCNT: It is a hallow cylindrically-shaped allotrope of carbon with high aspect ratio and its walls are shaped by multiple one-atom-thick sheets of carbon as indicated in the name. However, the multi-walls are interconnected together by relatively weak Vander Waals force of attraction [23]. Therefore, the mechanical properties of the MWCNT are highly dependent on the nature of weak Vander Waals force. The tensile strength of MWCNT has been investigated and the Young's modulus was found to be as much as 270 to 950 GPa which is sufficient enough to resist the high tensile load. Molecular dynamics simulation (MDS) has been employed to observe the resistance of MWCNT under bending deformations and found that the initial buckling was not dependent on the thickness of MWCNT and the result showed that Vander Waals force of attraction has little effect on bending deformations [27]. Several studies showed that addition of MWCNT into other materials as reinforcement materials increases the tensile, compressive, bending and mechanical properties of the composite [28].

2.1.4 Mechanics of Titanium Dioxide

Titanium dioxide (TiO_2) is one type of nanomaterial with excellent properties such as transparency, ultraviolet absorption etc. The discovery of TiO_2 in the science area leads to moving one step ahead on the change in the mechanical and electrical properties of the composite world. It is broadly used in the production of ceramics, application in coatings, in the paper-making industry, plastics manufacturing, fibre industries, inks manufacturing and so forth. This is because the addition of TiO_2 increases the tensile strength of resins [29, 30], enhances the mechanical properties of nanocomposites [31], and TiO_2 possesses excellent hardness and fracture toughness [32].

The other very important property of titanium dioxide is its ability to resist wear and scratch and abrasion resistance. Some efforts have been made to enhance the corrosion and abrasion resistance using titanium dioxide and found a permissible result [33]. When a material is subjected to tensile load owing to break due to less resistance of the applied load. Nowadays, titanium dioxide has been incorporated in many composite materials to enhance the resistance against heavy loads. One method to combat this breakage is by making nanocomposites using TiO_2. For instance, producing micro composites of polystyrene using TiO_2 filler help in increasing the tensile modulus of the composite [34]. In addition, coating of aramid fibre using TiO_2 nanoparticles increases the interfacial strength of the fibre by 40–67% while keeping the original bulk strength of the aramid fibre [35]. TiO_2 can also help to keep the strength of the cotton fabrics without losing several hours of photocatalytic decomposition [36].

3 Polymer Nanocomposites

3.1 Introduction

Polymer nanocomposites have a unique feature that it enhances the mechanical, thermal, gas and organic vapour barrier properties of intended materials [37]. No wonder how many scholars are researched so many times in this area. However, there are still challenges are lack of property-property relation models [38]. Indeed, the polymer part provides flexibility, lightweight, easy fabrication and assembly at a very low cost when compared. Therefore, polymer nanocomposites (PNCs) is a radical shift from the conventional manufacturing system [38]. Due to the additions of polymer matrix to the nanocomposites, not only the cost and flexibility, but also fundamental properties such as mechanical properties, biodegradability, thermal stability, flame resistant property and gas barrier properties can also be improved [39]. A representative of polymer nanocomposite can be shown as in Fig. 3.

In the following chapter, the mechanicals of polymer nanocomposites in various perspectives will be discussed.

As many research outputs indicate, polymer nanocomposites are the hot issues in recent years. From 1985 to 2022, the number of documents published on polymer nanocomposite materials has increased tremendously showing that the polymer nanocomposite era is growing rapidly (Fig. 4).

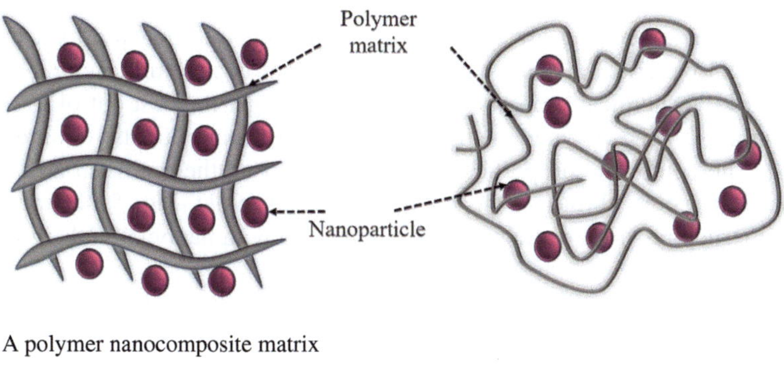

Fig. 3 A polymer nanocomposite matrix

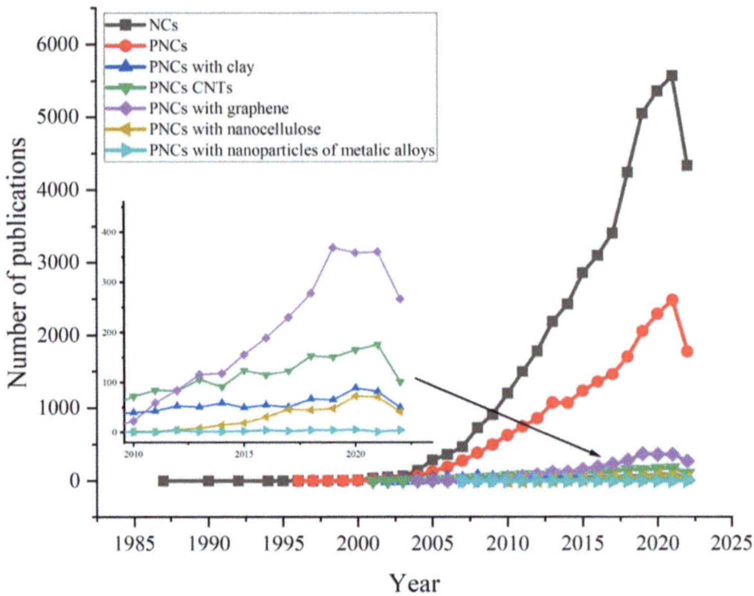

Fig. 4 The increasing trends of polymer nanocomposites per year on the topic of nanocomposites (NC), polymer nanocomposites (PNCs), polymer nanocomposites with clay (PNCs with clay), polymer nanocomposites with carbon nanotube (PNCs with CNTs), polymer nanocomposites (PNCs) with graphene, polymer nanocomposites (PNCs) with nano cellulose and polymer nanocomposites (PNCs) with the nanoparticles of the metallic alloys (as obtained from MEDLINE databases and accessed on 14/10/2022)

3.2 Structure of Polymer Nanocomposites

Composite manufacturing always plays an important role that the total behavior of individual materials is changed categorically that is structural modifications as well as physicochemical properties [40]. In order to get the best out of it, polymer nanocomposites, the polymer matrix should be distributed to the optimum position [8]. The composite properties therefore depend on the optimum arrangement and matrix forming ability of the polymers. For example, Fig. 5 illustrates three different structures of how the polymer is matrixed with nanomaterials in this case clay nanomaterials.

There are three ways to make polymer nanocomposites: in situ polymerization, solution blending, and melt blending. The kind of polymeric matrix, the nanofiller, and the desired qualities for the finished goods are taken into consideration while choosing a suitable process [41].

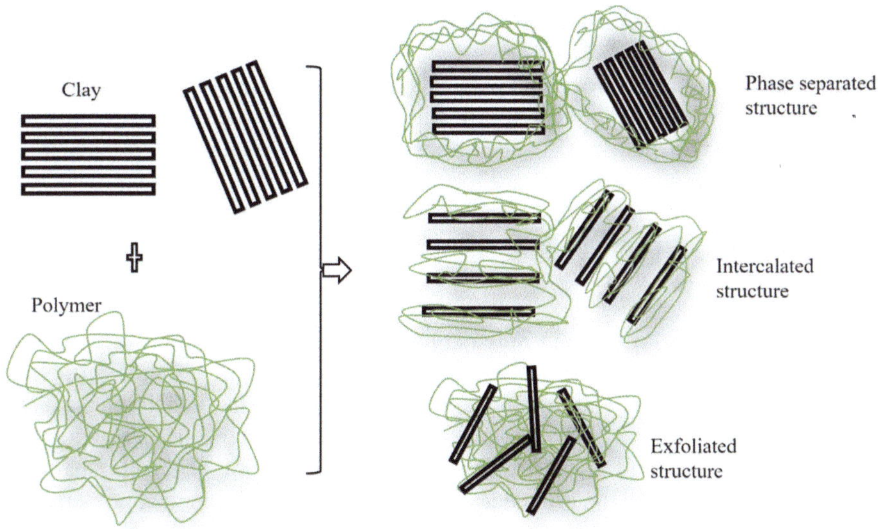

Fig. 5 Polymer nanocomposite structure (adapted from [8])

3.3 Mechanics of Polymer Nanocomposites (PNCs)

The mechanics of polymer nanocomposite is very crucial to get the desirable properties out it. The mechanics of the polymer nanocomposite materials is different from at the macroscale level [42]. Researchers have used various in situ mechanical tests to differentiate this. PNCs are hybrid materials that employed polymer material as a matrix and that of the nanomaterials as the filler. Depending of the synchronization and the types of polymers and to that of the properties of nanomaterials, the mechanical properties also varied. In general, the matrix can provide excellent shear properties with low densities while the filament (filler) can provide high strength, high stiffness and low density. The final composite (the polymer nanocomposite) possesses the combination of these that is high strength, high stiffness, excellent shear properties and finally low density. The mechanics of nanocomposites cannot be discussed in bulk and could be discussed separately as PNCs are defined as polymers in which very small nanofillers (from 1 to 100 nm range) consistently mixed by several other materials by %wt. Therefore, this sub-chapter aims to review the mechanics of polymer nanocomposites with different types of nanofillers with possible enhancement in mechanical strength, toughness, bending strength and other mechanical properties.

3.3.1 Mechanical Property of PNCs Based on Nanoclays

Polymer nanocomposites processed with nanoclays (hybrid organic–inorganic nanomaterial and serve as rheological modification and gas absorbers) provides a potential alternative to conventional nanocomposites due to their upgraded mechanical properties [43]. That is why there are lots of commercial products using nanoclays. The introduction of nanoclays to form polymer–clay-nanocomposites has shown an improved tensile strength of the composite materials [44]. Storage modulus is an important mechanical property in which energy gets back after force is applied on the material.

Addition of nanoclays during polymerisation to produce polymer nanocomposites improves the storage modulus up to ~170% [45]. Fatigue, the most important mechanical property of a material in response to cyclic loading which means materials resistance to fracture. Nano clay enhances the fatigue characteristics of polymer nanocomposites and the fatigue life depends on the dispersion of the clay particles [46]. Table 1 summarises the influence of nanoclays on the mechanical properties of polymer nanocomposites.

Overall, nanoclays plays an important role in shifting/improving the mechanical/viscoelastic behaviors of polymer nanocomposites. However, processing conditions play an important role on the stratification and diffusion of clay additives in the polymer nanocomposite matrix.

3.3.2 Mechanics of PNCs Based on Carbon Nano Tubes

Carbon nanotubes (CNTs) are ultrafine carbon-based fibres with extremely in size in both diameter and length wise that is with nanometres-sized diameter and micrometre sized length. It comprises of graphite sheet distributed in honeycomb lattices [57]. The carbon atoms in CNT have arranged in a planar-hexagonal system. The mechanics of CNTs could then depend on the mechanics of graphite carbon. CNTs exhibit exceptional elastic modulus and might reach up to 1 TPa (MWCNTs) [58]. That is why CNTs are the perfect candidate for composite manufacturing. Young's modulus has been projected to be 0.9 TPa [59] where this and other mechanical properties leads to the high consumption of CNTs for the manufacture of PNCs.

Superior mechanical properties have been achieved in the production of PNCs with CNTs [60]. However, these PNCs matrixed with CNTs as a nanofiller have faced some difficulties in obtaining the perfect interaction. At a level where the assumption of continuum level interactions is not valid, CNTs physically connect with polymeric matrix [61]. The mechanical properties of polyamide have been greatly improved when a c composite is made from CNT [62]. Table 2 summarises the improvement in some mechanical properties of PNCs featured with CNTs.

All these results indicated that CNTs have high strength, aspect ratio as well as high stiffness which means the addition of CNTs during the production of PNCs can be benefited from these high values.

Table 1 Influence of nanoclays on the mechanical properties of PNCs

Composites	Mechanical properties	Effect on mechanical properties	Refs.
PP/wood flour	Flexural modulus (fm), tensile modulus (tm), elongation at break (eb), and impact strength	fm, tm, and eb increases up to 3 [a]phc nanoclay and then decreases; impact strength decreases	Hemmasi et al. [47]
Wood polymer	Storage modulus, dynamic young's modulus	Both show significant improvement	Islam et al. [48]
PMMA	Elastic modulus, tensile strength, and elongation at break	Improved performance but decreased with excessive nanoclay	Fu and Naguib [49]
Bitumen	Stiffness	Improved stiffness	Jahromi and Khodaii [50]
Nylon 6	Friction and wear resistance	Superior friction and wear resistance	Srinath and Gnanamoorthy [51]
Epoxy	Compressive property	Depend on the degree of exfoliation of the clay nanoplatelets	Jumahat et al. [52]
PP/bagasse	Tensile strength, impact strength	Increased but decrease at 4% nanoclay, impact decreased	Nourbakhsh and Ashori [53]
Polyamide	Elastic modulus and stiffness	Both improved	Baniassadi et al. [54]
Bamboo/kenaf	Tensile, flexural, and impact	Enhanced mechanical properties	Chee et al. [55]
Montmorillonite/PP	Storage modulus, loss modulus	Significantly higher than virgin	Venkatesh et al. [56]

[a] Per hundred compounds

3.3.3 Mechanics of PNCs Based on Nanoparticles of Metallic Alloys

PNCs with nanoparticles have acknowledged considerable attention as they are used in a wide range of applications where mechanical performance is a priority requirement. Among the metallic nanoparticles, silver nanoparticles (AgNPs) and gold nanoparticles (AuNPs) have got quite conceivable consideration due to their unique physical properties [69]. However, there are still difficulties in achieving the desired properties as it is tough to get an even dispersion of metals on the polymer matrix [70]. The basic advantages of PNCs based on metallic alloys are having a high degree of strength and stiffens compared to the pure polymer, their mechanical properties are potentially superior. Therefore, adding metallic alloys to form PNCs is paramount important this is because there is an increasing demand of materials with superior mechanical properties.

Table 2 Influence of CNTs on the mechanical properties of PNCs

Composites	Mechanical properties	Effect on mechanical properties	Refs.
Polyamide 12	Flexural, impact, and tensile	All properties enhanced	Bai et al. [62]
Polystyrene	Young's modulus	Enhanced	Cruz-Chú et al. [63]
Epoxy resin	Tensile strength	Greatly increased	Gu et al. [64]
TiO_2	Tensile, elongation	Enhanced	Alghamdi and Rajeh [65]
Epoxy/amine	Flexure, quasi-static compression and strain rate	Better properties	Hosur et al. [66]
Epoxy-based	Elastic modulus	Enhanced	Wang et al. [67]
Fuzzy fibre-reinforced	Elastic modulus, Poisson's ratio	Depends on the off-axis angle	Hassanzadeh-Aghdam et al. [68]

3.3.4 Mechanical Property of PNCs Based on Graphene

Graphene has been explored to a greater extent where a special emphasis has been given in the area of nanoscience due to its account of mechanical properties after its discovery since 2007 by Geim and Novoselov [71]. Graphene-based PNCs have been given a special attention due to its load-bearing capability [72]. Potts [73] has reviewed a lot of research works on graphene-based PNCs and their mechanical properties. According to this review work, elastic modulus, tensile strength and other mechanical properties have been improved due to graphene fillers. Table 3 summarises the effect of graphene-based fillers on the mechanical properties of PNCs.

All the results and other review works [82] revealed that graphene and modified graphene-based PNCs have shown improved mechanical properties. In addition to its mechanical strength, graphene can also bring lightweight and tough materials when a composite is made along with the polymers [81].

4 Future Outlook

The pace of very fast findings now in nanomaterials and polymer nanocomposites is predictable to rush in the next years in the science arena. These findings will have a great impact on existing applications of nanotechnology where a great mechanical performance is required somewhere else in the industry. Therefore, potential applications with high mechanical performance can be expected in the areas of nanomaterials and polymer nanocomposites.

Table 3 Effect of graphene-based fillers on the mechanical property of PNCs

Matrix polymer	Filler type	Mechanical properties influenced	Refs.
Polyurethane	G–O	Tenfold increase in tensile stiffness	Kim et al. [74]
PVDF	FGS[a]	Storage modulus increased	Ansari and Giannelis [75]
Poly (vinyl chloride)	GNF[b]	58% increase in Youngs modulus, 130% increase in tensile strength	Vadukumpully et al. [76]
Poly (styrene)	G–O	Improved mechanical stiffness	Stankovich et al. [77]
Epoxy matrix	Graphene	Young's modulus improved	Giannopoulos and Kallivokas [78]
PVA	G–O	Enhanced modulus	Compton et al. [79]
Paraffin	Graphene	~11% increase in tensile strain	Wang et al. [80]
Chitosan	GNs	Enhanced tensile strength	Lu et al. [81]

[a] Functionalized graphene sheet; [b] graphene nanoflakes

5 Conclusions

Nanomaterials and polymer nanocomposites represent one of the most outstanding candidates in various application fields specifically when mechanical performance is a mandatory requirement. However, there are still many challenges in the composite manufacturing to reach their high level. For instance, the interaction of the nanomaterials with that of polymer matrix is not disappeared well and has faced stability issues. In addition, some nanomaterials have faced some problems regarding environmental toxicity. Furthermore, it is a very challenging task to produce polymer nanocomposites at a very large scale due to complex steps during fabrication. If the interaction of the nanofiller is controlled and chain flexibility optimized and finally if the size of the nanomaterials is controlled, nanomaterials and polymer nanomaterials will provide a significant contribution for various application areas.

Acknowledgements This work was assisted financially by Alexander Von Humboldt foundation at Albstadt-Sigmaringen University via Georg Forster scheme.

Conflicts of Interest The authors declare no conflict of interest.

References

1. Tadesse MG (2019) Quality inspection and evaluation of smart or functional textile fabric surface by skin contact mechanics. Högskolan i Borås
2. Gebeyehu EK, Sui X, Adamu BF, Beyene KA, Tadesse MG (2022) Cellulosic-based conductive hydrogels for electro-active tissues: a review summary. Gels 8(3):140
3. Tadesse MG, Kasaw E, Lübben JF (2023) Valorization of banana peel using carbonization: potential use in the sustainable manufacturing of flexible supercapacitors. Micromachines 14(2):330
4. Tadesse MG, Loghin C, Chen Y, Wang L, Catalin D, Nierstrasz V (2017) Effect of liquid immersion of PEDOT: PSS-coated polyester fabric on surface resistance and wettability. Smart Mater Struct 26(6):065016
5. Tadesse MG, Lübben JF (2023) Review on hydrogel-based flexible supercapacitors for wearable applications. Gels 9(2):106
6. Tadesse MG, Mengistie DA, Chen Y, Wang L, Loghin C, Nierstrasz V (2019) Electrically conductive highly elastic polyamide/lycra fabric treated with PEDOT: PSS and polyurethane. J Mater Sci 54(13):9591–9602
7. Gogotsi Y (2006) Nanomaterials handbook, 1st edn. CRC press
8. Passador F, Ruvolo-Filho A, Pessan L (2017) Nanocomposites of polymer matrices and lamellar clays. In: Nanostructures. Elsevier, pp 187–207
9. Rahimi MR, Mosleh S (2021) Intensification of sorption processes: active and passive mechanisms. Elsevier
10. Jeevanandam J, Barhoum A, Chan YS, Dufresne A, Danquah MK (2018) Review on nanoparticles and nanostructured materials: history, sources, toxicity and regulations. Beilstein J Nanotechnol 9(1):1050–1074
11. George S (2015) Nanomaterial properties: implications for safe medical applications of nanotechnology. In: Nanotechnology in endodontics. Springer, pp 45–69
12. Vajtai R (2013) Springer handbook of nanomaterials. Springer Science & Business Media
13. Du X, Chen R, Liu F (2017) Investigation of graphene nanosheets reinforced aluminum matrix composites. Dig J Nanomater Biostruct(DJNB) 12:37–45
14. Andrew RC, Mapasha RE, Ukpong AM, Chetty N (2012) Mechanical properties of graphene and boronitrene. Phys Rev B 85(12):125428
15. Ovid'Ko I (2013) Mechanical properties of graphene. Rev Adv Mater Sci 34(1):1–11
16. Papageorgiou DG, Kinloch IA, Young RJ (2017) Mechanical properties of graphene and graphene-based nanocomposites. Prog Mater Sci 90:75–127
17. Frank I, Tanenbaum DM, van der Zande AM, McEuen PL (2007) Mechanical properties of suspended graphene sheets. J Vacuum Sci Technol B: Microelectron Nanometer Struct Process, Measurement, and Phenomena 25(6):2558–2561
18. Sakhaee-Pour A (2009) Elastic properties of single-layered graphene sheet. Solid State Commun 149(1–2):91–95
19. Min K, Aluru NR (2011) Mechanical properties of graphene under shear deformation. Appl Phys Lett 98(1):013113
20. Yu M-F, Lourie O, Dyer MJ, Moloni K, Kelly TF, Ruoff RS (2000) Strength and breaking mechanism of multiwalled carbon nanotubes under tensile load. Science 287(5453):637–640
21. Qian D, Wagner a, Gregory J, Liu WK, Yu M-F, Ruoff RS (2002) Mechanics of carbon nanotubes. Appl Mech Rev 55(6):495–533
22. Harik V (2018) Mechanics of carbon nanotubes: fundamentals, modeling and safety. Academic Press
23. Rafiei S (2015) Foundations of nanotechnology, vol 3. Mechanics of carbon nanotubes. CRC Press
24. Yao Z, Zhu C-C, Cheng M, Liu J (2001) Mechanical properties of carbon nanotube by molecular dynamics simulation. Comput Mater Sci 22(3–4):180–184
25. Yu M-F, Files BS, Arepalli S, Ruoff RS (2000) Tensile loading of ropes of single wall carbon nanotubes and their mechanical properties. Phys Rev Lett 84(24):5552

26. To CW (2006) Bending and shear moduli of single-walled carbon nanotubes. Finite Elem Anal Des 42(5):404–413
27. Chang T, Hou J (2006) Molecular dynamics simulations on buckling of multiwalled carbon nanotubes under bending. J Appl Phys 100(11):114327
28. Kebede Kassa M, Babu Arumugam A (2022) Bending response analysis of a laminated, tapered, curved, composite panel made from an agglomerated and wavy MWCNT–glass fiber–polymer hybrid. Trans Can Soc Mech Eng 46(1):103–131
29. Shirkavand S, Moslehifard E (2014) Effect of TiO2 nanoparticles on tensile strength of dental acrylic resins. J Dental Res, Dental Clinics, Dental Prospects 8(4):197
30. Sodagar A, Bahador A, Khalil S, Shahroudi AS, Kassaee MZ (2013) The effect of TiO2 and SiO2 nanoparticles on flexural strength of poly (methyl methacrylate) acrylic resins. J Prosthodont Res 57(1):15–19
31. Hosoda T, Yamada T (2014) Effect of TiO2 on morphology and mechanical properties of PVDF/PMMA blend films prepared by melt casting process. J Appl Polym Sci 131(13)
32. Das R, Ambardekar V, Bandyopadhyay PP (2021) Titanium dioxide and its applications in mechanical, electrical, optical, and biomedical fields. Titanium Dioxide-Adv Appl
33. Barth N, Schilde C, Kwade A (2014) Influence of electrostatic particle interactions on the properties of particulate coatings of titanium dioxide. J Colloid Interface Sci 420:80–87
34. Selvin TP, Kuruvilla J, Sabu T (2004) Mechanical properties of titanium dioxide-filled polystyrene microcomposites. Mater Lett 58(3–4):281–289
35. Wang B, Duan Y, Zhang J (2016) Titanium dioxide nanoparticles-coated aramid fiber showing enhanced interfacial strength and UV resistance properties. Mater Des 103:330–338
36. Qi K, Wang X, Xin JH (2011) Photocatalytic self-cleaning textiles based on nanocrystalline titanium dioxide. Text Res J 81(1):101–110
37. Vaia RA, Krishnamoorti R (2001) Polymer nanocomposites: introduction. In. ACS Publications
38. Vaia RA, Giannelis EP (2001) Polymer nanocomposites: status and opportunities. MRS Bull 26(5):394–401
39. Karak N (2019) Fundamentals of nanomaterials and polymer nanocomposites. In: Nanomaterials and polymer nanocomposites. Elsevier, pp 1–45
40. Hussain CM, Mishra AK (2018) New polymer nanocomposites for environmental remediation. Elsevier
41. Mallakpour S, Naghdi M (2018) Polymer/SiO2 nanocomposites: production and applications. Prog Mater Sci 97:409–447
42. Roy S, Ryan J, Webster S, Nepal D (2017) A review of in situ mechanical characterization of polymer nanocomposites: prospect and challenges. Appl Mech Rev 69(5)
43. Rafiee R, Shahzadi R (2019) Mechanical properties of nanoclay and nanoclay reinforced polymers: a review. Polym Compos 40(2):431–445
44. Zare Y, Rhee KY (2021) Effects of interfacial shear strength on the operative aspects of interphase section and tensile strength of carbon-nanotube-filled system: a modeling study. Results Phys 26:104428
45. Koyilapu R, Subhadarshini S, Singha S, Jana T (2021) An in-situ RAFT polymerization technique for the preparation of poly (N-vinyl imidazole) modified Cloisite nanoclay to develop nanocomposite PEM. Polymer 212:123175
46. Vijayashakthivel A, Krishnamurthy R, Ravisankar B, Karunakaran C (2018) Influence of nanoclay addition on the fatigue behavior of polymer nanocomposites. Mater Today: Proc 5(2):7225–7229
47. Hemmasi AH, Khademi-Eslam H, Talaiepoor M, Kord B, Ghasemi I (2010) Effect of nanoclay on the mechanical and morphological properties of wood polymer nanocomposite. J Reinforced Plastics Compos 29(7):964–971
48. Islam MS, Hamdan S, Talib ZA, Ahmed AS, Rahman MR (2012) Tropical wood polymer nanocomposite (WPNC): the impact of nanoclay on dynamic mechanical thermal properties. Compos Sci Technol 72(16):1995–2001
49. Fu J, Naguib HE (2006) Effect of nanoclay on the mechanical properties of PMMA/clay nanocomposite foams. J Cell Plast 42(4):325–342

50. Jahromi SG, Khodaii A (2009) Effects of nanoclay on rheological properties of bitumen binder. Constr Build Mater 23(8):2894–2904
51. Srinath G, Gnanamoorthy R (2005) Effect of nanoclay reinforcement on tensile and tribo behaviour of Nylon 6. J Mater Sci 40(11):2897–2901
52. Jumahat A, Soutis C, Mahmud J, Ahmad N (2012) Compressive properties of nanoclay/epoxy nanocomposites. Procedia Eng 41:1607–1613
53. Nourbakhsh A, Ashori A (2009) Influence of nanoclay and coupling agent on the physical and mechanical properties of polypropylene/bagasse nanocomposite. J Appl Polym Sci 112(3):1386–1390
54. Baniassadi M, Laachachi A, Hassouna F, Addiego F, Muller R, Garmestani H, Ahzi S, Toniazzo V, Ruch D (2011) Mechanical and thermal behavior of nanoclay based polymer nanocomposites using statistical homogenization approach. Compos Sci Technol 71(16):1930–1935
55. Chee SS, Jawaid M, Alothman OY, Fouad H (2021) Effects of nanoclay on mechanical and dynamic mechanical properties of bamboo/kenaf reinforced epoxy hybrid composites. Polymers 13(3):395
56. Venkatesh G, Deb A, Karmarkar A, Chauhan SS (2012) Effect of nanoclay content and compatibilizer on viscoelastic properties of montmorillonite/polypropylene nanocomposites. Mater Des 37:285–291
57. Ebbesen TW (1994) Carbon nanotubes. Annu Rev Mater Sci 24(1):235–264
58. Lau K-T, Chipara M, Ling H-Y, Hui D (2004) On the effective elastic moduli of carbon nanotubes for nanocomposite structures. Compos B Eng 35(2):95–101
59. Demczyk BG, Wang YM, Cumings J, Hetman M, Han W, Zettl A, Ritchie R (2002) Direct mechanical measurement of the tensile strength and elastic modulus of multiwalled carbon nanotubes. Mater Sci Eng: A 334(1-2):173–178
60. de Oliveira AD, Beatrice CAG (2018) Polymer nanocomposites with different types of nanofiller. Nanocompos-Recent Evol 103–104
61. Pal G, Kumar S (2016) Modeling of carbon nanotubes and carbon nanotube–polymer composites. Prog Aerosp Sci 80:33–58
62. Bai J, Goodridge RD, Hague RJ, Song M (2013) Improving the mechanical properties of laser-sintered polyamide 12 through incorporation of carbon nanotubes. Polym Eng Sci 53(9):1937–1946
63. Cruz-Chú ER, Villegas-Rodríguez GJ, Jäger T, Valentini L, Pugno NM, Gkagkas K, Gräter F (2020) Mechanical characterization and induced crystallization in nanocomposites of thermoplastics and carbon nanotubes. npj Comput Mater 6(1):1–11
64. Gu H, Tadakamalla S, Zhang X, Huang Y, Jiang Y, Colorado HA, Luo Z, Wei S, Guo Z (2013) Epoxy resin nanosuspensions and reinforced nanocomposites from polyaniline stabilized multi-walled carbon nanotubes. J Mater Chem C 1(4):729–743
65. Alghamdi HM, Rajeh A (2021) Synthesis of carbon nanotubes/titanium dioxide and study of its effect on the optical, dielectric, and mechanical properties of polyvinyl alcohol/sodium alginate for energy storage devices. Int J Energy Res
66. Hosur MV, Rahman T, Brundidge-Young S, Jeelani S (2010) Mechanical and thermal properties of amine functionalized multi-walled carbon nanotubes epoxy-based nanocomposite. Compos Interfaces 17(2–3):197–215
67. Wang X, Wang J, Biswas S, Kim H, Nam I (2020) Mechanical, electrical, and piezoresistive sensing characteristics of epoxy-based composites incorporating hybridized networks of carbon nanotubes, graphene, carbon nanofibers, or graphite nanoplatelets. Sensors 20(7):2094
68. Hassanzadeh-Aghdam M, Mahmoodi M, Ansari R (2018) Micromechanics-based characterization of mechanical properties of fuzzy fiber-reinforced composites containing carbon nanotubes. Mech Mater 118:31–43
69. Tamayo L, Palza H, Bejarano J, Zapata PA (2019) Polymer composites with metal nanoparticles: synthesis, properties, and applications. In: Polymer composites with functionalized nanoparticles. Elsevier, pp 249–286
70. Soytaş SH, Oğuz O, Menceloğlu YZ (2019) Polymer nanocomposites with decorated metal oxides. In: Polymer composites with functionalized nanoparticles. Elsevier, pp 287–323

71. Geim AK, Novoselov KS (2007) The rise of graphene. Nat Mater 6(3):183–191
72. Lodhi RS, Kumar P, Achuthanunni A, Rahaman M, Das P (2022) Mechanical properties of polymer/graphene composites. In: Polymer nanocomposites containing graphene. Elsevier, pp 75–105
73. Potts JR, Dreyer DR, Bielawski CW, Ruoff RS (2011) Graphene-based polymer nanocomposites. Polymer 52(1):5–25
74. Kim H, Miura Y, Macosko CW (2010) Graphene/polyurethane nanocomposites for improved gas barrier and electrical conductivity. Chem Mater 22(11):3441–3450
75. Ansari S, Giannelis EP (2009) Functionalized graphene sheet—Poly (vinylidene fluoride) conductive nanocomposites. J Polym Sci, Part B: Polym Phys 47(9):888–897
76. Vadukumpully S, Paul J, Mahanta N, Valiyaveettil S (2011) Flexible conductive graphene/poly (vinyl chloride) composite thin films with high mechanical strength and thermal stability. Carbon 49(1):198–205
77. Stankovich S, Dikin DA, Dommett GH, Kohlhaas KM, Zimney EJ, Stach EA, Piner RD, Nguyen ST, Ruoff RS (2006) Graphene-based composite materials. Nature 442(7100):282–286
78. Giannopoulos GI, Kallivokas IG (2014) Mechanical properties of graphene based nanocomposites incorporating a hybrid interphase. Finite Elem Anal Des 90:31–40
79. Compton OC, Cranford SW, Putz KW, An Z, Brinson LC, Buehler MJ, Nguyen ST (2012) Tuning the mechanical properties of graphene oxide paper and its associated polymer nanocomposites by controlling cooperative intersheet hydrogen bonding. ACS Nano 6(3):2008–2019
80. Wang Y, Yang C, Mai Y-W, Zhang Y (2016) Effect of non-covalent functionalisation on thermal and mechanical properties of graphene-polymer nanocomposites. Carbon 102:311–318
81. Lu H, Chen Z, Ma C (2012) Bioinspired approaches for optimizing the strength and toughness of graphene-based polymer nanocomposites. J Mater Chem 22(32):16182–16190
82. Kumar A, Sharma K, Dixit AR (2019) A review of the mechanical and thermal properties of graphene and its hybrid polymer nanocomposites for structural applications. J Mater Sci 54(8):5992–6026

Dynamic Mechanical Behavior of Polymer Nanocomposites

Heitor Luiz Ornaghi Jr., Lídia Kunz Lazzari, Eduardo Fischer Kerche, and Roberta Motta Neves

Abstract Nanocomposites are a relatively new class in the composites field since deals with reinforcements in nanoscale. Nanomaterials have more superficial area and consequently have more interaction with matrices, enhancing materials properties. Among the techniques for characterize nanocomposites, dynamic mechanical analysis (DTMA) stands out as a very versatile technique since it can relate various mechanical responses versus temperature since polymeric materials have viscoelastic behavior. In this regard, DMTA explores the viscoelastic properties such as storage modulus (E'), loss modulus (E"), tandelta or damping factor (δ), of polymeric nanocomposites. DMTA is a powerful tool to extract information about interfacial bonding between the reinforcement and matrix, stiffness, polymeric chain movements, relaxation processes and, might predict the effects of time and temperature on the nanocomposites. This chapter details with the dynamic mechanical behavior of polymeric nanocomposites, showing the importance of this technique for further research related to polymeric nanocomposites materials, combining theory with relevant practical studies in the area.

Keywords Dynamic mechanical thermal analysis · Viscoelastic behavior · Nanofiller

H. L. Ornaghi Jr.
Mantova Indústria de Tubos Plásticos Ltda., Caxias do Sul, Brazil

L. K. Lazzari · E. F. Kerche
Ford Motor Company/Instituto Euvaldo Lodi, Camaçari, Bahia 42810-225, Brazil

R. M. Neves (✉)
Post Graduate Program in Engineering of Processes and Technologies/Caxias do Sul University, Caxias do Sul, Brazil
e-mail: robertamneves@gmail.com

© The Author(s), under exclusive license to Springer Nature Singapore Pte Ltd. 2023
H. Abdellaoui et al. (eds.), *Mechanics of Nanomaterials and Polymer Nanocomposites*, Smart Nanomaterials Technology,
https://doi.org/10.1007/978-981-99-2352-6_8

1 Dynamic Mechanical Thermal Analysis Fundaments

The dynamic mechanical thermal analysis (DMTA) is a technique employed to study the properties of polymeric materials (but not restricted to) in function of temperature, time, or frequency [29], being also used to characterize creep or stress relaxation at low stress/strains [2, 27, 41]. One of the main goals is to correlate the macroscopic properties (e.g., elastic modulus) with molecular relaxations caused by microscopic deformations via molecular motion [11].

It is important to have in mind that the viscoelastic nature of polymeric materials allows different curves depending on the time, temperature, or frequency used. As an example, Fig. 1 can be considered below of a hypothetic polymer where the stress vs. strain curves are given from distinct temperatures. It is noted that the elastic modulus slightly increases by decreasing the temperature due to the restriction imposed by the chain backbone at lower temperatures, which needs more force (energy) to strain the polymer [14].

The main difference is observed after the yield point (plastic region) where a more ductile characteristic is observed. Microscopically, by decreasing the temperature the number of conformational states of the backbone also decrease. With lower thermal energy imposed in the system, the polymeric chains are close together and hence more energy is required to deform the polymer [7]. By increasing the temperature, the number of conformational states at a given temperature increase (the segment of the polymer chain can occupy several spaces with given energy), and the chains are more apart each other. In this case, more free volume is available and more space among the polymer chain exists, being easier to deform the polymer. It is note to worthy that at given temperature, distinct elastic modulus can be obtained for different polymers due to distinct chemical structures—aromatic rings that tend to be more rigid and difficult for the molecular motion in the glassy state while oxygen tends to facilitate the rotation of the bond on the backbone [17, 24].

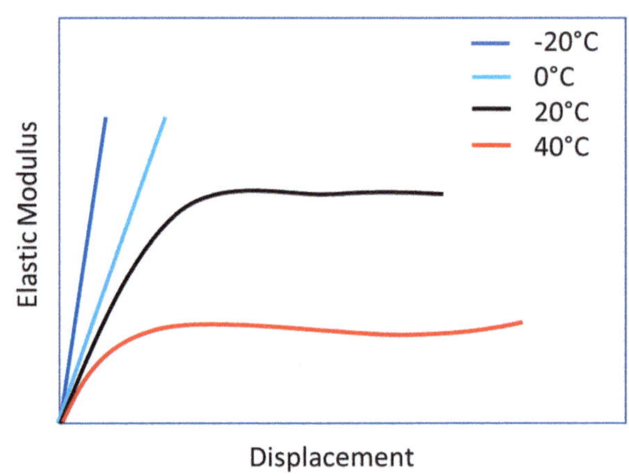

Fig. 1 Stress versus strain modulus in function of distinct temperatures for a distinct polymer

To better comprehend this behavior, the viscoelastic nature of the polymer as well as the fundaments of the DMTA are explained below. The principle of DMTA is based on an oscillate tension applied on be polymer given a shift (deformation) and a resulted amplitude. As the force is applied in the senoid form, the modulus can be expressed as the phase component (storage modulus), and the out-of-phase component (loss modulus). The reason loss/storage results in the tan δ (damping), which is a measurement of the glass transition of the polymer [28, 34].

Figure 2a, b shows the schematic representation of the in phase and out-of-phase features and the storage/loss characteristic. The spring component represents the storage component while the dashpot represents the viscous component. When the force is applied in the spring, all energy is recovered by removing the tension applied. On the other hand, if the force is removed from the dashpot, the energy is not totally recovered, and the recovering process is slower in comparison to the spring. Phenomenologically and considering the temperature for a better understanding, after imposed a tension, the polymer can behave in two different ways: when the deformation is imposed at lower temperatures, after removal of any strain, the polymer will return to its initial conformational state [13, 28].

In the case of higher temperatures (in the elastomeric region after T_g), the higher internal energy imposed by the temperature allows different conformational states and different polymeric segments diffuse each other (Brownian motion), retarding the motion and increasing the local viscosity. In addition, after removal the initial configuration is impossible to be achieved because the initial configurational state changed, and the polymeric chains ca do not return to their original states [14].

Regarding Fig. 2b, can be noted that the rubber and the tennis balls return to different positions after being dropped from the hand. The storage energy of the rubber ball is higher than the tennis ball and consequently the loss energy is lower. Phenomenological is a means that the ability to store energy after some stress/deformation imposed is recovered in a greater amount, returning almost to its original state/position. In other words, higher the ability to store energy of a polymeric material, lower will be the loss energy (dissipation) and properties as the elastic modulus tend to maintain the same in a distinct temperature range if no changes in this capacity are achieved [35, 36].

DMTA is a characterization method in which the viscoelastic parameters related to the behavior of the material, as the storage modulus, loss modulus, and damping factor are determined. Besides, it is possible to determine primary, secondary, and tertiary relaxations, as the glass transition temperature and other segmental movements. The great number of clamps allows the determination of these parameters under different forms, as compression, tensile of films, submersion mode, three-point bending, single- and dual-cantilever, among others [1, 18, 26]. All types of tests (creep, stress relaxation, flexure) in different clamps are done in accordance with ASTM D4092 [4]. One of the major advantages over other mechanical tests is the specimen size (usually 50 mm × 10 mm × 3 mm) that is considerably lower than tensile, or flexure tests and the temperature range tested. Also, the storage and loss moduli are obtained simultaneously. It worthy to mention that the dimensions will depend on the type of material to be analyzed (e.g., materials with low modulus

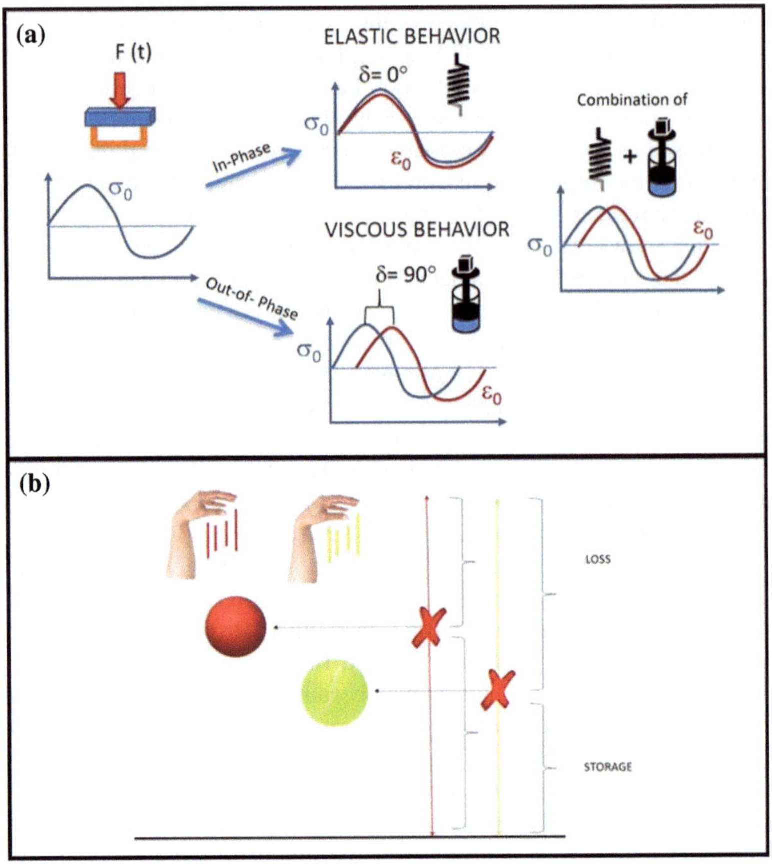

Fig. 2 **a** σ × ε when applied a force showing the δ for in phase and out-of-phase and **b** ball releasing showing the store capacity for a rubber ball and a tennis ball. Adapted from [38]—Accessed 10/05/2022

require more width specimens, while the opposite is valid for materials with higher modulus [3].

Briefly, all transition/events that provide significant molecular motion (not only in the backbone chain) will decrease the ability to storage energy and hence the modulus tends to decrease, as presented in Fig. 3. The transitions below T_α (T_g) have low influence in the elastic modulus and are related to ramification relaxations, side groups relaxations and motion of other small segments that involve at least four carbon atoms [12]. The more abrupt decay related to the main transition (T_g) is related to an abrupt expansion of the main backbone in a short temperature range, decreasing the capacity to store energy and dissipating a higher amount of energy as heat. All polymers have a glass transition temperature (T_g) in which an abrupt decay of the modulus (in order of 3–4 magnitude order) occurs. Since this transition is related to the amorphous portion of the polymeric chain, higher the amount of

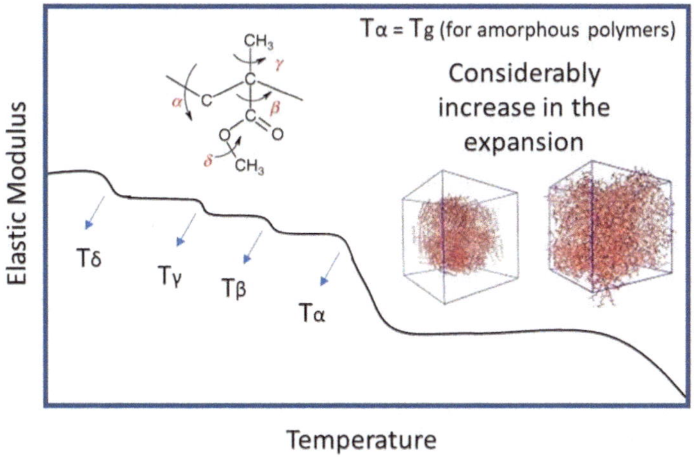

Fig. 3 Schematic representation of the DMTA curve and the main transitions of a poly (methyl methacrylate)

crystalline content, lower the tendency to decrease the magnitude of the modulus at T_g. If a polymer had no amorphous regions, no T_g would appears.

2 Types of Fillers

Polymer matrix composite materials offer many essential benefits over traditional polymers and the addition of fillers to polymers considerably increases their physical, mechanical and dynamic properties [5]. Some carbonaceous fillers used for the improvement of dynamical mechanical response of composites include multi wall carbon nanotubes (MWCNT), single wall carbon nanotubes (SWCNT), graphene (Gr). Furthermore, nano silica (NSi), nano clays and polyhedral oligomeric silsesquioxane (POSS) [37] are other conventional fillers used into polymeric matrices aiming to enhance viscoelastic response of such nano composites.

In recent years, the use of natural fibers (coir, rice husk, banana leaf, date palm, coconut shells, peanut shells and groundnut shells) has increased as reinforcements in composites due to their low cost, reuse of agricultural waste and the replacement of conventional materials. Several authors focus on natural fibers fillers with various polymer resins, but hybrid composites with synthetic fibers were also researched, with the objective of improving the mechanical and thermal properties of composites. In addition to these properties, natural fibers have low thermal conductivity, which makes them useful in the production of composites for thermal insulation. Epoxy resin is the most used resin for applications in the automotive and industrial sector, having a good bonding with natural fibers [31, 42].

The functionalization of microcrystalline cellulose (MCC) with 3-(aminopropyl) triethoxysilane (APTES), produces composites with epoxy resin, reported by Neves et al. [32] that increased the storage modulus in the glassy region and the loss modulus compared to their nonfunctionalized counterparts. These results demonstrate good dispersion of the filler in the epoxy resin and that the effect of the filler is more pronounced in the rubbery state and the chemical modification in the glassy state. Additionally, authors in [15] analyze the effect of fiber content and chemical treatments on the dynamic mechanical or viscoelastic behavior of PP/jute yarn composites. This study concluded that increasing of fiber content leads to increasing the storage and loss modulus of the composite. Neves et al. [32] studied thermal and dynamic mechanical behavior of epoxy composites reinforced with post-consumed yerba mate (YM) without chemical treatment. The authors concluded that the addition of yerba mate was strongly evidenced in the dynamic-mechanical properties, since with the addition of wt% of YM, all parameters showed significant increases.

Kamble et al. [21] developed a new type of nano/micro composites by utilizing cotton textile waste, reduced graphene oxide (rGO), and hemp fiber (HF) microparticles, along with the epoxy matrix. The addition of reinforcements provided an increase in the storage modulus at low concentrations of rGO (0.1–0.3wt%), indicating that the reinforcement effect of rGO increases the stiffness of the composites. however, at higher concentrations of rGO (0.5–1.0% by weight) the result was the opposite, with a decrease from the unfilled composite attributed to an agglomeration of rGO particles, which results in a reduced interface area with a matrix and ultimately leads to less stress transfer from the matrix to the nanoparticles. Attributed to the presence of an interface between rGO and epoxy matrix, which results in less energy dissipation, the tan delta values of composites with low concentrations of rGO are lower than unfilled composites. At higher concentrations, due to the agglomeration of rGO that results in a reduced interface between rGO and epoxy, the Tan delta values are higher than the unfilled composites.

In multiwalled carbon nanotubes (MWCNT)/glass fiber multiscale filler reinforced polypropylene composites [39] the incorporation of the optimal filler content of 3% by weight of MWCNT together with 20% by weight of glass fiber (H3), showed an improvement in the storage modulus of the hybrid composite, which demonstrates the effect of stiffening offered by the fillers to the matrix. For all composites, the loss modulus increases, reaches a maximum value and then decreases in the glass transition temperature range. For the H3 composite, the loss modulus beyond 30 °C was significantly higher compared to the others, demonstrating greater energy dissipation and superior mechanical properties. The lowering and widening of the tanδ peak were notable for the composite 'H3', this can be referred to the better interfacial adhesion between the fillers and the matrix. The T_g exhibited by the 'H3' composite was the highest due to the restrictions offered by the addition of MWCNT at an optimal content of 3% by weight. Figure 4 complies all the results regarding DMTA from [39].

The hybridization of carbon nanotubes (CNTs) with graphene oxide, reduces the problem of agglomeration of nanotubes in composites, and this integration in a hybrid structure generates multifunctional materials. Jyoti et al. [20], Jyoti

Fig. 4 The variation of **a** storage modulus **b** loss modulus **c** tan δ as a function of temperature at a frequency of 1 Hz and **d** Nomenclature. Adapted from [39]

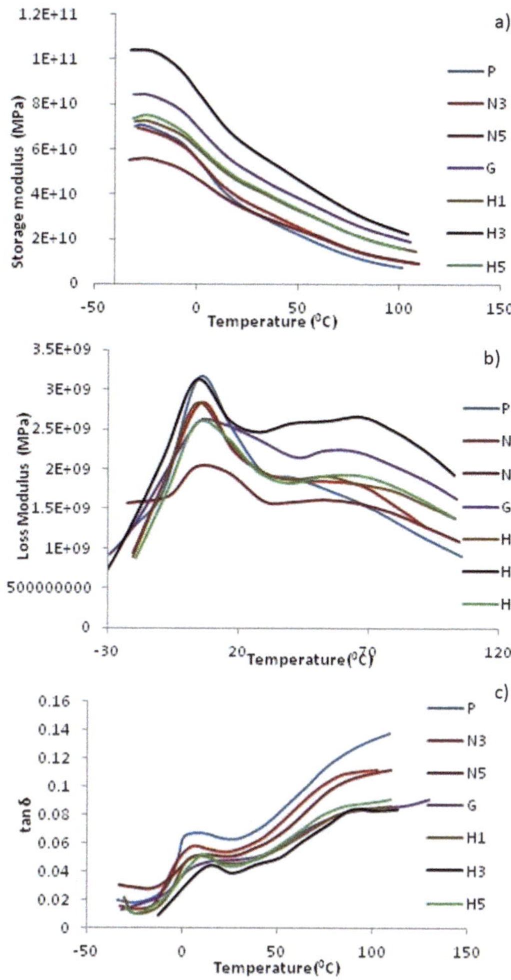

d)

Material	Polypropylene (P) (wt%)	Glass fiber (G) (wt%)	MWCNT (N) (wt%)
P	100	0	0
N1	99	0	1
N3	97	0	3
N5	95	0	5
G	80	20	0
H0.5	79.5	20	0.5
H1	79	20	1
H2	78	20	2
H3	77	20	3
H5	75	20	5

and Arya [19] reported the DMA analysis of the three different carbon nanofillers (MWCNTs, functionalized carbon nanotubes (FCNTs) and graphene oxide–carbon nanotubes (GCNTs)) acrylonitrile butadiene styrene (ABS) reinforced composites. The FCNTs-ABS composites showed a higher storage modulus than the corresponding MWCNTs-ABS and GCNTs-ABS, at temperatures below 90 °C, and as the temperature increases there is a significant decrease compared to MWCNTs-ABS and GCNTs-ABS. The addition of different types of fillers generates a greater number of chain segments, inhibiting the relaxation process within the composite, causing a broadening of the peak loss modulus.

In graphene and nano silica reinforced hybrid epoxy composites [5] the storage modulus decreases at high temperatures due to the loss of stiffness in the matrix due to its softening, and also with the use of nano silica concentrations below 8wt%. At concentrations above 8wt% of nanosilica, there is an increase in the storage module due to the better interfacial bond between the fillers and the matrix, showing that the elasticity of the composite is increasing. The authors observed that the value of the loss modulus increased up to T_g and then decreased with increasing temperature. A higher value of T_g indicates better thermal stability of the composite, indicating that there is a lower mobility of the molecules due to the inclusion of fillers.

3 Interaction of Nano Fillers with Thermoplastics and Thermoset Matrices

Regardless the use of thermoset or thermoplastic matrices, in a general way the use of micro and/or nano fillers increase elastic response, due to the restriction of molecular chain motions [23, 37]. On the other hand, many concerns need to be taken into account when these two different matrices are employed, aiming to fabricate nano composites. For instance, for thermoplastics it was proved that nanofillers had a tendency to disperse more in the amorphous region rather than in the crystalline one. Less energy is required for dispersion of nano fillers into amorphous chains due to the lower entanglement between polymeric chains itself and consequently a higher free volume is expected. Then, a thermoplastic with high crystallinity degree results in poor compatibility and adhesion toward nano or inorganic fillers. A lower entanglement means a lower interaction reinforcement-polymer matrix and lower nano composite stiffness.

For instance, in elastomers in general, it is expected that a high entanglement occurs between the polymer's macromolecules and nano fillers, due to the predominant amorphous phase in this system. A high entanglement means a better interaction between the filler and matrix and high stress is transferred from the matrix to reinforcement. On the other hand, other mechanisms occur, such as touch-touch of filler and polymer molecule. Figure 5 presents the main mechanism that occurs when a nano filler is inserted into a semi-crystalline polymer. In addition to the simple touching, that occurs between the filler and crystalline regions, the high aspect ratio

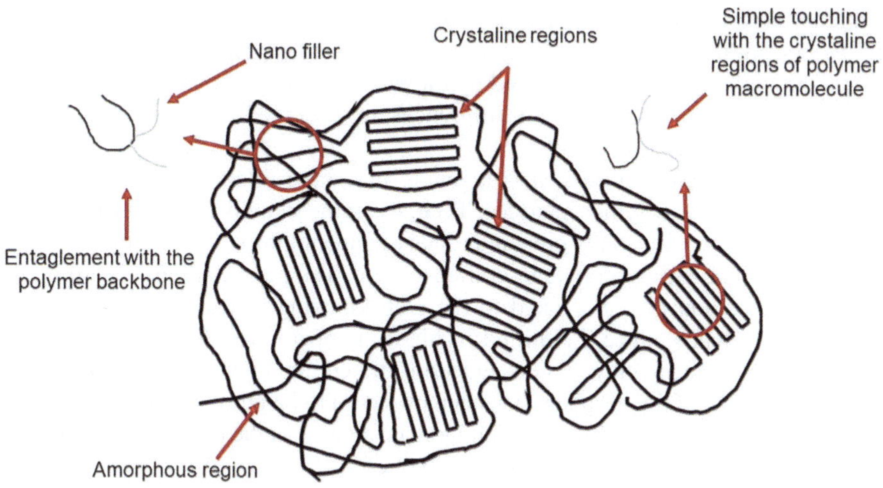

Fig. 5 Schematic representation of a semi-crystalline thermoplastic and the possible mechanism interactions between a nano reinforcement and the amorphous' polymer region. Adapted from [9]

of nano reinforcements enhances the entanglement of fillers with the polymer back bone [9].

Theoretically, it is expected that the insertion of nano fillers increases E′ and T_α due to the well-known excellent theoretical mechanical response of conventional nano materials. However, the aggregation of reinforcements will hinder their reinforcing effectiveness due to the low formation of interface between the filler-matrix which may also hinder the load transferring throw shear forces from the matrix to the reinforcement. Then, as the higher the crystallinity of the used thermoplastic matrix, the higher the tendency to form aggregates and lower the reinforcement effectiveness, which may also reflect on the quasi-static mechanical response. A high content of aggregates is taken as discontinuities on the polymer matrix and then, can generate premature failures when loads are applied [43].

Despite the improvements in glass transition temperature, if a highly dispersibility is reach on the overall nano composite, the incorporation of nano fillers usually does not modify the melt temperature of thermoplastics [30]. This behavior is again related to the fact that the nanofillers are better distributed on the amorphous region. For the same reason, more significant changes on storage modulus are observed above T_g, where the amorphous region's chain mobility takes place [8]. In the same way, secondary transitions are affected by the use of nano fillers. In particular, the intensity of β relaxation depends on various factors, such as crystalline fraction, orientation of the amorphous phase, and nano filler fraction [8].

Milani et al., investigated the synthesis, characterization and properties of isotactic PP/Gr nanosheet nano composites, prepared by in situ polymerization, using a metal-locene complex and methylaluminoxane as cocatalyst. The authors reported that the molecular characteristics of iPP, such as molecular weight, polydispersity and tacticity were not affected by the presence of nanoparticles, which was attributed to the non-influence of Gr on the main macromolecular properties of PP. Furthermore, the higher stiffness introduced by the presence of graphene slightly modifies the amorphous regions, and small differences are observed in the location and the intensity of the β process.

Regarding polymer blends, it is knowing that nano fillers may interact differently with different polymer matrices. Then, the same trend is reported for polymer blends. Dayma and Satapathy prepared ternary nanocomposites containing nano clay into a polyamide-6/polypropylene-grafted-maleic. The main findings regarding the viscoelastic response were related to the well knowing restriction of mobility of the polymer chains, since T_g shifted to higher temperatures. Furthermore, it was observed that the nano clay was dispersed/infiltrated into both the phases. However, an increase in T_g of the PP phase has been observed to be more uniform and consistent with the nano clay addition, which indicate an ease in the nano clay localization amidst the PP phase, compared to the PA-6 phase.

Figure 6 represents the interaction mechanism of nano fillers with polymer blends. As aforementioned, depending on the interaction of the filler-matrix system, the dispersion of nanoparticles may occur more in the continuous phase (high concentration, Fig. 5a) or in the dispersed phase (less concentration, Fig. 5b). If the nanoparticles interact better with the dispersed phase, a thicker layer of the nano filler may be expressive at the interface of the dispersed and matrix phases. This scenario coupled with a higher concentration of nanofillers may lead to agglomeration of nano reinforcements, causing stress concentration at the interface, and, then, decrease the overall composite stiffness [6].

For thermoset nano composites, the dispersion of nano reinforcements is so easier, compared to thermoplastic one, especially because the use of techniques such as sonication is possible for liquid resins and no preparation of master batches is required. Epoxy is from far the most used thermoset for the manufacturing of nano composites. Some properties, such as higher dimensional stability at high temperatures, as well as high stiffness aligned with low brittleness, compared to other thermosets (e.g., polyester and vinyl ester) make epoxy a thermoset used for advanced applications.

Differently from thermoplastic nano composites, when thermosets are reinforced by nano fillers, there are increments in storage modulus in both elastic and rubbery regions, due to the restricted mobility of the epoxy polymer chains. Furthermore, usually, when a filler with a high interaction with the polymer matrix is used, the cross-link density for thermoset composites may increase. Pistor et al. used POSS for the reinforcement of an epoxy matrix. A decrease on the E' in the rubbery region, loss modulus and composites T_g were reported. This behavior was explained by the fact that greater lengths are promoted by the POSS incorporation, leading to a more flexible system through an increase in the free volume.

It is noteworthy that many factors influence the overall dynamic-mechanical response of thermoset nano composites. A good dispersion may reflect an increased T_g, E' at both elastomeric and rubbery region and so on. Sonication is an alternative technique for the dispersion of nano fillers. However, the processing time, amplitude

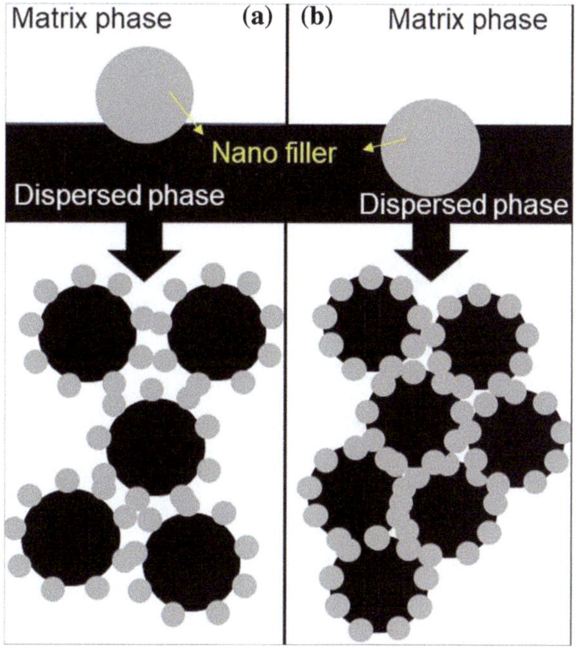

Fig. 6 Possible interactions of nano fillers with a polymer blend with selective interactions with **a** matrix phase, **b** dispersed phase

and pulse need to be optimized, aiming to ensure a good dispersion without causing a disruption of the macromolecular structure of thermoset oligomers [25].

Another alternative to increase the dispersibility of nano fillers is the use of coupling agents. Amino functionalization, using silane coupling agents is usually the most used chemical as coupling agent. For instance, Gojny and Schulte reported that the use of amino-functionalized CNTs may favor the formation of covalent bonds between the surface of carbon nanotubes and the epoxy matrix. This chemical linkage may reduce even more the polymer matrix mobility and cause a significant shift of T_g to higher temperatures. Furthermore, as the nanotube content increases, the T_g and E' of the composites increase, linearly. Figure 7 presents the possible reaction of amino functional groups on the CNT and Gr fillers. The epoxy ring opening may occur in a high velocity compared with the reaction with hydroxyl groups from untreated fillers and then, the interfacial characteristics of reinforcement-matrix is enhanced, which will reflect on the overall composite dynamical mechanical behavior [40].

4 Conclusion

This chapter allied theory about DMTA and its applications in the polymeric nanocomposite field. It was stressed that the DMTA offers precise information about interfacial bonding between the reinforcement and matrix, stiffness, polymeric chain

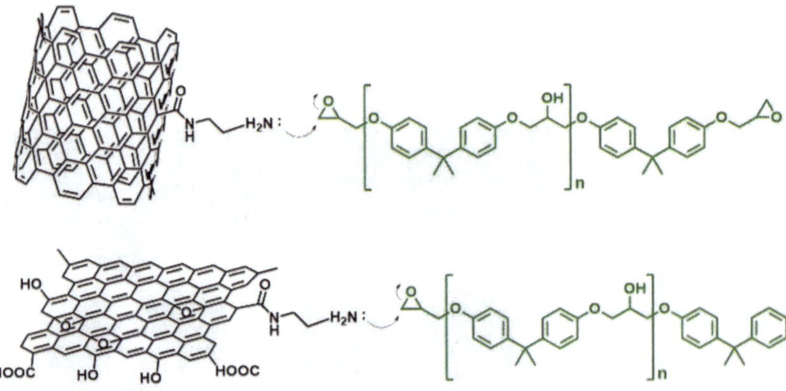

Fig. 7 Possible reactive mechanism of functionalized carbon nanotubes and a graphene nano sheet used as reinforcement for the improvement of dynamical mechanical behavior of an epoxy system Adapted from [22]

movements, and relaxation processes and might predict the effects of time and temperature on the nanocomposites. Furthermore, DMTA gives accurate E', E", T_g, and tandelta of polymers and their nanocomposites. Generally, independent of the polymer's nature (elastomer, thermoset, or thermoplastic), the incorporation of nanoparticles or nanofillers increased the storage modulus suggesting a matrix stiffening. Regarding E", in most of the cases, the nanofillers' addition enhances T_g values compared to the neat polymer, indicating, on some level, influence in the molecular mobility of the polymeric chains on the transition region. Finally, regarding tandelta, literature has been demonstrating that the addition of nanofillers decreases the value of damping, which relates to a reduction in the motion of polymer chains because of their large surface area. Furthermore, we stressed the influence of the nanoparticle surface chemical modification on the dynamic-mechanical behavior of its related nanocomposites, once it improved the dispersion and chemical interaction between the nanoparticle/matrix. After reading this chapter, we expect the researchers have some information that will help you, from the choice of the clamp to the interpretation of the results.

References

1. Akay M (1993) Aspects of dynamic mechanical analysis in polymeric composites. In: Composites science technology, vol 47
2. Almeida JH, Ornaghi HL Jr, Lorandi NP, Marinucci G, Amico SC (2018) On creep recovery stress relaxation of carbon fiber-reinforced epoxy filament wound composites. Polym Eng Sci 58(10):1837–1842.https://doi.org/10.1002/pen.24790
3. American Society for Testing Materials (2017) ASTM D4065–01: standard practice for plastics: dynamic mechanical properties: determination report of procedures

4. American Society for Testing Materials (2021) ASTM D4092 21—Standard terminology for plastics: dynamic mechanical properties
5. Anilkumar PR, Manjunatha LH, Venkategowda T (2022) Dynamic mechanical analysis of graphene nano silica reinforced hybrid epoxy composites under dual cantilever mode. Mater Today: Proc 54:402–408. https://doi.org/10.1016/j.matpr.2021.09.467
6. Baudouin AC, Auhl D, Tao F, Devaux J, Bailly C (2011) Polymer blend emulsion stabilization using carbon nanotubes interfacial confinement. Polymer 52(1):149–156. https://doi.org/10.1016/j.polymer.2010.11.004
7. Boyd RH (1985) Relaxation processes in crystalline polymers: molecular interpretation-a review
8. Dabrowska I, Fambri L, Pegoretti A, Slouf M, Vackova T, Kolarik J (2015) Spinning drawing physical properties of polypropylene nanocomposite fibers with fumed nanosilica. Express Polym Lett 9(3):277–290. https://doi.org/10.3144/expresspolymlett.2015.25
9. Dalmas F, Cavaillé JY, Gauthier C, Chazeau L, Dendievel R (2007) Viscoelastic behavior electrical properties of flexible nanofiber filled polymer nanocomposites. Influence of processing conditions. Compos Sci Tech 67:(5)829–839. https://doi.org/10.1016/j.compscitech.2006.01.030
10. Dayma N, Satapathy BK (2010) Morphological interpretations micromechanical properties of polyamide-6/polypropylene-grafted-maleic anhydride/nanoclay ternary nanocomposites. Mater Design 31(10):4693–4703. https://doi.org/10.1016/j.matdes.2010.05.024
11. Fatemi Khadar P, Maghsoud Z, Dashti A (2020) Synthesis characterization dynamic-mechanical properties of styrene-acrylate/nanoclay interpenetrating polymer network. Polym Composi 41(3):982–993. https://doi.org/10.1002/pc.25429
12. Feng J, Guo Z (2016) Temperature-frequency-dependent mechanical properties model of epoxy resin its composites. Compos Part B: Eng 85:161–169. https://doi.org/10.1016/j.compositesb.2015.09.040
14. Ferry JD (1980) Viscoelastic properties of polymers. In: Viscoelastic properties of polymers, 3rd edn. Wiley. https://doi.org/10.1149/1.2428174
15. George G, Tomlal EJ, Åkesson D, Skrifvars M, Nagarajan ER, Joseph K (2012) Viscoelastic behaviour of novel commingled biocomposites based on polypropylene/jute yarns. Compos Part A: App Sci Manuf 43(6):893–902. https://doi.org/10.1016/j.compositesa.2012.01.019
16. Gojny FH, Schulte K (2004) Functionalisation effect on the thermo-mechanical behaviour of multi-wall carbon nanotube/epoxy-composites. Compos Sci Tech 64:2303–2308. https://doi.org/10.1016/j.compscitech.2004.01.024
17. Huang CC, Du MX, Zhang BQ, Liu CY (2022) Glass transition temperatures of copolymers: molecular origins of deviation from the linear relation. Macromolecules 55(8):3189–3200. https://doi.org/10.1021/acs.macromol.1c02287
18. Jones DS (1999) Dynamic mechanical analysis of polymeric systems of pharmaceutical biomedical significance. Int J of Pharmaceutics 179:167–178. https://doi.org/10.1016/S0378-5173(98)00337-8
19. Jyoti J, Arya AK (2020) EMI shielding dynamic mechanical analysis of graphene oxide-carbon nanotube-acrylonitrile butadiene styrene hybrid composites. Polym Testing 91:106839. https://doi.org/10.1016/j.polymertesting.2020.106839
20. Jyoti J, Babal AS, Sharma S, Dhakate SR, Singh BP (2018) Significant improvement in static dynamic mechanical properties of graphene oxide–carbon nanotube acrylonitrile butadiene styrene hybrid composites. J Mater Sci 53(4):2520–2536. https://doi.org/10.1007/s10853-017-1592-6
21. Kamble Z, Behera BK, Mishra R, Behera PK (2021) Influence of cellulosic non-cellulosic particle fillers on mechanical dynamic mechanical thermogravimetric properties of waste cotton fibre reinforced green composites. Compos Part B: Eng 207:108595. https://doi.org/10.1016/j.compositesb.2020.108595
22. Kerche EF, Fonseca E, Schrekker HS, Amico SC (2022) Ionic liquid-functionalized reinforcements in epoxy-based composites: a systematic review. Polym Compos 43:5783–5801. https://doi.org/10.1002/pc.26956

23. Kerche EF, Silva VD, Fonseca E, Salles NA, Schrekker HS, Amico SC (2021) Epoxy-based composites reinforced with imidazolium ionic liquid-treated aramid pulp. Polymer 226:123787. https://doi.org/10.1016/j.polymer.2021.123787
24. Kunal K, Robertson CG, Pawlus S, Hahn SF, Sokolov AP (2008) Role of chemical structure in fragility of polymers: a qualitative picture. Macromolecules 41(19):7232–7238. https://doi.org/10.1021/ma801155c
25. Lam CK, Lau KT, Cheung HY, Ling HY (2005) Effect of ultrasound sonication in nanoclay clusters of nanoclay/epoxy composites. Mater Lett 59(11):1369–1372. https://doi.org/10.1016/j.matlet.2004.12.048
26. Lorandi NP, Cioffi MOH, Ornaghi HLJ (2016) Dynamic mechanical analysis (DMA) of polymeric composite materials. Scientia cum Industria 4(48):48–60
27. Lorandi NP, Cioffi MOH, Shigue C, Ornaghi Jr HL (2018) On the creep behavior of carbon/epoxy non-crimp fabric composites. Mater Res 21(3). https://doi.org/10.1590/1980-5373-mr-2017-0768
28. Menard KP (2008) Dynamic mechanical analysis—a practical introduction, 2nd edn. CRC Press
29. Menard KP, Menard N (2017) Dynamic mechanical analysis. Encyclopedia of analytical chemistry, pp 1–22.https://doi.org/10.1002/9780470027318.a2007.pub3
30. Milani MA, González D, Quijada R, Basso NRS, Cerrada ML, Azambuja DS, Gall GB (2013) Polypropylene/graphene nanosheet nanocomposites by in situ polymerization: synthesis characterization fundamental properties. Compos Sci Tech 84:1–7. https://doi.org/10.1016/j.compscitech.2013.05.001
31. Mohammed L, Ansari MNM, Pua G, Jawaid M, Islam MS (2015) A review on natural fiber reinforced polymer composite its applications. J Polym Sci. https://doi.org/10.1155/2015/243947
32. Neves RM, Vanzetto AB, Lazzari LK, Zattera AJ (2021) Thermal dynamic mechanical behavior of epoxy composites reinforced with post-consumed yerba mate. J Polym Sci 50438.https://doi.org/10.1002/app.50438
33. Neves RM, Zattera AJ, Amico SC (2021) Enhancing thermal dynamic-mechanical properties of epoxy reinforced by amino-functionalized microcrystalline cellulose. J Polym Sci 138:51329. https://doi.org/10.1002/app.51329
34. Ornaghi HL, Silva HSP, Zattera AJ, Amico, SC (2012) Dynamic mechanical properties of curaua composites. J App Polym Sci 125.https://doi.org/10.1002/app.36378
35. Ornaghi HL, Neves RM, Monticeli FM, Thomas S (2022) Modeling of dynamic mechanical curves of kenaf/polyester composites using surface response methodology. J App Polym Sci 139(18). https://doi.org/10.1002/app.52078
36. Ornaghi HL, Poletto M, Ornaghi FG, Zattera AJ, Amico SC (2016) Dynamic mechanical properties the dynamic fragility concept applied to vegetal fiber on vegetal composite materials. J Compos Mat 50(18):2469–2475. https://doi.org/10.1177/0021998315604208
37. Pistor V, Ornaghi FG, Ornaghi HL, Zattera AJ (2012) Dynamic mechanical characterization of epoxy/epoxycyclohexyl-POSS nanocomposites. Mat Sci Eng A 532:339–345. https://doi.org/10.1016/j.msea.2011.10.100
38. Polymer database (2022) Dynamical mechanical analysis. https://Polymerdatabase.com/Polymer%20physics/DMA.html
13. Queiroz HFM, Banea MD, Cavalcanti DKK, Silva Neto JS (2021). The effect of multiscale hybridization on the mechanical properties of natural fiber-reinforced composites. J App Polym Sci 138:(41). https://doi.org/10.1002/app.51213
39. Rasana N, Jayanarayanan K, Deeraj BDS, Joseph K (2019) The thermal degradation dynamic mechanical properties modeling of MWCNT/glass fiber multiscale filler reinforced polypropylene composites. Compos Sci Tech 169:249–259. https://doi.org/10.1016/j.compscitech.2018.11.027
40. Rozenberg BA (1986) Kinetics, thermodynamics and mechanism of reactions of epoxy oligomers with amines. In: Dušek K (ed) Epoxy resins and composites II. Advances in polymer science, vol 75. Springer, Berlin, Heidelberg. https://doi.org/10.1007/BFb0017916

41. Sarva SS, Deschanel S, Boyce MC, Chen W (2007) Stress-strain behavior of a polyurea and polyurethane from low to high strain rates. Polymer 48(8):2208–2213. https://doi.org/10.1016/j.polymer.2007.02.058
42. Sharma M, Sharma R, Sharma SC (2020) A review on fibres fillers on improving the mechanical behaviour of fibre reinforced polymer composites. Mater Today: Proc 46:6482–6489. https://doi.org/10.1016/j.matpr.2021.03.667
43. Teng CC, Ma CCM, Huang YW, Yuen SM, Weng CC, Chen CH, Su SF (2008) Effect of MWCNT content on rheological dynamic mechanical properties of multiwalled carbon nanotube/polypropylene composites. Compos Part A: Appl Sci Manuf 39 (12) :1869–1875. https://doi.org/10.1016/j.compositesa.2008.09.004

Fracture Toughness of Polymer Nanocomposites

Azzam Ahmed and Hashim Kabrein

Abstract Fracture toughness is an important mechanical property of materials that determines their ability to resist fracture under applied stress. In recent years, polymer nanocomposites have become increasingly popular due to their superior mechanical properties compared to conventional polymers. In this chapter, we discuss the fracture toughness of polymer nanocomposites, their mechanisms, and their potential applications. The potential applications of polymer nanocomposites with improved fracture toughness are numerous. They can be used in automotive, aerospace, and biomedical applications, among others. They can also be used for protective coatings, as well as for structural components that require high levels of toughness.

Keywords Polymer composites · Polymer nanocomposites · Mechanical characterization · Fracture toughness

1 Introduction

Composite materials are multiphase materials processed by two or more materials such as metal materials, ceramic, or polymer materials. Polymer nanocomposites are a class of materials composed of polymers and nanoscale fillers. They offer a combination of the mechanical properties of both components, leading to enhanced properties such as stiffness, strength, and toughness. The toughness of a nanocomposite is determined by the interaction between the polymer matrix and the nanofillers. The fracture toughness depends on the size, shape, and morphology of the nanofillers,

A. Ahmed (✉)
Department of the Textile Engineering, College of Engineering and Technology of Industries, Sudan University of Science and Technology, Khartoum, Sudan
e-mail: azzam.ko@hotmail.com

A. Ahmed · H. Kabrein
Safat College of Science and Technology, Khartoum, Sudan

H. Kabrein
Department of Mechanical Engineering, Faculty of Engineering, International University of Africa, Khartoum, Sudan

© The Author(s), under exclusive license to Springer Nature Singapore Pte Ltd. 2023
H. Abdellaoui et al. (eds.), *Mechanics of Nanomaterials and Polymer Nanocomposites*, Smart Nanomaterials Technology,
https://doi.org/10.1007/978-981-99-2352-6_9

as well as on the type of polymer matrix and the degree of dispersion. Polymer nanocomposites are composite materials in which polymer is the matrix continuous phase and nanometer-sized fillers are the dispersed phases. Among them, nano-sized dispersed phases include metals, semiconductors, inorganic salt nanoparticles, nanofibers, carbon nanotubes, graphene, etc. These nanosized materials are uniformly dispersed in polymer substrates through appropriate preparation methods to form polymer nanocomposite material. This type of material combines the advantages of different material components and nanomaterials to produce a synergistic effect. The comprehensive performance is better than that of the original composition materials, and it has high mechanical properties, electrical conductivity, heat insulation, and biocompatibility. The synthesis and functional research of polymer nanocomposites are in a stage of rapid innovation. Many researchers from all over the world are studying polymer nanocomposites and their applications, and have achieved significant results. At present, many materials have initially entered the commercial development stage.

Over the last years, four main manufacturing processes have been established for the successful incorporation of inorganic nanofillers into a polymer matrix: (i) direct mixing of polymer and filler (Fig. 1), (ii) intercalation based on the exfoliation (Fig. 2), (iii) sol–gel processes (Fig. 3), and (vi) in situ formation of nanofillers in the polymer matrix (Fig. 4). The direct mixing of polymer and filler is the most direct method for preparing polymer nanocomposites, which is suitable for nanoparticles of various forms as seen in Fig. 1.

The intercalation method (Fig. 2) is another important method for preparing polymer nanocomposites. The process is simpler than the sol–gel method (Fig. 3), and the source of raw materials is abundant and cheap. Many inorganic compounds,

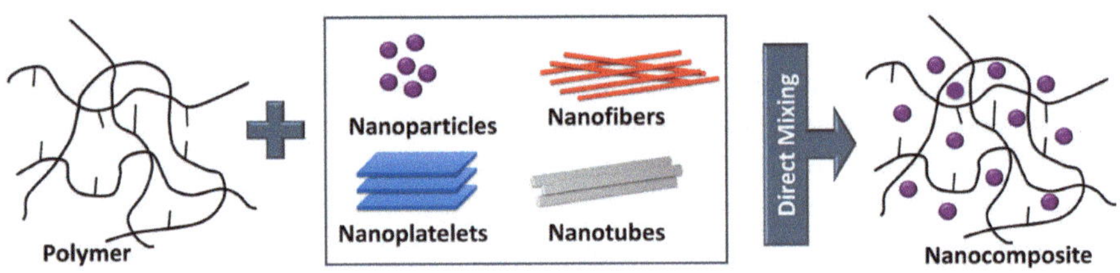

Fig. 1 Preparation of nanocomposites by direct mixing [1, 2]

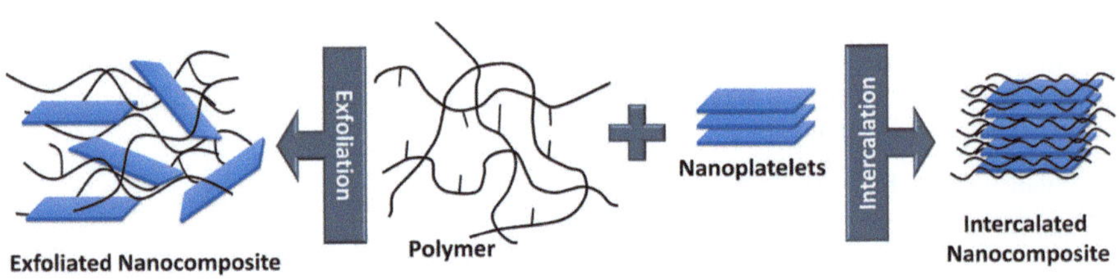

Fig. 2 Preparation of nanocomposites by exfoliation and intercalation [3]

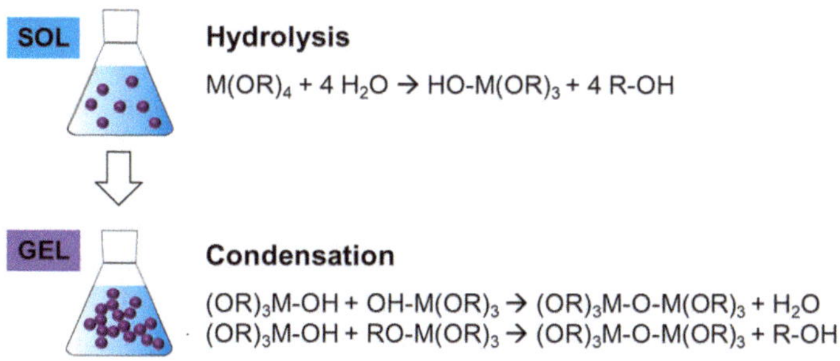

Fig. 3 Preparation of nanocomposites by sol–gel processes [4]

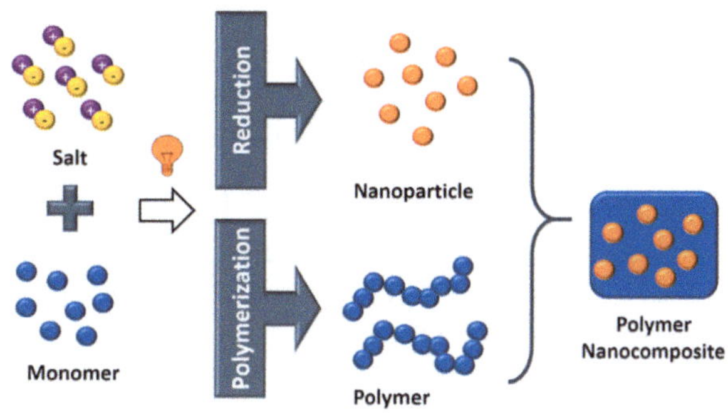

Fig. 4 Preparation of nanocomposites in situ polymerization [5]

such as silicate clays, phosphates, graphite, metal oxides, disulfides, and phosphorus trisulfide complexes, have a typical layered structure, which can be embedded in polymers to form polymer/inorganic nanocomposites material. According to the insertion system and method, it can be divided into several forms such as intercalation polymerization, solution intercalation, and melt intercalation.

The sol–gel composite method (Fig. 3) is mainly used to prepare inorganic–organic (polymer) nanocomposites and is also used for the preparation of nanoparticles, which belongs to the low-temperature wet chemical synthesis method.

The in situ polymerization method (Fig. 4) is to disperse the small molecular monomers into the silicate layer, and then intercalate into the silicate layer for in situ polymerization. In situ polymerization needs to meet two conditions: good dispersion of nanoparticles and polymerization of monomers.

2 Fracture Toughness Properties of Polymer Nanocomposites

Polymeric nanocomposites (PNCs) are commonly formed by an epoxy matrix reinforced with a nanosized filler. Due to its inherent characteristic of high crosslink density, an epoxy polymer is known to be a relatively brittle material [6]. At present, many researchers are trying to add suitable nanofillers to overcome their brittleness and improve the fracture toughness of epoxy composites [7, 8]. Several studies have investigated and enhanced the fracture toughness of nanocomposites prepared by plasma treatment and they found that its higher than that of pure epoxy resin, and the fracture toughness of nanocarbon fiber/epoxy resin composites will be improved [9–13]. Uniformly dispersed nanoparticles can simultaneously improve the stiffness and toughness of thermosetting polymers. Is known that the fracture toughness of nanocomposites is higher than that of pure resins, thus believe that the contribution of the nanoparticle–polymer interfacial influenced region plays a dominant role, and the interfacial influence region surrounding the nanoparticles begins to overlap with each other as the particle content increases, so that it is possible to form a three-dimensional network structure at the interface. Such a three-dimensional network interface structure controls the macroscopic mechanical properties of nanocomposites, including stiffness, hardness, and toughness. Rubber-enhanced polyamide nanofibers for a significant improvement of CFRP interlaminar fracture toughness have been investigated by Maccaferri et al. [14]. The use of inorganic nanoparticles and polymers to prepare polymer–inorganic nanocomposites is a new trend in polymer composites in recent years. The interfacial reaction between nanofiber and epoxy resin is the composite material that can greatly improve fracture toughness [15–17]. Nanofiber toughened film can significantly improve the fracture toughness of the product without increasing the weight and thickness, improve impact resistance, delamination resistance and fatigue resistance, shock absorption and energy absorption, improve resin toughness, improve the brittleness of carbon fiber composite materials, in military industry, aviation, automobile, high-end sports [18–21]. Some methods are used also to improve the fracture toughness of composites by electrospun nanofibers which can be used as interlayers between composite layers [15, 19, 22–24]. The effects of single-walled carbon nanotubes (SWCNTs) on the mechanical properties of nanocomposites with epoxy matrix were studied by various researcher, their results found that nanocomposites, in the presence of SWCNTs with greater volume of fractions, had a greater effect on fracture toughness of nanocomposites [25–30]. In general, nanocomposites with high levels of dispersion tend to be tougher than those with low levels of dispersion. This is due to the increased contact between the nanofillers and the polymer matrix, which increases the load transfer between them and enhances the material's resistance to fracture. Furthermore, the type and size of the nanofillers can also affect the fracture toughness. For example, nanofillers with higher aspect ratios, such as nanotubes and nanowires, can provide better load transfer than spherical particles and can therefore improve the fracture toughness of the nanocomposite. In addition to the type and size of the nanofillers, the interfacial adhesion between the polymer matrix and the nanofillers is also important for controlling the fracture toughness of nanocomposites. The interfacial adhesion is typically improved by the use of coupling agents, which can improve the load transfer between the components and reduce the number of defects in the interface. Effects of nanofiber on fracture toughness of the composites are presented in Table 1.

Table 1 Summary of published paper regarding the influence of nanofibers on fracture toughness response of composites

Nanofiber	Composites	Control	Value for composites	% Increase	Refs.
PPENK	CF/PPEK laminates	1238 J/m^2	1769 J/m^2	43	Zhang et al. [31]
ZnO	CF/PPEK laminates	1238 J/m^2	2618 J/m^2	112	Zhang et al. [31]
GO-GNFs	CFRPs	–	–	102.8	Kim and Park [32]
Polyvinyl butyral (PVB) nanofiber	Glass/phenolic prepreg UD laminated	798 J/m^2	1056 J/m^2	61	Barzoki et al. [33]
PVB nanofiber mat	Glass/phenolic composites	0.5814	0.7496	29	Kheirkhah Barzoki et al. [34]
Al_2O_3-PAN	CFRP	0.523	1.3	60	Razavi et al. [35]
PAN nanofibers	CFRP	0.523	1.09	52	Razavi et al. [35]
Cellulose nanofibers (CNFs)	CFRPs	–	–	25	Zhu et al. [36]
Nylon 6/6 electrospun nanofibers	Carbon fiber/epoxy laminates	0.68 kJ/m^2	0.80	15	Aljarrah and Abdelal [37]
Polyvinyl butyral (PVB) electrospun nanowebs	Phenolic resin/glass fiber composite	1142.3 J/cm^2	1855.92 J/cm^2	38	Ipackchi et al. [38]

3 Applications of Polymer Nanocomposites

The applications of polymer/nanocomposites are mainly in coatings, medical materials, electronic functional materials, packaging materials, catalysts, etc. and have broad application prospects. The development of polymer nanocomposites with excellent performance is the focus of our future work. Although the research on polymer-based nanocomposites has made great progress, due to its complex structure, small domain size, and surface effects. Due to the influence of factors such as quantum effects, the relationship between the structure and morphology of materials and the properties of materials needs to be further studied; the synthesis method still needs to break through the existing synthesis methods, especially in terms of broadening the scope of its application fields. It is an urgent problem to be solved at present. Polymeric nanocomposites (PNCs) are used widely in packaging, energy, safety, transportation, electromagnetic shielding, defense systems, sensors, catalysis, and the information industry [39–41]. The fracture toughness of polymer nanocomposites is an important property that determines their ability to resist fracture under applied stress. The toughness of a nanocomposite depends on the size, shape, and morphology of the nanofillers, as well as on the type of polymer matrix and the degree of dispersion. The potential applications of polymer nanocomposites with improved fracture toughness include automotive, aerospace, and biomedical applications, as well as protective coatings and structural components and other applications are presented in Table 2.

So far, polymer nanocomposites have been used in huge and high-performance applications (Fig. 5) such as energy storage [53, 54], environmental remediation [55, 56], electromagnetic (EM) absorption [57–59], sensing and actuation [60–62], transportation and safety [63, 64], defense systems [65], and information industry [66], to novel catalysts [67, 68].

Table 2 Recent polymer nanocomposite applications

Polymer nanocomposite material	Applications	Refs.
Boron nitride (BN) nanomaterials	Heat dissipation and thermal management	Mazumder et al. [42]
PVC/PVP/SrTiO3 polymer blend nanocomposites	Optoelectronic applications	Alshammari et al. [43]
ZnO/CuO nanocomposite-based carboxymethyl cellulose/polyethylene oxide polymer electrolytes	Energy storage applications	Hameed et al. [44]
Facile synthesis of rGO/Ag/PVA	Optoelectronic applications	Attia et al. [45]
Ni/ZnO nanohybrid-based polymer nanocomposites	Food packaging applications	Alghamdi et al. [46]
NiCoFe2O4 anchored polymer nanocomposites	Asymmetric supercapacitor application	Siva et al. [47]
g-C3N4/ZnO nanocomposites	Anti-fouling polymer membranes with dye and protein rejection superiority	Vatanpour et al. [48]
PVDF-Er2O3 polymer nanocomposites	Energy storage applications	Taha and Mahmoud [49]
BaTiO3@polymer/fluorinated polymers nanocomposites	Energy storage applications	Bouharras et al. [50]
P(VdF-HFP)-MMT-based nanocomposite gel polymer electrolytes	Energy storage devices	Borah and Deka [51]
Graphene/graphene oxide-based polymer nanocomposites	Corrosion applications	Kumar et al. [52]

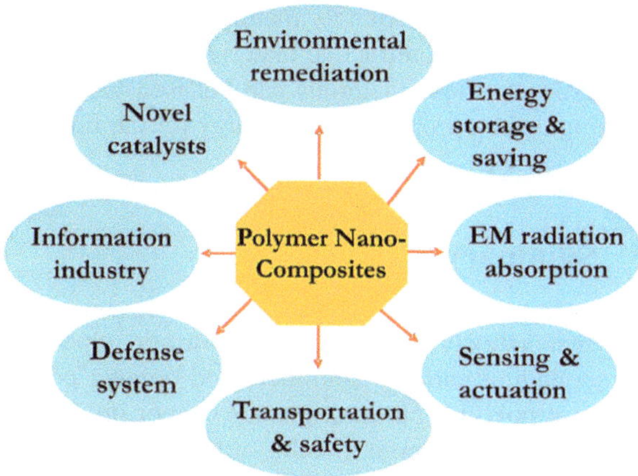

Fig. 5 Polymer nanocomposites for different applications [62]

4 Limitations of Polymer Nanocomposites

The limitations of polymer nanocomposites are varied depending on which materials will be used and nanocomposite preparation and final applications. Despite the successful use of these different methods for the preparation of polymer-based nanocomposites, information on various factors is still lacking, such as (i) the use of an appropriate method for a specific matrix-reinforcement combination or (ii) the maximum amount of reinforcements to give optimum property combinations and lower the cost of the processes. Therefore, it is still necessary to look into these aspects including the use of simulation and modeling techniques [3]. Polymer nanocomposites are materials composed of a polymer matrix filled with nanoparticles. These materials offer tremendous potential for a wide range of applications due to their unique properties, such as high thermal stability, superior mechanical strength, and improved electrical and optical properties. However, despite their numerous advantages, there are certain limitations associated with the use of polymer nanocomposites. One of the major drawbacks of polymer nanocomposites is the difficulty in obtaining uniform dispersion of the nanoparticles in the polymer matrix. In order to achieve a homogeneous distribution of nanoparticles, the particles must be dispersed to a very small size. This is a difficult task as the particles tend to agglomerate and form clusters, leading to poor dispersion. In addition, the nanoparticles often migrate to the surface of the matrix, reducing the effectiveness of the composite. Another limitation of polymer nanocomposites is the cost of production. The production process is quite complex and requires specialized equipment, such as high-pressure homogenizers and high-temperature reactors. Moreover, the synthesis of nanoparticles is often expensive, resulting in a high cost of the final product. Finally, the poor recyclability of polymer nanocomposites is a major limitation. Since the nanoparticles are often incompatible with the polymer matrix, the recycling process is often not possible. This results in a large amount of waste product, which is difficult to dispose of in an environmentally friendly manner. Despite these limitations, polymer nanocomposites are still an attractive material for a wide range of applications. The unique properties of these materials can be used to create new and innovative products, making them a promising technology for the future.

5 Summary

Polymer nanocomposite material is a new type of composite material formed by compounding various nanofibers and organic polymer materials in various ways. Because polymer nanocomposites cannot only reflect the characteristics of small nanoparticle size, large specific surface area, surface effect, quantum effect, etc. but also maintain various excellent properties of the polymer body itself, the polymer nanocomposites show the characteristics that conventional materials do not have. It possesses the characteristics and has broad application prospects. Fracture toughness of polymer composites has been taken full consideration when it is needed to design composite materials, however recently, with new applications of nanofibers and their superior mechanical properties when mixing with a resin using different ways, they found all the mechanical properties increased gradually with increase of nanofiber amount. Thus, the advantages of using these materials for enhancing the mechanical properties of the composites. The amount of the nanofiber by wt.% and volume fraction of the fiber has significant issues in the fracture toughness property. The challenges of these materials in near future are good preparation and interfacial bonding with resin as well as finding a suitable amount of these materials that can achieve unique mechanical properties. In conclusion, we have identified the key parameters influencing the fracture toughness of particle/polymer nanocomposites, the effect of using nanofiber to improve mechanical properties regarding fracture toughness of composites is obvious that range of increase of any mechanical property depends on which type of nanofiber is used and which amount by wt% is used, preparation method, and final applications.

References

1. Roy M, Nelson JK, Schadler L, Zou C, Fothergill JC (eds) (2005) The influence of physical and chemical linkage on the properties of nanocomposites. In: CEIDP'05 2005 annual report conference on electrical insulation and dielectric phenomena. IEEE
2. Reed CW (ed) (2007) Self-assembly of polymer nanocomposites for dielectrics and HV insulation. In: 2007 IEEE international conference on solid dielectrics. IEEE
3. Tronto J, Bordonal AC, Naal Z, Valim JB (2013) Conducting polymers/layered double hydroxides intercalated nanocomposites. Mater Sci-Adv Topics 3–30
4. Mera G, Gallei M, Bernard S, Ionescu E (2015) Ceramic nanocomposites from tailor-made preceramic polymers. Nanomaterials 5(2):468–540
5. Zhang H, Tang L-C, Zhang Z, Friedrich K, Sprenger S (2008) Fracture behaviours of in situ silica nanoparticle-filled epoxy at different temperatures. Polymer 49(17):3816–3825

6. Argon AS, Cohen RE (2003) Toughenability of polymers. Polymer 44(19):6013–6032
7. Domun N, Hadavinia H, Zhang T, Sainsbury T, Liaghat G, Vahid S (2015) Improving the fracture toughness and the strength of epoxy using nanomaterials–a review of the current status. Nanoscale 7(23):10294–10329
8. Tee ZY, Yeap SP, Hassan CS, Kiew PL (2022) Nano and non-nano fillers in enhancing mechanical properties of epoxy resins: a brief review. Polym-Plast Technol Mater 61(7):709–725
9. Chen Z, Dai XJ, Magniez K, Lamb PR, de Celis Leal DR, Fox BL, et al (2013) Improving the mechanical properties of epoxy using multiwalled carbon nanotubes functionalized by a novel plasma treatment. Compos Part A: Appl Sci Manuf 45:145–152
10. Huskić M, Bolka S, Vesel A, Mozetič M, Anžlovar A, Vizintin A et al (2018) One-step surface modification of graphene oxide and influence of its particle size on the properties of graphene oxide/epoxy resin nanocomposites. Eur Polym J 101:211–217
11. Mousavi SR, Estaji S, Paydayesh A, Arjmand M, Jafari SH, Nouranian S et al (2022) A review of recent progress in improving the fracture toughness of epoxy-based composites using carbonaceous nanofillers. Polym Compos 43(4):1871–1886
12. Domun N, Paton KR, Hadavinia H, Sainsbury T, Zhang T, Mohamud H (2017) Enhancement of fracture toughness of epoxy nanocomposites by combining nanotubes and nanosheets as fillers. Materials 10(10):1179
13. Domun N, Hadavinia H, Zhang T, Liaghat G, Vahid S, Spacie C et al (2017) Improving the fracture toughness properties of epoxy using graphene nanoplatelets at low filler content. Nanocomposites 3(3):85–96
14. Maccaferri E, Dalle Donne M, Mazzocchetti L, Benelli T, Brugo TM, Zucchelli A, et al (2022) Rubber-enhanced polyamide nanofibers for a significant improvement of CFRP interlaminar fracture toughness. Sci Rep 12(1):1–16
15. Ravindran AR, Ladani RB, Kinloch AJ, Wang C-H, Mouritz AP (2021) Improving the delamination resistance and impact damage tolerance of carbon fibre-epoxy composites using multi-scale fibre toughening. Compos A Appl Sci Manuf 150:106624
16. Beylergil B, Tanoğlu M, Aktaş E (2017) Enhancement of interlaminar fracture toughness of carbon fiber–epoxy composites using polyamide-6, 6 electrospun nanofibers. J Appl Polym Sci 134(35):45244
17. Shivakumar KN, Panduranga R, Sharpe M (eds) (2013) Interleaved polymer matrix composites-a review. In: 54th AIAA/ASME/ASCE/AHS/ASC structures, structural dynamics, and materials conference
18. Beckermann GW (2017) Nanofiber interleaving veils for improving the performance of composite laminates. Reinf Plast 61(5):289–293
19. Echeberria J, Ollo J, Bocanegra-Bernal M, Garcia-Reyes A, Domínguez-Rios C, Aguilar-Elguezabal A, et al (2010) Sinter and hot isostatic pressing (HIP) of multi-wall carbon nanotubes (MWCNTs) reinforced ZTA nanocomposite: microstructure and fracture toughness. Int J Refractory Metals Hard Mater 28(3):399–406
20. Park S-J, Park S-J (2020) Effect of ozone-treated single-walled carbon nanotubes on interfacial properties and fracture toughness of carbon fiber-reinforced epoxy composites. Compos A Appl Sci Manuf 137:105937
21. Tyson BM, Abu Al-Rub RK, Yazdanbakhsh A, Grasley Z (2011) Carbon nanotubes and carbon nanofibers for enhancing the mechanical properties of nanocomposite cementitious materials. J Mater Civil Eng 23(7):1028
22. Sun J, Gao L, Iwasa M, Nakayama T, Niihara K (2005) Failure investigation of carbon nanotube/3Y-TZP nanocomposites. Ceram Int 31(8):1131–1134
23. Fereidoon A, Mottahedin L, Latibari ST (2012) Investigation of fracture toughness parameters of epoxy nanocomposites for different crack angles. J Polym Eng 32(4–5):311–317
24. Zhang Y, Liu C, Jia H, Qiao Y, Dang X, Weng Z et al (2022) Effect of electrospun PPENK nanofiber loaded with ZnO nanowires on the mode II fracture toughness, flexural properties and ILSS of CF/poly(phthalazinone ether ketone) composites. Compos B Eng 244:110172
25. Kim S-H, Park S-J (2021) Effect of graphene oxide/graphitic nanofiber nanohybrids on interfacial properties and fracture toughness of carbon fibers-reinforced epoxy matrix composites. Compos B Eng 227:109387

26. Barzoki PK, Rezadoust AM, Latifi M (2018) Tunable effect of polyvinyl butyral nanofiber veil on fracture toughness of glass reinforced phenolic composites manufactured with out of autoclave method. Polym Testing 71:255–261
27. Kheirkhah Barzoki P, Rezadoust AM, Latifi M, Saghafi H (2018) The experimental and numerical study on the effect of PVB nanofiber mat thickness on interlaminar fracture toughness of glass/phenolic composites. Eng Fracture Mech 194:145–153
28. Razavi SMJ, Neisiany RE, Khorasani SN, Ramakrishna S, Berto F (2018) Effect of neat and reinforced polyacrylonitrile nanofibers incorporation on interlaminar fracture toughness of carbon/epoxy composite. Theor Appl Mech Lett 8(2):126–131
29. Zhu X, Li Y, Yu T, Zhang Z (2021) Enhancement of the interlaminar fracture toughness and damping properties of carbon fiber reinforced composites using cellulose nanofiber interleaves. Compos Commun 28:100940
30. Aljarrah MT, Abdelal NR (2019) Improvement of the mode I interlaminar fracture toughness of carbon fiber composite reinforced with electrospun nylon nanofiber. Compos B Eng 165:379–385
31. Ipackchi H, Rezadoust AM, Esfandeh M, Rezaei M (2020) Improvement of interlaminar fracture toughness of phenolic laminates interleaved with electrospun polyvinyl butyral nanofibers. Theoret Appl Fract Mech 105:102406
32. Sun X, Huang C, Wang L, Liang L, Cheng Y, Fei W et al (2021) Recent progress in graphene/polymer nanocomposites. Adv Mater 33(6):2001105
33. Chow W, Mohd Ishak Z (2020) Smart polymer nanocomposites: a review. Express Polym Lett 14(5)
34. Alateyah A, Dhakal H, Zhang Z (2013) Processing, properties, and applications of polymer nanocomposites based on layer silicates: a review. Adv Polym Technol 32(4)
35. Mazumder MRH, Mathews LD, Mateti S, Salim NV, Parameswaranpillai J, Govindaraj P et al (2022) Boron nitride based polymer nanocomposites for heat dissipation and thermal management applications. Appl Mater Today 29:101672
36. Alshammari AH, Alshammari M, Alshammari K, Allam NK, Taha TA (2023) PVC/PVP/SrTiO3 polymer blend nanocomposites as potential materials for optoelectronic applications. Results Phys 44:106173
37. Hameed ST, Qahtan TF, Abdelghany AM, Oraby AH (2023) ZnO/CuO nanocomposite-based carboxymethyl cellulose/polyethylene oxide polymer electrolytes for energy storage applications. J Market Res 22:531–540
38. Attia RM, Yousif NM, Balboul MR (2022) Facile synthesis of r GO/Ag/ PVA nanocomposites polymeric films by electron beam irradiation for optoelectronic applications. Opt Mater 126:112233
39. Alghamdi HM, Abutalib MM, Mannaa MA, Nur O, Abdelrazek EM, Rajeh A (2022) Modification and development of high bioactivities and environmentally safe polymer nanocomposites doped by Ni/ZnO nanohybrid for food packaging applications. J Market Res 19:3421–3432
40. Siva V, Murugan A, Shameem A, Thangarasu S, Bahadur SA (2022) A facile microwave-assisted combustion synthesis of NiCoFe2O4 anchored polymer nanocomposites as an efficient electrode material for asymmetric supercapacitor application. J Energy Storage 48:103965
41. Vatanpour V, Mousavi Khadem SS, Dehqan A, Paziresh S, Ganjali MR, Mehrpooya M, et al (2022) Application of g-C3N4/ZnO nanocomposites for fabrication of anti-fouling polymer membranes with dye and protein rejection superiority. J Membr Sci 660:120893
42. Taha TA, Mahmoud MH (2021) Synthesis and characterization of PVDF-Er2O3 polymer nanocomposites for energy storage applications. Mater Chem Phys 270:124827
43. Bouharras FE, Raihane M, Ameduri B (2020) Recent progress on core-shell structured BaTiO3@polymer/fluorinated polymers nanocomposites for high energy storage: synthesis, dielectric properties and applications. Prog Mater Sci 113:100670
44. Borah S, Deka M (2021) Study of electrical and electrochemical properties of P(VdF-HFP)-MMT based nanocomposite gel polymer electrolytes for application in energy storage devices. Mater Sci Eng, B 263:114822

45. Kumar SSA, Bashir S, Ramesh K, Ramesh S (2021) New perspectives on graphene/graphene oxide based polymer nanocomposites for corrosion applications: the relevance of the graphene/polymer barrier coatings. Prog Org Coat 154:106215
46. Xie L, Huang X, Huang Y, Yang K, Jiang P (2013) Core@ double-shell structured BaTiO3–polymer nanocomposites with high dielectric constant and low dielectric loss for energy storage application. The J Phys Chem C. 117(44):22525–22537
47. Siwal SS, Zhang Q, Devi N, Thakur VK (2020) Carbon-based polymer nanocomposite for high-performance energy storage applications. Polymers 12(3):505
48. Singh PP (2018) Environmental remediation by nanoadsorbents-based polymer nanocomposite. New polymer nanocomposites for environmental remediation: Elsevier, pp 223–241
49. Zhu J, Wei S, Chen M, Gu H, Rapole SB, Pallavkar S et al (2013) Magnetic nanocomposites for environmental remediation. Adv Powder Technol 24(2):459–467
50. Pai AR, Nair GG, Thomas PC, Thomas S (2021) High-performance conducting polymer nanocomposites for EMI shielding applications. CRC Press, Conducting polymers for advanced energy applications, pp 383–398
51. Liu S, Qin S, Jiang Y, Song P, Wang H (2021) Lightweight high-performance carbon-polymer nanocomposites for electromagnetic interference shielding. Compos A Appl Sci Manuf 145:106376
52. Yousefi N, Sun X, Lin X, Shen X, Jia J, Zhang B et al (2014) Highly aligned graphene/polymer nanocomposites with excellent dielectric properties for high-performance electromagnetic interference shielding. Adv Mater 26(31):5480–5487
53. Chen L, Liu C, Liu K, Meng C, Hu C, Wang J et al (2011) High-performance, low-voltage, and easy-operable bending actuator based on aligned carbon nanotube/polymer composites. ACS Nano 5(3):1588–1593
54. Bhanushali H, Amrutkar S, Mestry S, Mhaske S (2022) Shape memory polymer nanocomposite: a review on structure–property relationship. Polym Bull 79(6):3437–3493
55. Zhang D, Tong J, Xia B (2014) Humidity-sensing properties of chemically reduced graphene oxide/polymer nanocomposite film sensor based on layer-by-layer nano self-assembly. Sens Actuators, B Chem 197:66–72
56. Youssef AM, El-Sayed SM (2018) Bionanocomposites materials for food packaging applications: concepts and future outlook. Carbohydr Polym 193:19–27
57. Peng T, Xiao R, Rong Z, Liu H, Hu Q, Wang S, et al (2020) Polymer nanocomposite-based coatings for corrosion protection. Chem–An Asian J 15(23):3915–3941
58. Joshi M, Chatterjee U (2016) Polymer nanocomposite: an advanced material for aerospace applications. Elsevier, Advanced composite materials for aerospace engineering, pp 241–264
59. Bredeau S, Peeterbroeck S, Bonduel D, Alexandre M, Dubois P (2008) From carbon nanotube coatings to high-performance polymer nanocomposites. Polym Int 57(4):547–553
60. Huang G, Song P, Liu L, Han D, Ge C, Li R et al (2016) Fabrication of multifunctional graphene decorated with bromine and nano-Sb2O3 towards high-performance polymer nanocomposites. Carbon 98:689–701
61. Sun T, Garces JM (2002) High-performance polypropylene-clay nanocomposites by in-situ polymerization with metallocene/clay catalysts. Adv Mater 14(2):128–130
62. Yang C, Wei H, Guan L, Guo J, Wang Y, Yan X et al (2015) Polymer nanocomposites for energy storage, energy saving, and anticorrosion. J Mater Chem A. 3(29):14929–14941
63. Zheng N, Huang Y, Liu H-Y, Gao J, Mai Y-W (2017) Improvement of interlaminar fracture toughness in carbon fiber/epoxy composites with carbon nanotubes/polysulfone interleaves. Compos Sci Technol 140:8–15
64. Bortz DR, Merino C, Martin-Gullon I (2011) Carbon nanofibers enhance the fracture toughness and fatigue performance of a structural epoxy system. Compos Sci Technol 71(1):31–38
65. Song Y, Zheng N, Dong X, Gao J (2020) Flexible carboxylated CNT/PA66 nanofibrous mat interleaved carbon fiber/epoxy laminates with improved interlaminar fracture toughness and flexural properties. Ind Eng Chem Res 59(3):1151–1158
66. Nguyen DD, Vu CM, Choi HJ (2018) Improvement of the mode I interlaminar fracture toughness of glass fiber/epoxy composites using polystyrene electrospun nanofibres. Polym Bull 75(11):5089–5102

67. Manh CV, Choi HJ (2016) Enhancement of interlaminar fracture toughness of carbon fiber/epoxy composites using silk fibroin electrospun nanofibers. Polym-Plast Technol Eng 55(10):1048–1056
68. Ahmadloo E, Gharehaghaji A, Latifi M, Mohammadi N, Saghafi H (2017) How fracture toughness of epoxy-based nanocomposite is affected by PA66 electrospun nanofiber yarn. Eng Fract Mech 182:62–73

Finite Deformation of Polymer Nanocomposites

D. Balaji

Abstract Polymer nanocomposites are quite interesting to work with owing to their earth-friendly nature. Applications of polymer nanocomposites are getting wider recently. Consideration should be given to their behaviors under various environments. While exposing to various environments leads to deformation both larger and finite deformation. This chapter focuses on the finite deformation of polymer nanocomposites exposed to various circumstances including thermal, mechanical, and other deformations.

Keywords Finite deformation · Polymer nanocomposites · Crack propagation

1 Introduction

As performance limits are reached with conventional polymer composites that use micro- to macro-scale reinforcement additives [1, 2], the polymer matrix is increasingly being reinforced using nanoscale expansions. The produced polymer nanocomposites combine the unique characteristics of the nanoscale with the beneficial qualities of polymer, such as low weight and high ductility. High rigidity and low weight, epoxy composites strengthened with boehmite nanoparticles (BNPs) have recently been regarded as one of the most desirable composites for structural components [3–5]. Boehmite is an orthorhombic cell but has a bulk modulus of about 93 GPa. Scanning electron microscopy studies of BNP deployment in epoxy reveal that the nanoparticles typically develop aggregates there in the 100–200 nm range [6, 7], which may affect the mechanical qualities of epoxy resins.

As a relatively novel material, nanocomposites necessitate careful forethought to predict damage and fracture progression when exposed to exterior and morphologies elements such as applied load, heat, and particle weight fraction. Distinct physiological and experiential concepts have been used to develop various constitutive

D. Balaji (✉)
Department of Mechanical Engineering, KPR Institute of Engineering and Technology, Coimbatore, Tamilnadu 641407, India
e-mail: balaji.ntu@gmail.com

© The Author(s), under exclusive license to Springer Nature Singapore Pte Ltd. 2023

H. Abdellaoui et al. (eds.), *Mechanics of Nanomaterials and Polymer Nanocomposites*, Smart Nanomaterials Technology,
https://doi.org/10.1007/978-981-99-2352-6_10

models [8–13]. The microstructure of semi-crystalline polymers should be considered in composite-type constitutive equations, as proposed by Boyce et al. [14]. Based on their inherent rigidity, the crystalline materials are modeled as nanofillers, while the amorphous materials play the role of the matrix, Qi and Boyce established a viscoelastic-plastic basic prototype [15]. Based on the concept that epoxies are crosslinked networks with superimposing free chains, Li et al. published a strain-dependent viscous modeling approach of elastomers. They used theoretical (via molecular dynamics simulations) and practical (through actual testing) methods to dial in the parameters of the material model. To learn more about the nonlinear rate-dependence of amorphous glassy polymers, built and evaluated phenomenology poroelastic fundamental formulas for epoxy resins [16]. For thermal processing in concrete substances subject to finite strain, N'Guyen et al. [17] constructed a thermodynamic framework using Helmholtz free energy. To investigate the physical characteristics of BNP/epoxy nanocomposites, Fankhänel and coworkers developed a layered multiscale model. We used molecular simulations within the framework of the multi-resolution prototype to investigate the interphase properties of BNPs within a urethane mixture [18]. Typical nanocomposite finite element FE studies were utilized to homogenize the elastic characteristics of the nanocomposites, and then the interfacial properties were prerendered to the continuous level. Few researchers proposed a multiscale approach when it comes to laying the groundwork for a viscoelastic stress foundation scheme for BNP/epoxy nanocomposites [18, 19]. The model accurately represented critical aspects of a strain connection in these nanocomposites, as demonstrated by experimental–numerical validation. Properties such as nonlinear hyperelasticity, rate dependence, and elongation softening are all present in nanocomposites. The thermo-visco-elastic nature of BNP/epoxy nanocomposites was studied by Unger et al. [20] using a modified multiscale approach.

Using mathematical approaches to foresee the emergence and progression of fractures in nanocomposites is also crucial. As shown in some research works, crazes and linear cracks are formed in strained polymeric substances and are, thus, associated with the initiation and progression of cracks. Microvoids and micro-cracks (minimal breakage) eventually combine to develop cracks (complete damage) under increasing stresses. Over time, the damage tends to soften and become increasingly confined, with the affected area shrinking. As such, there is an inherent lattice dependency for strain relaxation concerns [21–23] in FE models predicated on a localized modal description of damage. To verify a well of the finite volume issues and get insight into the processes at play during a fracture. By allowing for interactions between close material sites, regularized solutions for depreciation and failure are some of the most successful [24–26]. Specifically, the solutions enable a link between damage accumulation and failure mechanics by pairing an equation of the kind characteristic of diffusion with a momentum balance equation with nonlocal components. These techniques effectively handle a sharp crack as though it was a diffusely damaged strip by incorporating an indicator that regulates connections among material spots [27, 28].

As a result of the normalization theories, many other models have been created, including the phase-field model (PFM) [29–31] and its offspring, the gradient-enhanced damage model (GEDM) [24–28]. PFM extends Griffith's theory by providing variational rupture designs that seek to minimize the energy stored. This includes the accumulated mass power, the job performed by foreign influences, and the surface chemistry. In discontinuous crack modeling [32–36], PFMs are promising for predicting fracture initiation, propagation, and even branching with complex patterns. Before now, PFMs have primarily been utilized to study ductile fracture and brittle and quasi-brittle fracture [37, 38]. Clay/epoxy nanocomposites' [39–42] longitudinal strain and crack hardness can be predicted with the help of a PFM developed by Msekh et al. Using the model, researchers analyzed how nanocomposites' surface energy dissipates following a fracture. For clay and epoxy nanocomposites in which the granules are regarded as linear elastic materials embedded in a hyperelastic matrix, the authors also constructed a PFM to account for the interphase areas among the polymer matrix and the clay nanofillers [43]. Goswami and their co-researchers [44, 45] recently introduced a neural network technique in which crack trajectories are predicted by attempting to minimize the system's variational energy for phase-field modeling of a crack in composite fracture. The simulation results show that the suggested method's projected fracture route is consistent.

Phase-field formulations have also been used to investigate the rate-dependent fracture of solids [36, 46–48]. Part played by viscous energy dissipation in the fracture process in viscoelastic materials with minor deformation, Shen and colleagues [49] presented a PFM. Rubber's rate-dependent PFM was devised and verified empirically by Loew and colleagues. The model's variables have been established by subjecting the material to uniaxial tensile testing and subsequently to double tensile testing. Also, the length scale parameter was estimated using digital image correlation to evaluate local strain at the fracture tip. Yin and Kaliske [44] used a negative viscoelastic simulation into PFM to examine the rate-dependent fracture behavior of elastomers. In contrast to other models proposed before [46], the model will not use lost energy to propel fracture development. To study the elastic breakage of elastomers, few researchers developed a PFM using a quantitatively physics-based micromechanical framework that characterizes a reorganization of a polymeric system over time [48].

Nanocomposites fabricated from thermosetting polymers are susceptible to damage because of the intricate microstructure, interconnected polymer matrices, and nanofillers [20, 49, 50]. Even though nano-modified polymer nanocomposites depend on either heat or the deformation rate for breakdown, this problem needs to be discussed in the literature when a PFM is used to deform the material. Also, no one has yet attempted to calibrate a phase-field breakage theory, which calls for highly intricate experimental setups. This investigation has two primary objectives. The fracturing history of nanocomposites is first explored using phase-field modeling [16], and a viscous constitutive law is developed and incorporated into the simulation. We calculate the free specific energy by decomposing the volume into positive and negative components using a volumetric-deviatoric decomposition.

1.1 Thermo-viscoelastic Properties

It is easier to anticipate the failure processes of polymer nanocomposites by accurately capturing several factors, including damage mechanisms, temperature, and percentage material characteristics. To assess the thermo-viscoelastic characteristics of boehmite nanoparticle/epoxy nanocomposites, this study details the development and application of a limited displacement phase-field fracture framework. The rate-dependent fracture history is defined by an additive decomposition of the regenerative power into an equilibrium, non-equilibrium, and subsequently a volumetric component with different definitions under tensile and compressive deformation. Researchers use the Guth-Gold and altered Kitagawa models to analyze the influence of heat and particulate percentage on the fracture behavior of nanocomposites. The viability of the model is determined via the contrast of values obtained from compact-tension tests with empirical observations. We have used both experiments and simulations to verify the model's capacity to make accurate predictions [51].

Boundary conditions and geometric features are shown in Fig. 1a. A horizontal notch, going from the specimen's outer surface through its center, has been cut into its left side. As the bottom of the sample stays put, the top is shifted. Tensile and shear loads are delivered at a rate of mm/min, with a constant relocation increase of 106 mm. Mechanical properties at 296 K in plane strain were used in the ensuing simulations. In regions where fracture propagation is anticipated, the mesh is subsequently refined. This yields a discretization for a tensile test with 12,509 elements and an exact dimension of 0.003 mm across the long direction of the middle strip of the specimen. As a vital zone in the shear test, this specimen's bottom right diagonal strip employs 21,045 elements with enhanced meshes to attain exact element sizes. The fracture patterns from the two accidents are shown in Images 1b and 1c. A linear crack pattern is expected in a pure shear condition, while a horizontal crack pattern is expected in a tensile scenario. The shear test in Fig. 1c demonstrates that compression cracking can be prevented by employing free energy decomposition.

An epoxy nanocomposite reinforced with BNPs and a finite deformation phase-field fracture model is described. Modeling has shown that the nanoparticle content

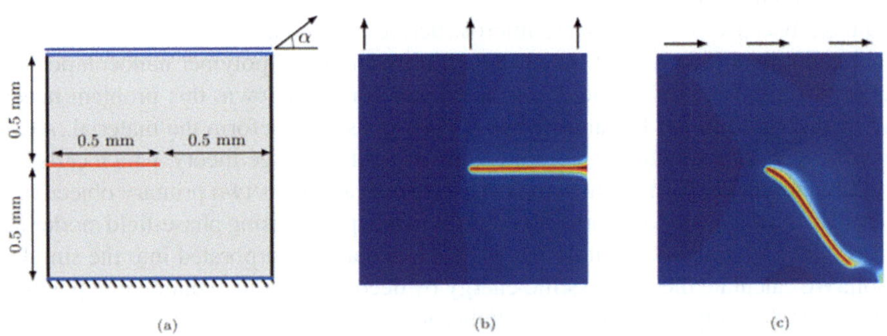

Fig. 1 **a** Notched sample, **b** unidirectional tension (α = 90), **c** shear deformation (α = 0) [28]

affects the rate and temperature dependence of breakage development in the nanocomposites. For this goal, we have employed phase-field modeling using the viscoelastic constitutive law given in [16]. As few researchers described, a favorable decomposition of free energy was implemented by summing the equilibrium, non-equilibrium, and bulk modulus components of free energy. Both the Guth-Gold model and a revised Kitagawa model have been applied to evaluate the effect of the concentration of nanoparticles and heat on fracturing development and recorded force–displacement curves. It has been shown that the suggested PFM is useful for FE assessment of CT tests. Hence, it is being recommended for use. By comparing numerical results with empirical observations, we confirm the PFM's ability to predict the fracture trend of BNP/epoxy nanocomposites at different nanoparticle concentrations [33, 42].

The effectiveness of the PFM can be assessed by comparing future mathematical predictions at varied heats and deformation rates with experimental data. With the help of a correlation between mathematical analysis and empirical studies, the temperature dependency of the speed of power release may be accurately identified. Furthermore, finite deformation at micro and sub-micro scales causes significant alterations to the physicochemical and chemical connections between particles and epoxy matrices. These adjustments would affect the material's viscosity and the rate at which it releases energy. The intricate interactions at the nanoparticle-epoxy interphase make it challenging to decipher the underlying mechanisms responsible for these shifts. An elementary starting point for the model is accounting for fluctuations in the nanoparticle volume percentage. Connecting coarse-grained models [52, 53] with phase-field modeling provides a molecular-scale characterization of polymer nanocomposites, which is crucial for a comprehensive knowledge of the microstructure's influence on the physicochemical characteristics. Inadequate dispersion in the epoxy matrix is caused by nanoparticle aggregation, which degrades the material's characteristics owing to considerably reduced interfacial interactions between the nanoparticles and the matrix. Nanocomposites enhanced by epoxy and surface-modified BNPs may exhibit improved fracture properties [51]. Still, more research is needed to know how modifying BNPs' interfaces may decrease accumulation and enhance their contact properties.

2 Finite Deformation Gradient—An Atomistically-Informed Multiscale Method

Nanoparticle/polymer nanocomposites' experimental data on dynamic behavior needs to be analyzed from multiple angles, such as nonlocal features of the damage mechanism, like nonlinear viscoelasticity. In this work, we provide the theoretical foundations, as well as the numerical implementation, of a new and improved degradation model for boehmite nanoparticle (BNP) and epoxy nanocomposites.

The specifications of the nonlinear creative description can be determined using a framework that relies on chemical investigations and simulation tests. Molecular simulations are used to parameterize the argon viscoelasticity paradigm, and data from compact-tension trials are used to infer the nonlocal parameters associated with damage. We validate the nonlocal constitutive equations included in a regressive Finite element analysis by matching computational results of a compact-tension evaluation of BNP/epoxy composites to data. The experimentation confirms the predictive power of the modeling framework. All nanoparticle-enhanced thermosetting polymers can be used with the proposed method [54].

Adopting a finite deformation GD framework accounts for time scales of interest in damage and failure process analysis. Considering nonlinear based materials, nonlinear viscoelasticity, and the existence of nanoparticles, this model has been developed to predict the degradation and breakage mechanisms in nanocomposites. The nonlocal constitutive characterization is assessed using MD calculations and experimental tests. The percentage elastic mixed rate-independent material behavior of epoxy resin has, therefore, been characterized using MD simulation-based methodologies [36, 55, 56] (Fig. 2).

For this reason, epoxy resin molecular tensile models were tested at a range of strain rates to ascertain its material properties using the Argon notion of viscoelasticity. Likewise, MD simulations have examined the equilibrium behavior of such epoxy resin under shear loading. Using the results of the simulations, the necessary hyperelastic material properties for the constitutive model may be predicted. CT tests

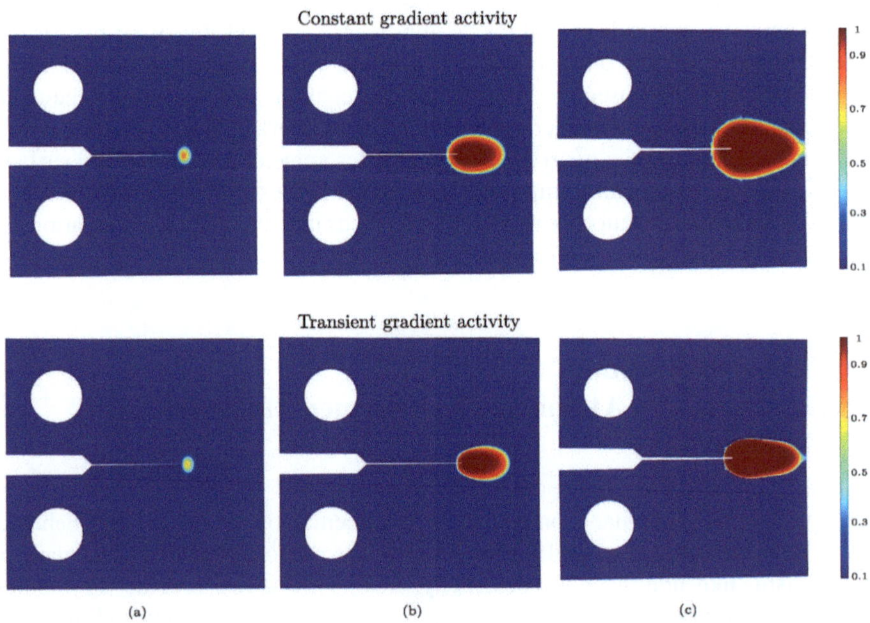

Fig. 2 Damage contour [36]

were used to establish experimental force–displacement curves, which allowed us to identify destructive and nonlocal factors and those collected at the atomic level. By combining molecular dynamics with empirical tests within a single framework, scientists can learn more about the nanocomposites' failure properties and spend less time on parameter identification. We have demonstrated the model's utility through its application to the Finite Element Analysis of CT Testing. Predicting the degradation and breakdown of BNP with epoxy nanocomposites is possible with this model because it faithfully depicts the fundamental physical processes at play. The consistency among experiments and simulations during the loading progression is promising. The present research may have far-reaching consequences for nanocomposite modeling as the method may be utilized to model diverse nanoparticles. The impact of temperature on the properties of nanocomposites is a field that needs more research. This potential enhancement could lead to a better understanding of the thermo-viscoelastic characteristics of the materials. An empirical assessment of local stress areas using CT testing can also be used to determine the nanolocal characteristics, allowing for a more precise comparison. Moreover, the existing modeling framework could be supplemented with coarse-grained dynamics [57–60] to broaden the length and time scales of the examined polymer systems. Now that the GD model has been calibrated and validated, we can train neural network models that can extract the relevant material features and even provide direct solutions to the corresponding threshold-valued issues [36, 61].

2.1 *Crack Progress*

Using a computational technique relying on the eXtended Finite Element Method, we can determine the extent to which cracks and crack propagation alter the electrical resistivity of a material (XFEM). Two phases are necessary to solve the problem of nearly frequent inspection using the finite element tool ANSYS. The XFEM is initially employed to ascertain the composite domain's strain response after damage. The strain state is transferred to the electrical properties of the piezoresistive gadgets positioned in the field, and the result is the electrical impedance among the broken frame's two electrodes. We can tell if a fracture occurs and, if so, how large it is by comparing the measured electrical characteristics of the damaged plate to that of the unharmed one. Also, the crack's progression can be followed by monitoring the variations in electric resistance. There are several numerical examples provided to show how effective this computational framework is [60, 62] (Fig. 3).

Because of their length and twisted shape, nanotubes typically fracture rather than unwind. As a result, the nanotube is pulled out of the polymers, creating friction and, eventually, cracks, releasing the stored energy. Toughening mechanisms discovered in this research [61, 63] include nanotube pull-out, delaminating, and polymeric void creation. (Check out Fig. One example of how this can be seen as a form of toughening is in developing sword drawing skills.) The MWCNT sheath can be broken like a sword in a scabbard by applying tension to the material, which

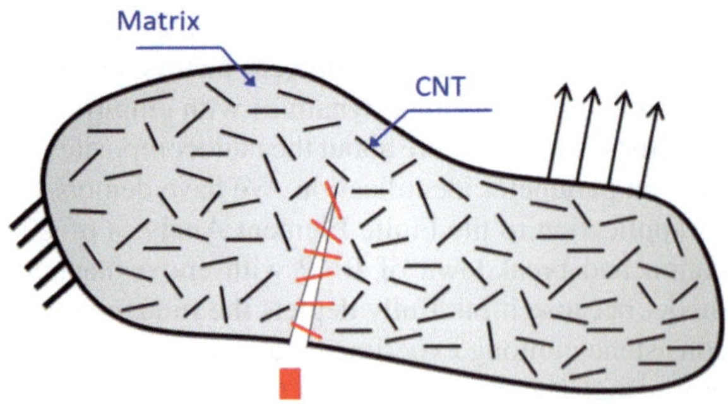

Fig. 3 Fracture mechanisms [62]

causes the outer shell to crack and release the inner surfaces. However, since no shells were in the crevices, the sword-in-scabbard pull-out method can be discounted. Even small amounts of carbon nanotubes (CNT) can dramatically improve the mechanical characteristics of the producing CNT-reinforced composite (CNTRC). Additionally, a specific number of well-dispersed CNT creates conductivity properties in the normally non-conductive polymeric matrix. Because of this latter quality, self-sensing CNTRC has been developed to establish a relationship between electrical resistivity changes and mechanical strain adjustments.

The development of numerical tools is necessary to estimate the electromechanical behavior of these novel materials when crack-type damage is present. Introduce a computational framework based on XFEM to model the impact of crack formation on the physical and electrical behavior of CNTRC materials; in other words, how damage compromises the component's structural integrity and capacity to self-sense. A two-stage method has been used to study crack growth with the use of the commercialized finite element program ANSYS: Firstly, the mechanical issue is resolved using the extended FEM, so that the tension condition in the fractured area is calculated at every element. Second, using the estimated strain field, we must update the elements' piezoresistive characteristics. This leads to the formulation of a problem with non-homogeneous electrical conductivity, which is then solved by utilizing the coupled-field elements in ANSYS.

Suggested numerical scheme is used to investigate a variety of fixed-crack and crack-growth scenarios. This is done to describe the impact cracking has on the electrical current flow as determined in the plates beneath the conductors' configuration. The analysis took into account several factors, including the magnetic conductivity of the fracture, the crack's orientation, and the crack's severity. The sensor efficiency improves as the crack permittivity decreases—when the crack face is electrically insulating, the piezoresistive effect has little to no effect. That is to say, the influence of the crack's alteration of the stress area on the piezo resistivity is outweighed by the electric field discontinuity generated by its presence—variations in the width and orientation of the fracture are readily apparent in electric resistance tests. Mechanical failure isn't the only thing cracked CNTRC parts affect; it also changes their ability to sense strain. Using the suggested numerical XFEM scheme, simulated crack propagation tracking can be carried out successfully, which can be used to connect fractures' existence further and intensity using the electric observations in CNTRC plates [64].

2.2 Elastic Properties

An issue with identifying the elastic continuum model in atomistic simulations of graphene polymer nanocomposites. To calculate an approximation of the local stiffness tensor everywhere in a polymer nanocomposite with graphene laminate, the Atomistic Local Identific Ation of Stiffness (ALIAS) approach was created. Continuum results are consistent with a universal defective contact of zero thickness representing graphene at a large scale. The identification process also uncovered a 1 nm thick interphase on both sides of the graphene, which is 1.5% stronger than the polymer bulk matrix. The elastic characteristics of nanocomposites with a sandwich microstructure are investigated using the specified continuum model. Since graphene plane slip is exceedingly flexible, this investigation at the continuum scale demonstrates a dampening effect. The interface softening is more significant than the interphase stiffening. Furthermore, the continuum model predicts that nanocomposites' stiffness will rise due to graphene cracks [65].

2.3 Shape Memory

Polymers that can revert to an initial, predetermined shape after being subjected to a stimulus are described as "shape memory polymers" (SMPs). The addition of nanoclay filler has enhanced the material's thermomechanical characteristics and broadened its potential for industrial uses. Characterizing the thermomechanical flow of the material using standard experimental equipment is challenging because of the small size of the refill and the different microstructure layers at various scale levels [66–69]. It is also challenging to provide a single numerical model that can reproduce the material's thermomechanical behavior by considering all the influences of the material's underlying atomic- and molecular-scale structure. Under a finite deformation, the material flow and the aggregate calculated lattice constants are studied as a function of the padding weight fraction. Elastic constants and stress–strain curves are in reasonable agreement with analytical predictions. In addition, the findings shed light on the material's behavior and serve as a springboard for more advanced modeling strategies [70–72].

2.4 Heterogeneous Elastic Structural Elements

Combining finite deformation with five-constant (Murnaghan) regressive flexibility, one obtains, under certain conditions, the unique dynamic double diffraction equations (DDE) for a portion of strain rate in rod and shell, with variable coefficients. From the results of numerical simulations and discussions of experimental data, we conclude the tremendous impact of diversity on the dynamic characteristics of strain particles in solids. We have demonstrated how strain solitons can be amplified or decayed depending on the elasticity change in both the polymer rod and the polymer shell. These results may be applicable in NDT issues and solid integrity issues under elastic modulus pulse stress [73].

3 Deformation Mechanisms

Comprehensive research into the importance of nanomanipulation of architecture in polymeric composites to their performance has been hampered by a lack of an appropriate experimental paradigm. This dissertation studied the effects of varying structural parameters on LBL-produced polymer-montmorillonite (MTM) clay nanocomposites. To anticipate the stress behavior of the nanocomposites at lower strain, a continuum-based parametric framework was constructed. Through LBL's systematic control over the nanostructure, we could probe the significance of factors including nanomaterials bulk proportion, particle level spacing, nanoparticle layer stratification, and the contact among the polymer and nanoparticles. By changing the distance between the MTM layers, we created a variety of polyurethane (PU)-MTM nanocomposites with a broad range of MTM nanoparticle volume fractions. Increases in MTM volume fraction resulted in greater yielding strength and stiffness in the nanocomposites. At large nanoparticle volume fractions, ductile behavior gave way to brittle behavior, and there was discovered to be a critical nanoparticle dispersion below which brittle behavior predominated the reaction of the nanocomposites. It was thought that a nanoparticle-stratified layer would provide an extra slip mechanism, leading to improved ductility.

Polyacrylic acid (PAA) was incorporated using an exponential (e)-LBL technique, which changed the contact between the polymer and the nanoparticle layers. The nanocomposites' stiffness and strength were improved because of a more robust interface. The nanocomposite volume was supposed to be filled with effective particles made up of MTM layers and a modified PU prophase area near the MTM layers to create the constitutive model. The behavior of the bulk polymer was captured using a hyperelastic model. The model's practical particle component was a linear viscoelastic spring, a viscoplastic dash-pot, and a nonlinear spring element, which caught the material's early elastic behavior, yield strength, and strain-hardening response. The model predicted all the essential aspects of the uni-axial stress–strain parametric response of a variety of PU-MTM nanocomposites, proving its usefulness [74].

3.1 Elastic Response

There is a fundamental flaw in fiber-reinforced polymeric composites in that they fail too soon because of poor mechanical characteristics in the transverse direction of the fiber. To combat this issue, researchers are developing nano-reinforced laminated composites by growing carbon nanotubes on the surface of fiber filaments. This will enhance the matrix-dominated composite properties. Because of the carbon nanotubes, the interface area between the fiber and the polymeric matrix is increased, increasing the fiber's effective diameter. The elastic characteristics of nano-reinforced composites were predicted numerically in recent work. The practical mechanical attributes of the nano-reinforced laminated composite were determined using a multiscale modeling method and the Finite Element Method. Interface modeling between nanotubes and polymer matrix was performed using the cohesive zone method. To anticipate and correlate the elastic characteristics of nano-reinforced laminated composites, iso-strain and iso-stress models were used to predict the elastic moduli, shear moduli, and Poisson's ratios.

The elastic moduli of the nano-reinforced laminated composite were calculated numerically and then compared to experimental values by many researchers. When it comes to the matrix-dominated features of fiber-reinforced polymeric composites, a unique CNT nano-reinforced laminated composite (NRLC) was developed. This article offers the most critical findings from our investigation of its mechanical response. It was an experimental program coupled with a numerical strategy based on a multiscale modeling technique. It produces more accurate findings than perfect bonding models, indicating that multiscale modeling may be utilized efficiently to analyze nano-reinforced laminated composites. The NRLC FEM results indicate that a nanocomposite matrix material may enhance the matrix-dominated features of traditional fiber-reinforced composites. Evidence showed that by including around 10% CNT volume content in the NRLC, the elastic modulus of the NRLC in the transverse direction to the fiber upsurges by a factor of several times over the value of the polymer binder. The study's findings also show that the transverse elastic modulus is more easily anticipated than the longitudinal modulus or Poisson's ratios [75].

3.2 Glassy Polymers

The mechanical characteristics of agitated polymers are the compromised product of damage and toughening. Firstly, it is shown that the entropy interval of planar crazes is (0, 2), and it is proposed that the craze variable be used to characterize the combined impact of destruction and hardening by the multifractal. Then, the effect of elastic and plastic deformation is removed by defining a new method for calculating fractal dimensions via area conversion with finite deformation and by adjusting the related craze variable. Hence, the constitutive model of fractal crazing was developed.

Thirdly, PMMA's stress barrier of crazing drops linearly with time and tends to remain constant, while its geometric size is quadratic with strain and quadratic with time. Additionally, the damage equation's parameters are calculated, demonstrating that the frenzy factor can account for the fractal discontinuities and strain consistency of frenzy in addition to the three-way, as mentioned above, split between craze and microcrack. Crazy polymer research is given a fractal twist. To begin, the fractal dimension interval is calculated in the context of statistical fractal research (0, 2). The second finding is that the finite distortion leads the fractal dimension to fluctuate and decreases its scale range. It is found that during the limited distortion, the traditional fractal dimension cannot describe the genuine damage degree of crazy polymer, so a modified fractal dimension notion is provided by using area transformation [76].

The fractal dimension defines a novel damage variable. The level of cavitation and fibrilization in a polymer are reflected in the multifractal of crazes. The notion of frenzy constant is to characterize the impact of craze destruction and toughening on materials, building on the harm attribute and hardening function created by the fractal dimension. Both the spiral discontinuities and strain continuation of craze, as well as the three distinctions between craze and tiny crack noted in the introduction, may now be explained by the craze variable. Fractal dimension was determined using the box-counting technique, and creep trends of PMMA were measured throughout a range of stress and time to do so. Observations reveal that PMMA's stress threshold for crazing due to creep falls rapidly with duration and stabilizes at a constant value. Crawl crazes have a fractal dimension that grows exponentially with time and linearly with stress. If the stress level is high enough, many crazes will form, and the craze's fractal dimension will increase by one step. According to the data, the spiral aspect of the creep craze is shown to be related to both stress and time. Using the Kachanov-Rabotnov effective stress theory, a nonlinear microscopic cracks model of the crazed polymer has been constructed for use under finite deformation conditions. The PMMA creep experiment serves as the basis for the model's parameters. It is broadly like the experimental results, allowing for approximations of the three phases of creep growth [77, 78].

4 Conclusion

Polymer nanocomposites are assessed for finite deformation under various conditions. To assess the thermo-viscoelastic characteristics of boehmite nanoparticle with epoxy is studied in detail to the development and application of a limited displacement phase-field fracture framework. The theoretical foundations, as well as the numerical implementation, of a new and improved degradation model for BNP with epoxy nanocomposites. The specifications of the nonlinear creative description can be determined using a framework that relies on chemical investigations and simulation tests. Mechanical failure isn't the only thing cracked CNTRC parts affect. It also changes their ability to sense strain. Using the suggested numerical XFEM scheme, simulated crack propagation tracking can be carried out successfully, which can be used to connect fractures' existence further and intensity using the electric observations in CNTRC plates. Since graphene plane slip is exceedingly flexible, this investigation at the continuum scale demonstrates a dampening effect. The interface softening is more significant than the interphase stiffening.

Furthermore, the continuum model predicts that nanocomposites' stiffness will rise due to graphene cracks. Under a finite deformation, the material flow and the aggregate calculated lattice constants are studied as a function of the padding weight fraction. In addition, the findings shed light on the material's behavior and serve as a springboard for more advanced modeling strategies. How strain solutions can be amplified or decayed depending on the elasticity change in both the polymer rod and the polymer shell. These results may be applicable in NDT issues and solids integrity issues under elastic modulus pulse stress. The model predicted all the essential aspects of the uni-axial stress–strain parametric response of a variety of PU-MTM nanocomposites, proving its usefulness. Study's findings also show that the transverse elastic modulus is more easily anticipated than the longitudinal modulus or Poisson's ratios. The PMMA creep experiment serves as the basis for the model's parameters. It is broadly similar to the experimental outcomes, allowing for approximations of the three phases of creep growth.

References

1. Spitalsky Z, Tasis D, Papagelis K, Galiotis C (2010) Carbon nanotube-polymer composites: chemistry, processing, mechanical and electrical properties. Prog Polym Sci 35(3):357–401
2. Ramesh M, RajeshKumar L, Bhuvaneshwari V (2021) Bamboo fiber reinforced composites. In: Jawaid M, Mavinkere Rangappa S, Siengchin S (eds) Bamboo fiber composites, composites science and technology. Springer, Singapore, pp 1–13. https://doi.org/10.1007/978-981-15-8489-3_1
3. Li Y, Wang S, Arash B, Wang Q (2016) A study on tribology of nitrile-butadiene rubber composites by incorporating carbon nanotubes: molecular dynamics simulations. Carbon 100:145–150
4. Ramesh M, Kumar LR (2020) Bioadhesives. In: Inamuddin R, Boddula MI, Ahamed, Asiri AM (eds) Green adhesives. Wiley, Germany, pp 145–167. https://doi.org/10.1002/9781119655053
5. Ramesh M, Maniraj J, RajeshKumar L (2021) Biocomposites for energy storage. In: Khan A, Rangappa SM, Siengchin S, Asiri AM (eds) Biobased composites: processing, characterization, properties, and applications. Wiley Online Library, Germany, pp 123–142. https://doi.org/10.1002/9781119641803.ch9
6. Chen W, Wu S, Lei Y, Liao Z, Guo B, Liang X, Jia D (2011) Interfacial structure and performance of rubber/boehmite nanocomposites modified by methacrylic acid. Polymer 52(19):4387–4395
7. Ramesh M, Rajeshkumar L, Balaji D, Bhuvaneswari V (2021) Green composite using agricultural waste reinforcement. In: Thomas S, Balakrishnan P (eds) Green composites. materials horizons: from nature to nanomaterials. Springer, Singapore, pp 21–34. https://doi.org/10.1007/978-981-15-9643-8_2
8. Jux M, Fankhänel J, Daum B, Mahrholz T, Sinapius M, Rolfes R (2018) Mechanical properties of epoxy/boehmite nanocomposites in dependency of mass fraction and surface modification-an experimental and numerical approach. Polymer 141:34–45

9. Ramesh M, Rajeshkumar L (2021) Technological advances in analyzing of soil chemistry. In: Ahamed MI, Boddula R, Altalhi T (eds) Applied soil chemistry. Wiley-Scrivener Publishing LLC, USA, pp 61–78. https://doi.org/10.1002/9781119711520.ch4
10. Ramesh M, Rajeshkumar L, Balaji D (2021) Aerogels for insulation applications. In: Inamuddin AMI (ed) Aerogels II: preparation, properties and applications. Materials Research Foundations, United States, vol 98, pp 57–76. https://doi.org/10.21741/9781644901298-4
11. Ramesh M, Rajeshkumar L, Balaji D (2021) Mechanical and dynamic properties of ramie fiber reinforced composites. In: Nagarajan R, Thiagamani SMK, Krishnasamy S, Siengchin S (eds) Mechanical and dynamic properties of biocomposites. Wiley, Germany, pp 275–322
12. Ramesh M, Rajeshkumar L (2021). Case-studies on green corrosion inhibitors. In: Inamuddin AMI, Luqman M, Ahamed MI, Altalhi T (eds) Sustainable corrosion inhibitors, materials research foundations, vol 107, pp 204–221. https://doi.org/10.21741/9781644901496-9
13. Tunega D, Pašalić H, Gerzabek MH, Lischka H (2011) Theoretical study of boehmite's structural, mechanical, and spectroscopic properties (γ-AlOOH). J Phys: Condens Matter 23(40):404201
14. Boyce MC, Socrate S, Llana PG (2000) Constitutive model for the finite deformation stress-strain behavior of poly (ethylene terephthalate) above the glass transition. Polymer 41(6):2183–2201
15. Qi HJ, Boyce MC (2005) Stress-strain behavior of thermoplastic polyurethanes. Mech Mater 37(8):817–839
16. Li Y, Tang S, Kröger M, Liu WK (2016) Molecular simulation guided constitutive modeling on finite strain viscoelasticity of elastomers. J Mech Phys Solids 88:204–226
17. N'guyen TA Lejeunes S, Eyheramendy D, Boukamel A (2016) A thermodynamical framework for the thermo-chemo-mechanical couplings in soft materials at finite strain. Mech Mater 95:158–171
18. Fankhänel J, Arash B, Rolfes R (2019) Elastic interphase properties of nanoparticle/epoxy nanocomposites: a molecular dynamics study. Compos B Eng 176:107211
19. Khorasani MGZ, Silbernagl D, Szymoniak P, Hodoroaba VD, Sturm H (2019) The effect of boehmite nanoparticles (γ-AlOOH) on nanomechanical and thermomechanical properties correlated to crosslinking density of epoxy. Polymer 164:174–182
20. Unger R, Arash B, Exner W, Rolfes R (2020) Effect of temperature on the viscoelastic damage behaviour of nanoparticle/epoxy nanocomposites: constitutive modelling and experimental validation. Polymer 191:122265
21. Bardella L (2001) A phenomenological constitutive law for the nonlinear viscoelastic behaviour of epoxy resins in the glassy state. Eur J Mech-A/Solids 20(6):907–924
22. Ramesh M, Balaji, D, Rajeshkumar L, Bhuvaneswari V, Saravanakumar R, Anish Khan, Abdullah M Asiri (2021) Tribological behavior of glass/sisal fiber reinforced polyester composites. In: Jawaid M, Khan A (eds) Vegetable fiber composites and their technological applications. Composites science and technology. Springer, Singapore, pp 445–459. https://doi.org/10.1007/978-981-16-1854-3_20
23. Ramesh M, Rajeshkumar L, Balaji D, Priyadharshini M (2021) Properties and characterization techniques for waterborne polyurethanes. In: Inamuddin AMI, Boddula R, Khan A (eds) Sustainable production and applications of waterborne polyurethanes. Advances in science, technology & innovation (IEREK Interdisciplinary Series for Sustainable Development). Springer, Cham, pp 109–123. https://doi.org/10.1007/978-3-030-72869-4_6
24. Rajeshkumar L (2021) Biodegradable polymer blends and composites from renewable Resources. In: Sanjay MR, Parameswaranpillai J, Siengchin S, Ramesh M (eds) Biodegradable polymer blends and composites. Woodhead publishing. Elsevier, pp 527–549. https://doi.org/10.1016/B978-0-12-823791-5.00015-6
25. Nguyen VD, Lani F, Pardoen T, Morelle XP, Noels L (2016) A large strain hyperelastic viscoelastic-viscoplastic-damage constitutive model based on a multi-mechanism nonlocal damage continuum for amorphous glassy polymers. Int J Solids Struct 96:192–216

26. Deepa C, Rajeshkumar L, Ramesh M (2021) Thermal properties of kenaf fiber-based hybrid composites. In: Senthilkumar K, Senthil Muthu Kumar T, Chandrasekar M, Rajini N, Siengchin S (eds) Natural fiber-reinforced composites: thermal properties and applications. Wiley, USA, pp 167–181. https://doi.org/10.1002/9783527831562.ch10
27. Arash B, Exner W, Rolfes R (2019) A viscoelastic damage model for nanoparticle/epoxy nanocomposites at finite strain: a multiscale approach. J Mech Phys Solids 128:162–180
28. Arash B, Exner W, Rolfes R (2021) Viscoelastic damage behavior of fiber reinforced nanoparticle-filled epoxy nanocomposites: multiscale modeling and experimental validation. In: Acting principles of nano-scaled matrix additives for composite structures. Springer, Cham, pp 377–410
29. Chowdhury KA, Talreja R, Benzerga AA (2008) Effects of manufacturing-induced voids on local failure in polymer-based composites
30. Ramesh M, Rajeshkumar L, Balaji D, Bhuvaneswari V (2021) Keratin-based biofibers and their composites. In: Sanjay MR, Madhu P, Parameswaranpillai J, Siengchin S, Gorbatyuk S (eds) Advances in bio-based fiber: moving towards a green society. Woodhead publishing. Elsevier, pp 315–334
31. Ramesh M, Rajeshkumar L, Balaji D, Bhuvaneswari V (2022) Influence of moisture absorption on mechanical properties of biocomposites reinforced surface modified natural fibers. In: Muthukumar C, Krishnasamy S, Thiagamani SMK, Siengchin S (eds) Aging effects on natural fiber-reinforced polymer composites. Composites science and technology. Springer, Singapore, pp 17–34. https://doi.org/10.1007/978-981-16-8360-2_2
32. Vandoren B, Simone A (2018) Modeling and simulation of quasi-brittle failure with continuous anisotropic stress-based gradient-enhanced damage models. Comput Methods Appl Mech Eng 332:644–685
33. Ramesh M, Balaji D. Rajeshkumar L, Bhuvaneswari,V (2022) Manufacturing methods of elastomer blends and composites. In: Sanjay MR, Jyotishkumar P, Siengchin S, Ozbakkaloglu T (eds) Elastomer blends and compositess—principles, characterization, advances, and applications. Elsevier, pp 11–25
34. Sathishkumar TP, Satheeshkumar S, Rajeshkumar L (2022. Jute fibers, their composites, and applications. In: Rangappa SM, Parameswaranpillai J, Siengchin S, Ozbakkaloglu T, Wang H (eds) Plant fibers, their composites, and applications. Elsevier, pp 253–282. https://doi.org/10.1016/B978-0-12-824528-6.00020-5
35. Poh LH, Sun G (2017) Localizing gradient damage model with decreasing interactions. Int J Numer Meth Eng 110(6):503–522
36. Arash B, Unger R, Exner W, Rolfes R (2021) A finite deformation gradient-enhanced damage model for nanoparticle/polymer nanocomposites: an atomistically-informed multiscale approach. Compos Struct 258:113211
37. Francfort GA, Marigo JJ (1998) Revisiting brittle fracture as an energy minimization problem. J Mech Phys Solids 46(8):1319–1342
38. Ramesh M, Rajeshkumar L, Balaji D, Sivalingam S (2022) Chapter 15 Mechanical and viscoelastic properties of wool composites. In: Thomas S, Jose S (eds) Wool fiber reinforced polymer composites. Elsevier, pp 299–318
39. Ambati M, Kruse R, DeLorenzis L (2016) A phase-field model for ductile fracture at finite strains and its experimental verification. Comput Mech 57(1):149–167
40. Ramesh M, Rajeshkumar LN, Srinivasan N, Kumar DV, Balaji D (2022) Influence of filler material on properties of fiber-reinforced polymer composites: a review. e-Polymers 22(1):898–916. https://doi.org/10.1515/epoly-2022-0080
41. Sathishkumar TP, de-Prado-Gil J, Martínez-García R, Rajeshkumar L, Rajeshkumar G, MavinkereRangappa S, Siengchin S, Alosaimi AM, Hussein MA (2022) Redeemable environmental damage by recycling of industrial discarded and virgin glass fiber mats in hybrid composites—an exploratory investigation. Polym Compos https://doi.org/10.1002/pc.27047
42. Zhou S, Zhuang X, Zhu H, Rabczuk T (2018) Phase field modelling of crack propagation, branching and coalescence in rocks. Theoret Appl Fract Mech 96:174–192

43. Zhou S, Zhuang X, Rabczuk T (2019) Phase-field modeling of fluid-driven dynamic cracking in porous media. Comput Methods Appl Mech Eng 350:169–198
44. Yin B, Kaliske M (2020) Fracture simulation of viscoelastic polymers by the phase-field method. Comput Mech 65(2):293–309
45. Miehe C, Hofacker M, Welschinger F (2010) A phase field model for rate-independent crack propagation: robust algorithmic implementation based on operator splits. Comput Methods Appl Mech Eng 199(45–48):2765–2778
46. Bourdin B, Francfort GA, Marigo JJ (2000) Numerical experiments in revisited brittle fracture. J Mech Phys Solids 48(4):797–826
47. Ramesh M, Tamil Selvan M, Rajeshkumar L, Deepa C, Awais Ahmad (2022) Influence of Vachellia Nilotica subsp. indica tree trunk bark nano-powder on properties of milkweed plant fiber reinforced epoxy composites, J Nat Fibers https://doi.org/10.1080/15440478.2022.2106341
48. Wu JY (2017) A unified phase-field theory for the mechanics of damage and quasi-brittle failure. J Mech Phys Solids 103:72–99
49. Shen R, Waisman H, Guo L (2019) Fracture of viscoelastic solids modeled with a modified phase field method. Comput Methods Appl Mech Eng 346:862–890
50. Balaji Devarajan, Rajeshkumar L., Bhuvaneswari V, Sanjay MR, Suchart Siengchin G (2022) Additive manufacturing of jute fiber reinforced polymer composites—a concise review of material forms and methods. Polym Compos 43(10):6735–6748. https://doi.org/10.1002/pc.26789
51. Msekh MA, Silani M, Jamshidian M, Areias P, Zhuang X, Zi G, He P, Rabczuk T (2016) Predictions of J integral and tensile strength of clay/epoxy nanocomposites material using phase field model. Compos B Eng 93:97–114
52. Msekh MA, Cuong NH, Zi G, Areias P, Zhuang X, Rabczuk T (2018) Fracture properties prediction of clay/epoxy nanocomposites with interphase zones using a phase field model. Eng Fract Mech 188:287–299
53. Goswami S, Anitescu C, Chakraborty S, Rabczuk T (2020) Transfer learning enhanced physics informed neural network for phase-field modeling of fracture. Theoret Appl Fract Mech 106:102447
54. Goswami S, Anitescu C, Rabczuk T (2020) Adaptive fourth-order phase field analysis using deep energy minimization. Theoret Appl Fract Mech 107:102527
55. Loew PJ, Peters B, Beex LA (2019) Rate-dependent phase-field damage modeling of rubber and its experimental parameter identification. J Mech Phys Solids 127:266–294
56. Brighenti R, Rabczuk T, Zhuang X (2021) Phase field approach for simulating failure of viscoelastic elastomers. Eur J Mech-A/Solids 85:104092
57. Dastgerdi JN, Anbarlooie B, Marzban S, Marquis G (2015) Mechanical and real microstructure behavior analysis of particulate-reinforced nanocomposite considering debonding damage based on cohesive finite element method. Compos Struct 122:518–525
58. Hussein A, Kim B (2019) Micromechanics based FEM study on the mechanical properties and damage of epoxy reinforced with graphene based nanoplatelets. Compos Struct 215:266–277
59. Arash B, Exner W, Rolfes R (2021) A finite deformation phase-field fracture model for the thermo-viscoelastic analysis of polymer nanocomposites. Comput Methods Appl Mech Eng 381:113821
60. Arash B, Park HS, Rabczuk T (2015) Tensile fracture behavior of short carbon nanotube reinforced polymer composites: a coarse-grained model. Compos Struct 134:981–988
61. Arash B, Park HS, Rabczuk T (2016) Coarse-grained model of the J-integral of carbon nanotube reinforced polymer composites. Carbon 96:1084–1092
62. Rodríguez-Tembleque L, Vargas J, García-Macías E, Buroni FC Sáez A (2022) XFEM crack growth virtual monitoring in self-sensing CNT reinforced polymer nanocomposite plates using ANSYS. Compos Struct 284:115137
63. Balokas G, Czichon S, Rolfes R (2018) Neural network assisted multiscale analysis for the elastic properties prediction of 3D braided composites under uncertainty. Compos Struct 183:550–562

64. Samaniego E, Anitescu C, Goswami S, Nguyen-Thanh VM, Guo H, Hamdia K, Zhuang X, Rabczuk T (2020) An energy approach to the solution of partial differential equations in computational mechanics via machine learning: concepts, implementation and applications. Comput Methods Appl Mech Eng 362:112790
65. Gojny FH, Wichmann MH, Fiedler B, Schulte K (2005) Influence of different carbon nanotubes on the mechanical properties of epoxy matrix composites–a comparative study. Compos Sci Technol 65(15–16):2300–2313
66. Ramesh M, Bhuvaneswari V, Balaji D, Rajeshkumar L (2022) Self-healable conductive and polymeric composite materials. In: Inamuddin AMI, Altalhi T, Adnan SM (eds) Aerospace polymeric materials. Wiley-Scrivener LLC, pp 231–258. https://doi.org/10.1002/9781119905264.ch10
67. Deepa C, Balaji D, Bhuvaneswari V, Rajeshkumar L, Ramesh M, Priyadharshini M (2022) Deep learning for the selection of multiple analogs. In: Inamuddin AMI, Altalhi T, Cruz JN, Refat MSED (eds) Drug design using machine learning. Wiley-Scrivener LLC, pp 117–142. https://doi.org/10.1002/9781394167258.ch4
68. Ramesh M, Rajeshkumar L, Balaji D, Bhuvaneswari V, Sivalingam S (2021) Self-healable conductive materials. In: Inamuddin AMI, Ahamed MI, Boddula R, Altalhi TA (ed) Self-healing smart materials. Wiley, United States, pp 297–320. https://doi.org/10.1002/9781119710219.ch11
69. Ramesh M, Rajeshkumar L, Saravanakumar R (2021) Mechanically-induced self-healable materials. In: Inamuddin AMI, Ahamed MI, Boddula R, Altalhi TA (ed) Self-healing smart materials. Wiley, United States, pp 379–404. https://doi.org/10.1002/9781119710219.ch15
70. Hsieh TH, Kinloch AJ, Taylor AC, Kinloch IA (2011) The effect of carbon nanotubes on the fracture toughness and fatigue performance of a thermosetting epoxy polymer. J Mater Sci 46(23):7525–7535
71. Bhuvaneswari V, Priyadharshini M, Deepa C, Balaji D, Rajeshkumar L, Ramesh M (2021) Deep learning for material synthesis and manufacturing systems: a review. Mater Today: Proc 46(9):3263–3269. https://doi.org/10.1016/j.matpr.2020.11.351
72. Ramesh M, Kumar LR, Khan A, Asiri AM (2020) Self-healing polymer composites and its chemistry. In: Khan A, Jawaid M, Asiri AMA (eds) Self-healing composite materials. Elsevier, pp 415–427. https://doi.org/10.1016/b978-0-12-817354-1.00022-3
73. Lu X, Detrez F, Yvonnet J, Bai J (2021) Identification of elastic properties of interphase and interface in graphene-polymer nanocomposites by atomistic simulations. Compos Sci Technol 213:108943
74. Salman M, Guski V, Schmauder S (2022) Two-scale modeling of nano-clay-filled shape memory polymers. J Micromech Mol Phys 1–10
75. Samsonov AM, Semenova IV, Garbuzov FE (2017) Nonlinear guided bulk waves in heterogeneous elastic structural elements. Int J Nonlinear Mech 94:343–350
76. Kaushik AK (2010) Deformation mechanisms in polymer-clay nanocomposites. The University of Michigan
77. Kulkarni M, Carnahan D, Kulkarni K, Qian D, Abot JL (2010) Elastic response of a carbon nanotube fiber reinforced polymeric composite: a numerical and experimental study. Compos B Eng 41(5):414–421
78. Li Y, Sun X, Zhang S, Han S (2022) A fractal crazing constitutive model of glassy polymers considering damage and toughening. Eng Fract Mech 267:108354

Micromechanics of Nanomaterials Based Polymer Nanocomposites

V. Bhuvaneswari

Abstract In today's materialistic world, there is constant progress toward discovering and inventing new materials. Scientists have broadened their focus on the nanoscale since the groundbreaking discovery of 2D materials. Researchers from all over the world are interested in nanocomposites because of the exciting possibilities they present for the synthesis of new nanomaterials. In addition, there has been a rise in interest in biocomposite materials as a solution to the environmental impacts of global warming and other related issues. The micromechanics of polymer nanocomposites is, therefore, the primary topic of this article. Hybrid combinations of carbon nanotube (CNT), graphene, and a few other materials and their behavior at the nanoscale.

Keywords Micromechanics · Polymer nanocomposites

1 Introduction

The elastic modulus [1, 2], the strength [3, 4], and the fracture toughness [5] of composites, among others, have all been the subject of numerous exploratory and computation studies aimed at measuring and predicting these properties. The connectivity characteristics distinguish polymer matrix composites among the reinforcement, and the matrices significantly impact the effective physical properties [6, 7]. In addition, numerous nanocomposites have been developed in light of recent progress in the customizable polymerization of nano-sized particulate [8–10]. The interface properties of a nanocomposite have a more significant impact on its effective physical properties than they would have on those of a conventional strengthened matrix of the identical weight percentage. Several previous studies [11–15] have used FEA to foretell the composites' actual properties. However, three-dimensional FEA calls for many components; specifically, a few small parts must be employed nearer to

V. Bhuvaneswari (✉)
Department of Mechanical Engineering, KPR Institute of Engineering and Technology, Coimbatore, Tamilnadu 641407, India
e-mail: bhuvanashankar82@gmail.com

© The Author(s), under exclusive license to Springer Nature Singapore Pte Ltd. 2023

H. Abdellaoui et al. (eds.), *Mechanics of Nanomaterials and Polymer Nanocomposites*, Smart Nanomaterials Technology,
https://doi.org/10.1007/978-981-99-2352-6_11

the boundary to take the pressure attention into account, calling for robust computational resources. The interface framework of the nanocomposite can't be modeled in FEA, and its interfacial bond parameters can't be determined with any accuracy. The use of molecular dynamics (MD) experiments to foretell the nanocomposite's beneficial physical properties [16–19] has thus emerged as an alternative method in the last few decades. Once the inter-atomic ability is described, the functionality can be modeled additional precisely with no unnecessary appropriate factors, making atomic-level simulations preferable to finite element analyzes. Since many atoms must be simulated in MD simulations, the process is also computationally intensive [20].

Micromechanics-based homogenization methods and MD simulations are used together in multiscale frameworks [21–24] to circumvent the restrictions of brute-force MD simulations. Once the hypothetical estimate is validated in opposition to MD simulations for only several issues, the homogenization method could be used to anticipate the characteristics of the composite across wide varieties of bulk concentration and direct allocation of strengthening because it provides prediction equations of the efficient physical characteristics of a reinforced composite. Since it allows for a closed-form solution, the Mori–Tanaka approach [25] is one of the most popular mean-field homogenization techniques. Studies have used an interfacial framework [26] or an interface spring framework [27–29] to predict the nanocomposite's advantageous qualities by considering the interface's properties. We could expect the unknown but required interaction variables for theoretical modeling by comparing the MD simulation outcomes with the hypothetical forecasts that were adapted employing interfacial factors. After obtaining the interface parameters, the efficient rigidity of the composite is expected for different design parameters.

The linear spring model [27] is one of the earliest models to account for interfaces and is also one of the simplest mathematically. In the linear spring model, the rigidity of the spring at the reinforcement-matrix interface is assumed to be the determining factor in the severity of the imperfection. The two surfaces have completely debonded if there is no spring stiffness at the interface. Based on the interface spring model, the composite's elastic modulus focuses on the Mori–Tanaka remedy whenever the rigidity of the included interaction is endless. As a result, the linear spring model cannot predict a modulus larger than the Mori–Tanaka solution. Yet a large body of experimental research [30–32] demonstrates that specific nanocomposites' appropriate rigidity is more significant than the Mori–Tanaka solution. Often these established multiscale research on polymer matrix nanocomposites has used the interfacial prototype by implying interfacial is stronger and more durable than the matrix [24], which is necessary to explain the experimental results. An increase in the interphase fraction causes the efficient elastic properties of the composite to rise monotonically with a decrease in particle size at a fixed volume fraction. Several experimental and computational studies [33–35] have shown that the efficient modules of some nanocomposites reduce with decreasing grain size due to interfacial defects brought on by interfacial sharpness, synthetic rust, or lattice mismatch.

Research into multiscale models has some limitations, one of which is the inability to systematically test these models' efficacy by selecting specimens randomly from their boundaries. Changing the stiffness or dispersed phase variables to fit the test outcomes prevents a comprehensive investigation of the interface models' applications. The influence of adsorption adherence rigidity and energy within multi-resolution modeling structures that contribute to the interfacial characteristics has yet to be examined. In addition, the limitations of the sequential spring prototype have yet to be explained physically. This article studies how well homogenization methods collaborate for this polymer matrix nanocomposite. We discuss why every interaction prototype must be used when modeling nanocomposites and how this criterion depends on the rigidity of the interfacial bond [28]. Normalization hypothesis, interaction designs, and molecular dynamics simulations are all used to model the system at multiple scales. We analyze the efficient properties of the Nylon6-SiC nanoparticle matrix and predict its elastic modulus of elasticity by considering a wide range of interfacial bond stiffness (energy). We explain why efficient moduli obtained from MD simulations depend on the particle size of the interfacial bond energy (stiffness). Different models should be used for the numerous interface relationship power generation ranges. Because of the theoretical improvements made in our earlier research [29], we adopt the interface spring model to account for imperfections at the interface. Our MD simulations show that in low-bond energy regimes, interfacial damage emerges, and in high-bond energy regimes, the interphase becomes stiffer. We then describe the previous polymer matrix nanocomposites experimental results and describe the amalgamation effect as a function of the interfacial bond stiffness [36].

2 Shape Memory

A monoclinic micromechanics framework is established to foretell the physical modulus of shape memory polymer nanocomposites (SMPN) with silica nanoparticles. Model includes a transition region, or interphase zone, representing an area of the shape memory polymer matrix perturbed close to the SiO2 nanoparticles. When a layer of constricted polymer is present, SMPN ductile characteristics are highly heat-dependent. The elastic modulus of a shape memory polymer nanocomposite reduces, and the normalized modulus increases in a nonlinear fashion with increasing temperature [37–39]. Poisson's ratio falls with rising temperature in a nonlinear manner. Nanoparticle diameter in an interphase zone considerably disturbs the mechanical properties of SMPN. A size reduction of the nanoparticles substantially enhances the elastic modulus. Poisson's ratio also exhibits a nonlinear decrease with decreasing nanoparticle diameter. Additionally, as the temperature is raised, the significance of nanoparticle diameter grows. The elastic modulus of the SMPN and SiO_2 nanoparticles is shown to increase nonlinearly as the volume fraction of SiO_2 nanoparticles increases. In contrast, the Poisson's ratio is shown to decrease. In conclusion, it is demonstrated that the normalized modulus of a SMPN can be increased by increasing the interphase thickness [40–42].

3 Carbon Nanotube (CNT)

3.1 CNT

Electro-mechanical properties of polymer nanocomposites containing carbon nanotubes (CNTs) can vary greatly depending on factors such as the volume concentration of CNTs, the diffusion of CNTs, the orientation of CNTs, and the characteristics of the polymer itself. The electrical percolation phenomenon is intriguing because it causes the conductance of the CNT/polymer nanocomposite to increase by several orders of magnitude as the volume fraction of CNTs increases. Estimates and distributions are obtained and analyzed for the electrical characteristics and piezoresistive multipliers of this same CNT/polymer nanocomposite as the CNT volume concentration grows, and the blocking possibilities are varied. By expressing the percolation transition in aspects of the volume concentration of CNTs, it is clear that hurdle possibilities are crucial in deciding wiring precipitation. Results show that the electro-mechanical characteristics of CNTs are enhanced in the alignment direction compared to the perpendicular direction, allowing for an examination of the impact of CNT orientation. After transforming the estimated piezoresistive coefficients into gauge factors, we discuss them in the sense of innovative data. The methodology employed in this study entails generating a set of algorithmic implementations to a magnitude of 5 m $\times$ 5 m that are semi-random about the weight percentage of CNTs and the spawning of some of these CNTs. The outcomes of this FEA are then used to generate estimates of robotic rigidity, conductivity, and piezoresistive indices. Examining the dispersion of these characteristics can be done by using the multipliers of variability, deviation, and sample variance. These measures of variability go from minimal to maximum and back down to low as the CNT weight percentage rises, trying to capture the occurrence of electronic precipitation [43, 44] (Fig. 1).

FEA and a cloaking impedance design were used to estimate the electrical characteristics of CNT-reinforced nanocomposite materials processes containing oriented CNTs up to a CNT volume concentration of 0.0504. We calculated the mechanical stiffness Cij, the piezoresistive coefficient ij, and the initial electrical conductivity ij for each CNT volume fraction by averaging the outcomes of numerous tiny-magnitude nano composite realizations. The current work utilized a wolfram script to generate a variety of microscale realizations of CNTs. An in-house finite element program was used to examine these manifestations. The impact of CNT distribution was studied by looking at multiple CNT microscale realizations with varying CNT placements. To discuss how CNT alignment affected the results, researchers compared the properties measured in the synchronization path to those measured in the transverse direction. By changing the fence probable by three orders of magnitude, we could analyze the impact of the electrical tunneling effect. Using varying barrier potentials, we observed and reported electrical percolation in both the orientation and circumferential directions. Findings from comparing the estimates for electro-mechanical factors with other description attempts discovered in the discussion showed a high degree of agreement. A quick rundown of the most critical findings

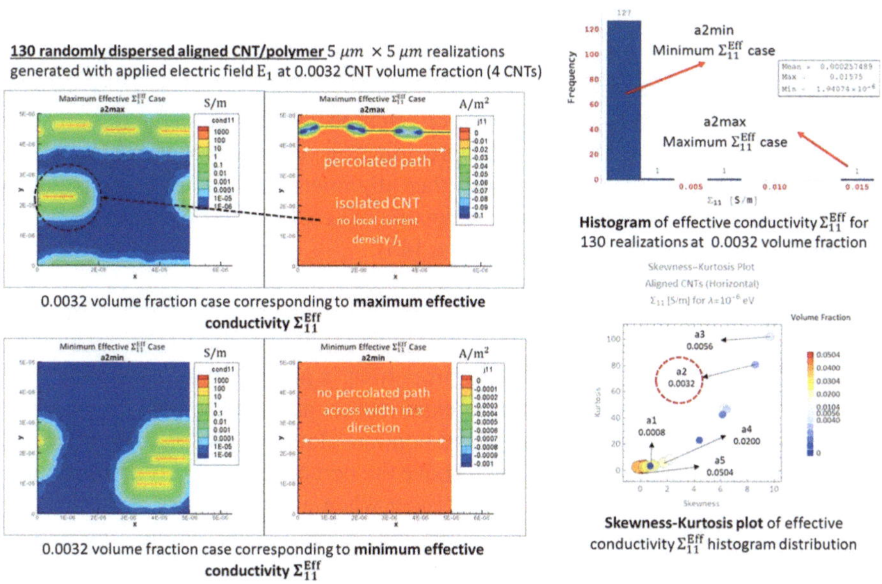

Fig. 1 CNT polymer nanocomposite electro-mechanical properties [43]

from the modeling efforts is provided. Alignment-directed mechanical reinforcement is always greater than transverse-directed mechanical reinforcement regardless of the percentage of CNTs present. This was discovered regarding predictions for the engineering constants E1 and E2 and the mechanical stiffness components C11 and C22.

The primary conductivity elements 11 and 22 went from having values closer to those of pure polymer to having qualities larger than 101S/m as the CNT relative density was increased. Consistently, it was found that the improvement in conductivity along the synchronization path was more significant than that along the transverse direction. We used a threshold coefficient of modification indicator to locate the percolation transition zones. Compliant with experimental efforts found in literature, PTR every time took place at lesser volume fraction variations in the synchronization path related to the oblique position. Analysis of bar graphs and skewness-kurtosis plots confirmed the onset and cessation of electrical percolation. An increase in barrier potential values caused the transition of the PTR to maximum weight concentration. Compared to the stiffness and conductivity components, the estimation of piezoresistivity evolution was more challenging to follow. It was discovered that the piezoresistivity of carbon nanotubes (CNTs) is affected by several variables, including their volume fraction, tendency to percolate, propensity to form redundant paths, and inherent piezoresistivity. Aligned CNT/polymer nanocomposites were compared to experimental efforts reported in the literature. It was discovered that the piezoresistive effect is more pronounced inside the synchronization path than in the transverse direction. Aligned CNT/polymer nanocomposites' electro-mechanical properties were estimated and found to be in acceptable agreement with both these parameterization attempts discovered in the research and those formed concerning the electrical precipitation of the nanocomposite. Model presented herein expanded to examine other alignment conditions, such as poorly oriented and dispersed CNT/polymer nanocomposites. In addition, future work will account for the influence of CNT waviness [43].

The infiltration criterion of nanoparticles is crossed, and the elastic deformation yield strength and an entire network of polymer/CNT nanocomposites are predicted using an innovative technique relying on a micromechanics model. To account for the absorption edge and other network properties, a new network parameter (NP) is introduced. Literature-based sample experiments are used to test the accuracy of the model's predictions. Using 3D and contour plots, we estimate the effects of control factors on the modulus and NP. Above the percolation threshold, the model's predictions match those of experiments. Even when a nanoparticle network is not present, the model consistently overestimates the modulus of the sample. Correlations between modulus and other valuable parameters like CNT volume concentration, infiltration limit, CNT aspect ratio, and CNT modulus are presented in the proposed model. Furthermore, the aspect ratio of CNTs and the percolation threshold are directly and inversely related to NP [45].

3.2 CNT—Polymer Media

Micromechanical analysis using an extended Takayanagi equation predicts high conductivity in polymer nanocomposites containing carbon nanotubes (PCNT). The complex model assumes that carbon nanotube (CNT) net, metabolic stage, molecules tunneling, and particle wet ability in polymer medium all play essential roles in conductivity. Interphase's effect on networks can be seen in the efficient padding accumulation and the onset of precipitation. The model also thoroughly investigates the tunneling properties and wettability using the existing equations. The predictions generally agree with the scientific data of the pieces used to approve the suggested model. Increased conductivity is achieved through various factors, including thick interphase, low CNT curliness, a thin Carbon nanotube, high CNT conduction, and extensive network size. In addition, at the largest tunneling radius ($r = 15$ nm) and the lowest tunneling distance (=3 nm), the conductivity dramatically increases to 16 S/m, whereas a shielded specimen is seen at >5 nm. Therefore, the conductivity of the nanocomposite is mainly under the control of the tunneling dimensions [46].

3.3 CNT—Interface–Treated

When describing the elastoplastic behavior of CNT/polymer nanocomposites, a two-step iterative micromechanics preparation is used in conjunction with finite element modeling and a weakened interface model. The validation is accomplished by contrasting the predicted outcomes under the ideal bonding assumption with the observed facts. Considering the weakened interface, the micromechanics method is extended to use an equivalent fiber to represent the nanotube, the polymer it is embedded in, and the interfacial interactions. The sliding parameter is identified as the most crucial factor in the developed method by comparing the model's results to those of a subatomic systemic mechanics-finite element multiscale approach under different loading conditions. Finally, using some examples, systematically investigate how the included force influences the elastoplastic nature of these nanocomposites. According to the findings, interfacial bonding characteristics are essential in improving the host polymer's mechanical behavior, and as such, they warrant extensive investigation [47].

4 Montmorillonite Nanoparticles

Various researchers used different bioceramics as reinforcement in polymer nanocomposites to improve their properties [48, 49]. Using Montmorillonite (MMT) nanoparticles as reinforcement, a polymer composite's mechanical and thermal properties are studied. The ABAQUS finite element commercial software is used for the computations, and the 3D representative volume elements (RVE) method is employed because of the low cost of analyzes. The matrix and nanoparticle materials are made of low-density polyethylene (LDPE) and metal–organic framework material (MMT), respectively. Mechanical and thermal properties like Moduli, heat expansion coefficient, shear strain, along with the heat transfer coefficient are investigated using nanoparticles of varying dimensional forms and weight fractions. Finite element modeling is carried out in 2 ways, excellent attachment and cohesive zone, because they both address the properties of the intercellular area among the matrix and nanoparticle. The findings are confirmed by cross-referencing them with published experimental results, with good agreement. The GA method presents a prediction function for Young's modulus. It can also be determined using the theoretical models of the Kerner and Paul approaches. It was determined that when MMT nanoparticles are included, the Young's and shear modules grow in size. Increased MMT nanoparticles added to a polymer matrix nanocomposite also result in lower heat expansion and heat transfer coefficients [50].

5 Graphene

The dynamic and coalescence nature of polyethylene-graphene nanocomposites benefit from the presence of oxygen dynamic sectors in mono-layer graphene oxide (GO), as predicted by molecular dynamics (MD) simulations and a mean-field micromechanics model. Once the graphene deflection is accounted for packing of the molecular cells of nanocomposites, the negative influences on the latitudes of Young's and in-plane shear moduli are pronounced. This is because the electrophile and alkoxy reduce the functional attributes of mono-layer graphene. However, it is anticipated that the nanocomposites will have a significantly elevated longitudinal shear modulus. The accomplishments of the included area and the interfacial tightening impact on the flexibility of nanocomposites are once more affirmed by comparing the MD simulated outcomes with the predictions of the double-inclusion (D-I) model. Finally, we show that single-layer GO's interplay with nearby PE matrix through particle virial stress leads to innovative development of shear and plane stress [51, 52] (Fig. 2).

The researcher conducted a functionalist molecular modeling calculation and an average micromechanics assessment to determine how the existence of single-layer graphene in PE nanocomposites affects the compressive and connection angle attitudes. Both oxygen dynamic sectors destroy the sp2 complexation of the intriguing planar nanocarbon arrangement, which is responsible for graphene's high elastic constants. Results in latitudinal strain and in-plane snipping demonstrate that anisotropy nanocomposites' strength and elastic modulus are oxygen functional group dependent. Compared to the stretchy deterioration of nanocomposites due to the deprivation of graphene, the presence of oxygen dynamic sectors was found

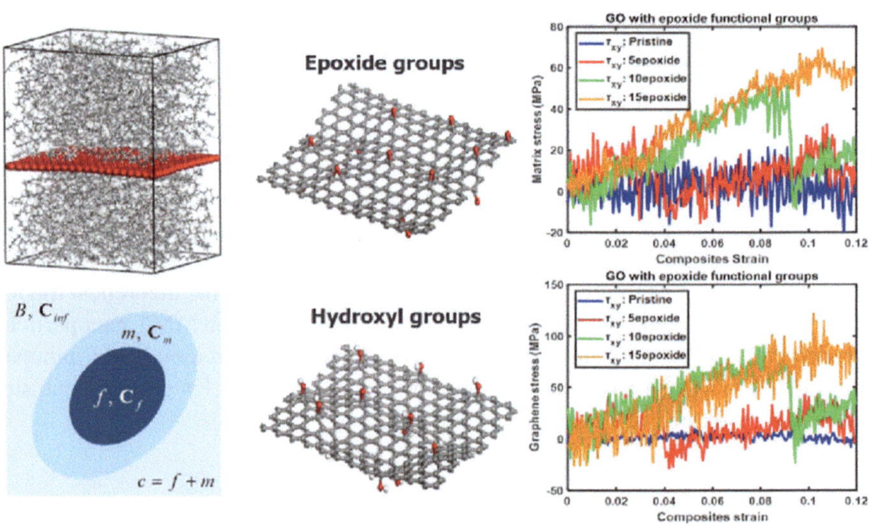

Fig. 2 Graphene polymer nanocomposite's mechanical and interfacial behavior [51]

to significantly increase the longitudinal compressive modulus, which includes the interfacial shear strain distribution capability. The fact that increasing the surface roughness of graphene by attaching oxygen dynamic sectors may be responsible for its unexpected improvement is a plausible explanation. We compare MD simulation results with D-I models and find that the interphase region forming closer to the graphene and GO has a significant impact on the circumferential strain of the nanocomposites. It was also shown that transverse loading causes significant transverse stress in nanocomposites containing graphene and GO for the initial period. The circumferential elastic constant of graphene and GO are solid and considered equivalent to the latitudinal elastic constant of both substances by evaluating the circumferential pressures in these materials beneath the longitudinal hoisting situation of nanocomposites.

Shear stress transition in the PE matrix and GO can shed light on the enhancement due to sheer load transition at the interface made possible by incorporating oxygen functional groups. When longitudinal shear stress was applied to the nanocomposites, pristine graphene displayed full interparticle sliding. At the same time, GOs exhibited various shear stress transfers because of the physical interlinking at the interplay via the oxygen-responsive groups. A comprehensive understanding of the effects of intrinsic and extrinsic faults on the properties of the materials, in addition to one's interaction characteristics with other organisms of particles, is necessary for developing innovative uses for graphene and other carbon isotopes. Achieving a maximum amount of graphene improvement is not a view, as this is determined by the required implementation requirements of the graphene. Optimizing the microstructure of the reinforcement to fulfill both prerequisites at an affordable price is the biggest challenge in creating institutional composites with outstanding characteristics and interfacial load transfer abilities. Our results show that graphene degradation can enhance the interfacial characteristics of nanocomposites, which opens up new opportunities for the use of constructed deficiencies or surface functionalizations. Because of the high cost and complexity of graphene synthesis, research into low-cost, high-performing graphene-reinforced nanocomposites is warranted. In addition, nanocarbon engineering is poised to become a prominent research field due to the need for more straightforward and applicable methods of capitalizing on inevitable functionalization and hybridization [52].

6 Graphene Oxide

Oxygen-functionalized graphene oxide (GO)-epoxy nanocomposites are investigated for their hygroelasticity. With water molecules either evenly dispersed or concentrated at the interface, two distinct molecular models of transversely isotropic nanocomposites are built. The stress–strain diagrams and CMEs (expansion coefficients due to moisture) are calculated. Interfacial decohesion tests reveal how the infiltrating moisture degrades the GO—epoxy interface. A new set of closed-form remedies for the efficient deformable rigidity and CME of multiphase fibers with interfacial imperfections are derived using the micromechanics model. Once nanocomposites absorb water, their hygroelastic properties deteriorate in a predictable manner independent of the local moisture distribution. We used MD simulations and a Thermo–elastic multi-inclusion model to investigate the hygroelastic behavior of GO/epoxy nanocomposites, a material whose aging characteristics are difficult to access experimentally. The hygroelastic characteristics of nanocomposites could be accurately predicted utilizing the interlayer adherence calculated from decohesion tests carried out using MD simulations. The results, which were not affected by the moisture concentration, showed significant volumetric swelling and a decrease in young's modulus and interfacial bond strength. It was also discovered that the GO/epoxy interface is inherently weak, even without wetness absorption. The suggested hygroscopic multi-inclusion model [53] can be used with composites that contain integrated, encased, or operationally appraised elements with a wide range of powder ratios, quantity fragments, and interlayer lumps and bumps.

Graphene-based materials with a high graphene intensity have attracted a lot of attention because of their potential as multifunctional, concept electrocatalysts that can hold electrical energy and bear structural loads. Graphene-based substances have a vast electrochemically energetic exterior, rising electrical properties, good stiffness, and strength. Graphene-based polymeric composites with the minimal intensity of conductors fillers have been the focus of most published lab testing of electrical conductivity for percolation. Reduced graphene oxide (rGO) and aramid nanofiber (ANF) nanocrystals with rGO concentrations up to 100 wt% are fabricated and characterized in this study. A deformation and experiential framework of electrical conductivity were also developed, taking into account the effects of rGO's waviness and conductivity, the volume fraction of ANF, the arbitrarily aligned rGO and ANFs, the included thickness, and the interphase conductivity. Experiments have shown that the in-plane conductivity drops precipitously when ANFs are incorporated into rGO nanocomposite films. For instance, reducing the ANF load distribution from 25 wt% to 0 wt% in rGO/ANF nanocomposite films increased the innovative in-plane permeability by 30. The proposed model explains this exponential connection. According to the model, interphase thickness and conductivity had a more significant impact than waviness. When waviness was reduced from its highest expected values to its minimum, the efficient in-plane permeability was altered by 20%. By increasing the interphase thickness from 0 to 0.5 nm, the effective in-plane conductivity dropped from 0.09 S m1 to 0.01 S m1. When the thickness and conductivity of the interphase changed in response to the volume fraction of rGO, the simulated results matched the experimental data. The crystalline structure of the composite, along with the interfacial framework and conductivity, are significantly impacted by the incorporation of ANFs [54–57].

7 CNT-Filled Polymer

Using the Ouali model for the modulus of composites as a starting point and developed a new model for the conductivity of polymer systems (PCNT) based on carbon nanotubes (CNTs) that accounts for the size of the CNTs, their conductivity, the concentration of the CNTs, the volume fraction of the CNTs that are attached, the depth of the interphase, and the characteristics of the tunnels. Efficiency in filler loading and the onset of percolation are linked to the volume fraction of the conductive net consisting of CNT and interphase. The established model also accounts for the tunneling distance and resistivity due to the polymer sheet. A state-of-the-art model is evaluated against the empirical data gathered from many samples. Consistently positive results between experimental data, the model's predictions, and the intuitive possessions of total factors on the nanocomposite's conductivity demonstrate the model's soundness. Although calculations show that increasing the CNT accumulation and interfacial depth can boost the conductivity of the polymer host, doing so with fewer values does not increase conductivity [57, 58] (Fig. 3).

7.1 Graphene-Based Nanoplatelets

To better understand how adding reduced graphene oxide (rGOnp) and theoretically optimal graphene nanoplatelets (Gnp) affects the mechanical characteristics and impact process of graphene-based nanocomposites, a mathematical micromechanics study is conducted. In this case, we focus on the effects of fragile crack formation and functionality delamination. The results demonstrated that the nanoplatelets/matrix modulus mismatch, volume percentage, connectivity strength, and alignment all play a role in the competition between the two mechanisms. Compared to rGOnp, Gnp had a better modulus but weaker strength and higher fracture strain. Because of Gnp's significant modulus mismatch, matrix cracking, stress concentration, and coalescing were all exacerbated under mild loading conditions. Results showed that interface debonding could be reduced by applying more robust interface properties, resulting in increased strength and lower fracture strain. Additionally, the alignment of nanoplatelets increases modulus, strength, and fracture strain all at once. This is because bio-inspired nacre-like structures have better load transfer and activate toughening mechanisms [59].

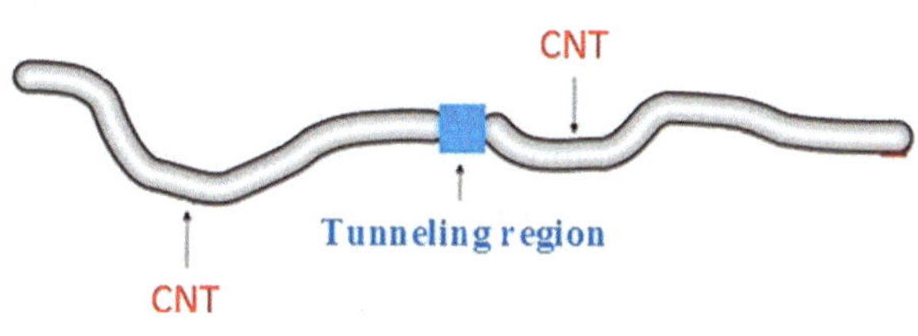

Fig. 3 CNT filled polymer system [57]

The graphene/epoxy nanocomposites were reported to have widely varying mechanical properties [60]. This is because graphene-based materials' mechanical and functional properties can vary greatly depending on the synthesis route taken. The primary focus was on contrasting two extreme graphene-based materials, Gnp and rGOnp. The graphene nanoplatelets are probably of the rGOnp type, which may help to explain the reported moderate mechanical properties. The nanocomposite modulus is proportional to the square of the nanoplatelet modulus. However, the sacrifice in durability and fracture resistance is significant. This is because the edges of the nanoplatelets experience a greater concentration of stress due to the more substantial modulus mismatch and serve as crack propagation locations. The possibility that these deformations will combine into larger cracks, failing lower loads, increases with the magnitude of the modulus mismatch. Increasing the volume percentage results in a greater concentration of these potential crack initiation sites, reducing the material's load-bearing capacity. In addition, it was demonstrated that the interface debonding vs. matrix cracking competition is influenced by the modulus mismatch. Compared to Gnp, rGOnp appears to suffer from more excellent interface debonding.

Furthermore, the results demonstrated that a more robust interface, as predicted for rGOnp due to dynamic groups, leads to significantly enhanced resilience and rupture stress, albeit still inadequate to the neat epoxy [61]. One possible explanation is that the power of the stress accumulation at the nanoplatelets' corners is dissipated during the matrix plastic deformation. Evidence from fractography studies confirmed the existence of excessive plastic deformation. Epoxy's plastic toughening mechanism may be controlled by its atypical chemical composition and curing times. The findings only account for brittle failure, so additional research into damaged plasticity is required [62]. It was discovered that the damage mechanism was modified depending on the nanoplatelet's orientation. Platelets with a more significant misalignment acted as stress concentration sites, demonstrating a greater propensity for interface debonding. Aligned particles, on the other hand, exhibited the optimal mix of characteristics. When compared to matrix cracking, the impact of interface debonding was minimal. Instead, the aligned particles impeded the crack's expected propagation, significantly boosting the nanocomposite's toughness. The familiarity between this toughening mechanism and that seen in genetic frameworks like nacre is intriguing. The powerful possibility of graphene-predicated nanocomposites can be unlocked by further research into the realization of this micro-structural via innovative analysis techniques.

Pure graphene (Gnp) has theoretically maximum properties compared to rGOnp, a reduced form of graphene oxide. Matrix fracturing and interaction delamination were found to be competing for mechanisms that control mechanical properties and failure. Each mode is activated by interface properties, modulus mismatch, volume fraction, and orientation. While Gnp did show increased modulus, this improvement came at the expense of decreased strength and increased fracture strain. However, rGOnp demonstrated moderate modulus enhancement alongside improved strength and fracture strain. Because the latter has a lower modulus, stresses aren't concentrated as much. As a result of their alignment, particles with better load transfer perform better. The configuration also displayed the toughening mechanism for crack resistance that has been reported for nacre-like structures [63].

8 CNT—Polyimide

There is an investigation into the thermoelastic behavior of polyimide nanocomposites enhanced to CNTs that are randomly oriented. A unit cell micromechanics model is used alongside an appropriate representative volume element. Nanocomposite modeling considers the interphase region formed by the non-bonded interaction among the CNT and the surrounding polyimide matrix. The thermal expansion coefficient of polyimide nanocomposite is studied about several different parameters, including the alignment versus random orientation of CNTs within the polyimide matrix, the volume fraction versus diameter of CNTs, the size versus adhesion exponent, and the substrate characteristics of the interfacial area. The results show that the CTE increases nonlinearly with decreasing CNT diameter when interphase is present but has no effect on the thermoelastic characteristics of polyimide nanocomposite when interphase is ignored. It is also demonstrated that an increase in the interphase's thickness, CTE, and stiffness leads to an increase in the CTE of a polyimide nanocomposite comprising isotropic CNTs. The presented model yields results that agree with experimental data to a high degree [64].

The unit cell micromechanical framework was proposed to foretell the thermoelastic characteristics of polyimide nanocomposites enhanced to 3D isotropic CNTs. Three distinct regions, or phases, make up the RVE of the model: the carbon nanotube (CNT), the polyimide matrix, and the interphase region formed by the non-bonded vdW interplay among the CNT and the surrounding polyimide. When comparing the results from the proposed SUC method to experimental data, indeed robust correlation was noted. Compared to the latitudinal CTE of aligned CNT-reinforced nanocomposite, the CTE of the nanocomposite that included 3D isotropic CNTs was greater. Transverse CTE increases to about 1% as the CNT volume fraction rises but then decreases. The CTE of the polyimide nanocomposite was shown to increase with decreasing CNT diameter. Increases in thickness, CTE, and stiffness of interphase all lead to an increase in the CTE of polyimide nanocomposites enclosing 3D isotropic CNTs. In addition, it was found that the interphase Poisson's ratio has no significant impact on the thermoelastic reaction of polyimide nanocomposite [65–67].

9 EGaIn—Polymer

The demand for smart systems with sensing and self-monitoring capabilities is on the rise. Researchers have focused much of their attention on soft matter composites as a material system in recent years. Their mechanical pliancy makes them useful for many projects, such as IoT devices, wearable biosensors, and human–computer interaction (IoT). The functional departments in these composites are micro- and nano-droplets of indium alloy wheels that are fluid at room air temp. Standard non-toxic LMs with high electrical and thermal conductivity include eutectic gallium indium (EGaIn) and eutectic gallium indium tin (Galinstan). We introduce a micromechanics model to foretell the elastic and functional behaviors of EGaInpolymer composites [68, 69]. As the solid gallium oxide layer shapes all over the liquid inclusions, this Eshelby incorporation model is more accurate. Even though this oxide interphase has little effect on the elasticity of LM nanocomposites, it does have a noticeable effect on composites with larger diameters (>30 microns). The developed model is applied to composites with various filler capacity fragments and polymer matrices, and it is also used to investigate the core structure of LM inclusions. This model is also used to predict the dielectric characteristics and heat capacity of EGaInpolymer composites. There is remarkable concordance between the model, the finite element analysis, and the existing experimental data. Finally, we discussed how LM composites might be used in future AI [70].

10 Conclusion

This article discusses the micromechanics of polymer nanocomposites at various hybrid combinations. The outcome of multiple combinations is discussed as follows. Research into nanocarbon engineering will likely focus on developing more accessible and practical methods for capitalizing on functionalization and hybridization. Aligned CNT/polymer nanocomposites and their electrical percolation properties in electro-mechanical CNT modulus, aspect ratio, percolation threshold, and volume fraction have logical relationships with modulus. To increase conductivity, CNTs should be flat, thin, have high conduction, and be part of an extensive network. Moreover, at the largest tunneling diameter ($d = 30$ nm) and the smallest tunneling distance (=3 nm), the conductivity increases dramatically to 16 S/m, whereas an insulated sample is observed at >5 nm. Therefore, the conductivity of the nanocomposite can be manipulated mainly by changing the tunneling dimensions. To improve the mechanical behavior of the host polymer, it is essential to investigate the interfacial bonding characteristics. Adding nanoparticles of Montmorillonite (MMT) to a polymer matrix reduces the coefficients of heat flow and heat exchange in the composite. Research into nanocarbon engineering will likely focus on developing more accessible and practical methods for capitalizing on functionalization and hybridization. The microstructure of the composite, interphase structure and conductivity all play a role. The conductivity of the polymer host cannot be improved by reducing the CNT concentration or the interphase thickness to small values. Greater efficiency is seen when particles are aligned, as they are better able to transfer force. Additionally, this arrangement demonstrated the same toughening mechanism for crack resistance observed in nacre-like structures. The modeling results show excellent agreement with finite element analysis and available experimental results. The potential applications of LM composites are in emerging intelligent systems.

References

1. Zakaria MR, Kudus MHA, Akil HM, Thirmizir MZM (2017) Comparative study of graphene nanoparticle and multiwall carbon nanotube filled epoxy nanocomposites based on mechanical, thermal and dielectric properties. Compos B Eng 119:57–66
2. Sathishkumar TP, de-Prado-Gil J, Martínez-García R, Rajeshkumar L, Rajeshkumar G, Mavinkere Rangappa S, Siengchin S, Alosaimi AM, Hussein MA (2022) Redeemable environmental damage by recycling of industrial discarded and virgin glass fiber mats in hybrid composites—an exploratory investigation. Polym Compos 44(1):318–324. https://doi.org/10.1002/pc.27047
3. Chen Y, Zhou S, Yang H, Wu L (2005) Structure and properties of polyurethane/nanosilica composites. J Appl Polym Sci 95(5):1032–1039
4. Ramesh M, Tamil Selvan M, Rajeshkumar L, Deepa C, Ahmad A (2022) Influence of Vachellia nilotica Subsp. indica tree trunk bark nano-powder on properties of milkweed plant fiber reinforced epoxy composites. J Nat Fibers 19(16):13776–13789. https://doi.org/10.1080/15440478.2022.2106341
5. Fu SY, Feng XQ, Lauke B, Mai YW (2008) Effects of particle size, particle/matrix interface adhesion and particle loading on mechanical properties of particulate–polymer composites. Compos B Eng 39(6):933–961
6. Priyadharshini M, Balaji D, Bhuvaneswari V (2022) Fiber reinforced composite manufacturing with the aid of artificial intelligence—a state-of-the-art review. Archives of Comput Methods Eng 29:5511–5524. https://doi.org/10.1007/s11831-022-09775-y
7. Luo H, Zhou X, Ellingford C, Zhang Y, Chen S, Zhou K, Zhang D, Bowen CR, Wan C (2019) Interface design for high energy density polymer nanocomposites. Chem Soc Rev 48(16):4424–4465
8. Toda I, Tsuruoka T, Matsui J, Murashima T, Nawafune H, Akamatsu K (2014) In situ synthesis of metal/polymer nanocomposite thin films on glass substrates by using highly cross-linked polymer matrices with tailorable ion exchange capabilities. RSC Adv 4(9):4723–4726
9. Devarajan B, LakshmiNarasimhan R, Venkateswaran B, Mavinkere Rangappa S, Siengchin S (2022) Additive manufacturing of jute fiber reinforced polymer composites: a concise review of material forms and methods. Polym Compos 43(10):6735–6748. https://doi.org/10.1002/pc.26789
10. Koo J, Kim H, Kim KY, Jang YR, Lee JS, Yoon SW, Suh BJ, Yu T, Bang J, Yoon K, Yuan G (2016) Controlling the magnetic properties of polymer–iron oxide nanoparticle composite thin films via spatial particle orientation. RSC Adv 6(61):55842–55847
11. Lee S, Jung J, Ryu S (2020) Micromechanics-based prediction of the effective properties of piezoelectric composite having interfacial imperfections. Compos Struct 240:112076
12. Mohankumar D, Rajeshkumar L, Muthukumaran N, Ramesh M, Aravinth P, Anith R, Balaji R (2022) Effect of fiber orientation on tribological behaviour of Typha angustifolia natural fiber reinforced composites. Mater Today: Proc 62(4):1958–1964. https://doi.org/10.1016/j.matpr.2022.02
13. Ramesh M, Rajeshkumar L, Balaji D, Bhuvaneswari V (2022) Influence of moisture absorption on mechanical properties of biocomposites reinforced surface modified natural fibers. In: Muthukumar C, Krishnasamy S, Thiagamani SMK, Siengchin S (ed) Aging effects on natural fiber-reinforced polymer composites. composites science and technology. Springer, Singapore, p 17–34. https://doi.org/10.1007/978-981-16-8360-2_2

14. Ramesh M, Rajeshkumar L, Sasikala G, Balaji D, Saravanakumar A, Bhuvaneswari V, Bhoopathi R (2022) A critical review on wood-based polymer composites: processing, properties, and prospects. Polymers 14(3):589. https://doi.org/10.3390/polym14030589
15. Lee S, Kim Y, Lee J, Ryu S (2019) Applicability of the interface spring model for micromechanical analyses with interfacial imperfections to predict the modified exterior Eshelby tensor and effective modulus. Math Mech Solids 24(9):2944–2960
16. Su CH, Chen HL, Ju SP, Chen HY, Shih CW, Pan CT, You TD (2020) The mechanical behaviors of polyethylene/silver nanoparticle composites: an insight from molecular dynamics study. Sci Rep 10(1):1–14
17. Ramesh M, Rajeshkumar L, Balaji D, Bhuvaneswari V (2021) Keratin-based biofibers and their composites. In: Sanjay MR, Madhu P, Parameswaranpillai J, Siengchin S, Gorbatyuk S (eds) Advances in bio-based fiber: moving towards a green society. Woodhead Publishing, Elsevier, USA, pp 315–334
18. Ramesh M, Rajeshkumar L, Bhuvaneswari V (2021) Leaf fibres as reinforcements in green composites: a review on processing, properties and applications. Emergent Mater 5(3):833–857. https://doi.org/10.1007/s42247-021-00310-6
19. Moon J, Yang S, Cho M (2017) Interfacial strengthening between graphene and polymer through Stone-Thrower-Wales defects: Ab initio and molecular dynamics simulations. Carbon 118:66–77
20. Zepeda-Ruiz LA, Stukowski A, Oppelstrup T, Bulatov VV (2017) Probing the limits of metal plasticity with molecular dynamics simulations. Nature 550(7677):492–495
21. Yang BJ, Shin H, Lee HK, Kim H (2013) A combined molecular dynamics/micromechanics/finite element approach for multiscale constitutive modeling of nanocomposites with interface effects. Appl Phys Lett 103(24):241903
22. Sahayaraj AF, Muthukrishnan M, Ramesh M, Rajeshkumar L (2021) Effect of hybridization on properties of tamarind (Tamarindus indica L.) seed nano-powder incorporated jute-hemp fibers reinforced epoxy composites. Polym Compos 42(12):6611–6620
23. Ramesh M, Balaji D, Rajeshkumar L, Bhuvaneswari V, Saravanakumar R, Anish K, Asiri AM (2021) Tribological behavior of glass/sisal fiber reinforced polyester composites. In: Jawaid M, Khan A (eds) Vegetable fiber composites and their technological applications. composites science and technology. Springer, Singapore, pp 445–459. https://doi.org/10.1007/978-981-16-1854-3_20
24. Yang S, Cho M (2008) Scale bridging method to characterize mechanical properties of nanoparticle/polymer nanocomposites. Appl Phys Lett 93(4):043111
25. Benveniste Y (1987) A new approach to the application of Mori-Tanaka's theory in composite materials. Mech Mater 6(2):147–157
26. Dunn ML, Ledbetter H (1995) Elastic moduli of composites reinforced by multiphase particles
27. Lee S, Ryu S (2018) Theoretical study of the effective modulus of a composite considering the orientation distribution of the fillers and the interfacial damage. Eur J Mech-A/Solids 72:79–87
28. Ramesh M, Deepa C, Rajeshkumar L, Tamil Selvan M, Balaji D (2021) Influence of fiber surface treatment on the tribological properties of Calotropis gigantea plant fiber reinforced polymer composites. Polym Compos 42(9):4308–4317
29. Lee S, Lee J, Ryu S (2019) Modified Eshelby tensor for an anisotropic matrix with interfacial damage. Math Mech Solids 24(6):1749–1762
30. Lin CH, Muliana A (2019) Micromechanics modeling of hysteretic responses of piezoelectric composites. In: Creep and fatigue in polymer matrix composites. Woodhead Publishing, pp 121–155
31. Ramesh M, Deepa C, Niranjana K, Rajeshkumar L, Bhoopathi R, Balaji D (2021) Influence of Haritaki (Terminalia chebula) nano-powder on thermo-mechanical, water absorption and morphological properties of Tindora (Coccinia grandis) tendrils fiber reinforced epoxy composites. J Nat Fibers 19(13):6452–6468
32. Ramesh M, Rajeshkumar L, Balaji D (2021) Influence of process parameters on the properties of additively manufactured fiber-reinforced polymer composite materials: a review. J Mater Eng Perform 30(7):4792–4807

33. Ramesh M, Rajeshkumar L, Deepa C, Tamil Selvan M, Kushvaha V, Asrofi M (2021) Impact of silane treatment on characterization of ipomoea staphylina plant fiber reinforced epoxy composites. J Nat Fibers 19(13):5888–5899
34. Bhuvaneswari V, Priyadharshini M, Deepa C, Balaji D, Rajeshkumar L, Ramesh M (2021) Deep learning for material synthesis and manufacturing systems: a review. Mater Today: Proc 46:3263–3269
35. Zhao FM, Takeda N (2000). Effect of interfacial adhesion and statistical fiber strength on tensile strength of unidirectional glass fiber/epoxy composites. Part I: experiment results. Compos Part A: Appl Sci Manuf 31(11):1203–1214
36. Lee S, Jung J, Ryu S (2021) Applicability of interface spring and interphase models in micromechanics for predicting effective stiffness of polymer-matrix nanocomposite. Extreme Mech Lett 49:101489
37. Ramesh M, Rajeshkumar L, Balaji D, Bhuvaneswari V, Sivalingam S (2021) Self-healable conductive materials. In: Inamuddin, Ahamed MI, Boddula R, Altalhi TA (eds) Self-healing smart materials. Wiley, United States, pp 297–320. https://doi.org/10.1002/9781119710219.ch11
38. Ramesh M, Rajeshkumar L, Saravanakumar R (2021) Mechanically-induced self-healable materials. In: Inamuddin, Ahamed MI, Boddula R, Altalhi TA (ed) Self-healing smart materials. Wiley, United States, pp 379–404. https://doi.org/10.1002/9781119710219.ch15
39. Ramesh M, Bhuvaneswari V, Balaji D, Rajeshkumar L (2022) Self-healable conductive and polymeric composite materials. In: Inamuddin, Altalhi T, Adnan SM (eds) Aerospace polymeric materials. Wiley-Scrivener LLC, pp 231–258. https://doi.org/10.1002/9781119905264.ch10
40. Hassanzadeh-Aghdam MK, Mahmoodi MJ (2018) Micromechanics-based characterization of elastic properties of shape memory polymer nanocomposites containing SiO2 nanoparticles. J Intell Mater Syst Struct 29(11):2392–2405
41. Deepa C, Balaji D, Bhuvaneswari V, Rajeshkumar L, Ramesh M, Priyadharshini M (2022) Deep learning for the selection of multiple analogs. In: Inamuddin, Altalhi T, Cruz JN, Refat MSED (eds) Drug design using machine learning. Wiley-Scrivener LLC, pp 117–142. https://doi.org/10.1002/9781394167258.ch4
42. Ramesh M, RajeshKumar L, Bhuvaneshwari V (2021) Bamboo fiber reinforced composites. In: Jawaid M, Mavinkere Rangappa S, Siengchin S (eds) Bamboo fiber composites. Composites science and technology. Springer, Singapore. https://doi.org/10.1007/978-981-15-8489-3_1
43. Talamadupula KK, Seidel GD (2021) Statistical analysis of effective electro-mechanical properties and percolation behavior of aligned carbon nanotube/polymer nanocomposites via computational micromechanics. Comput Mater Sci 197:110616
44. Ramesh M, Rajeshkumar L, Bhoopathi R (2021) Carbon substrates: a review on fabrication, properties and applications. Carbon Lett 31(4):557–580
45. Zare Y, Rhee KY (2021) Micromechanics modeling of electrical conductivity for polymer nanocomposites by network portion, interphase depth, tunneling properties and wettability of filler by polymer media. Fibers Polym 22(5):1343–1351
46. Karimi M, Ghajar R, Montazeri A (2018) A novel interface-treated micromechanics approach for accurate and efficient modeling of CNT/polymer composites. Compos Struct 201:528–539
47. Yas MH, ShahraniKorani H, ZareJouneghani F (2020) Studying the mechanical and thermal properties of polymer nanocomposites reinforced with montmorillonite nanoparticles using micromechanics method. J Solid Mech 12(1):90–101
48. Bhuvaneswari V, Rajeshkumar L, Ross KNS (2021) Influence of bioceramic reinforcement on tribological behaviour of aluminium alloy metal matrix composites: experimental study and analysis. J Market Res 15:2802–2819
49. Bhuvaneswari V, Rajeshkumar L, Saravanakumar R, Balaji D (2022) Synthesis and characterization of bioceramics reinforced aluminium matrix composites. Archives of Metall Mater 67(4):1217–1226. https://doi.org/10.24425/amm.2022.141045
50. Yang S, Kwon S, Lee MY, Cho M (2019) Molecular dynamics and micromechanics study of hygroelastic behavior in graphene oxide-epoxy nanocomposites. Compos B Eng 164:425–436

51. Yang S, Shin H, Cho M (2021) Contribution of oxygen functional groups in graphene to the mechanical and interfacial behaviour of nanocomposites: molecular dynamics and micromechanics study. Int J Mech Sci 189:105972
52. Deepa C, Rajeshkumar L, Ramesh M (2022) Preparation, synthesis, properties and characterization of graphene-based 2D nano-materials for biosensors and bioelectronics. J Market Res 19:2657–2694. https://doi.org/10.1016/j.jmrt.2022.06.023
53. Zhou T, Boyd JG, Loufakis D, Lutkenhaus JL, Lagoudas DC (2019) Fabrication, characterization and micromechanics modeling of the electrical conductivity of reduced graphene oxide/aramid nanofiber nanocomposites. Smart Mater Struct 28(9):094001
54. Wang E, Tehrani MS, Zare Y, Rhee KY (2018) A new methodology based on micromechanics model to predict the tensile modulus and network formation in polymer/CNT nanocomposites. Colloids Surf, A 550:20–26
55. Ramesh M, Rajeshkumar L (2018) Wood flour filled thermoset composites. In: Asiri AM, Khan A, Khan I, Bhawani SA (eds) Thermoset composites: preparation, properties and applications, Materials Research Foundations, vol 38, pp 33–65. https://doi.org/10.21741/9781945291876-2
56. Ramesh M, Deepa C, Kumar LR, Sanjay MR, Siengchin S (2020) Life-cycle and environmental impact assessments on processing of plant fibres and its bio-composites: a critical review. J Ind Text 51(4):5518S-5542S
57. Zare Y, Rhee KY(2021) Micromechanics simulation of electrical conductivity for carbon-nanotube-filled polymer system by adjusting Ouali model. The Eur Phys J Plus, 136(8):852
58. Hussein A, Kim B (2019) Micromechanics based FEM study on the mechanical properties and damage of epoxy reinforced with graphene based nanoplatelets. Compos Struct 215:266–277
59. Kim J, Kim J, Song S, Zhang S, Cha J, Kim K, Yoon H, Jung Y, Paik KW, Jeon S (2017) Strength dependence of epoxy composites on the average filler size of non-oxidized graphene flake. Carbon 113:379–386
60. Hussein A, Sarkar S, Lee K, Kim B (2017) Cryogenic fracture behavior of epoxy reinforced by a novel graphene oxide/poly (p-phenylenediamine) hybrid. Compos B Eng 129:133–142
61. Hussein A, Sarkar S, Oh D, Lee K, Kim B (2016) Epoxy/p-phenylenediamine functionalized graphene oxide composites and evaluation of their fracture toughness and tensile properties. J Appl Polym Sci 133(34)
62. Hussein A, Sarkar S, Kim B (2016) Low temperature reduction of graphene oxide using hot-plate for nanocomposites applications. J Mater Sci Technol 32(5):411–418
63. Tang LC, Wan YJ, Yan D, Pei YB, Zhao L, Li YB, Wu LB, Jiang JX, Lai GQ (2013) The effect of graphene dispersion on the mechanical properties of graphene/epoxy composites. Carbon 60:16–27
64. Wan YJ, Tang LC, Gong LX, Yan D, Li YB, Wu LB, Jiang JX, Lai GQ (2014) Grafting of epoxy chains onto graphene oxide for epoxy composites with improved mechanical and thermal properties. Carbon 69:467–480
65. Prolongo SG, Moriche R, Jiménez-Suárez A, Sánchez M, Ureña A (2014) Advantages and disadvantages of the addition of graphene nanoplatelets to epoxy resins. Eur Polymer J 61:206–214
66. Balaji D, Ramesh M, Kannan T, Deepan S, Bhuvaneswari V, Rajeshkumar L (2021) Experimental investigation on mechanical properties of banana/snake grass fiber reinforced hybrid composites. Mater Today: Proc 42:350–355
67. Ramesh M, Deepa C, Selvan MT, Rajeshkumar L, Balaji D, Bhuvaneswari V (2021) Mechanical and water absorption properties of Calotropis gigantea plant fibers reinforced polymer composites. Mater Today: Proc 46:3367–3372

68. Danzi F, Fanteria D, Panettieri E, Palermo M (2017) A numerical micro-mechanical study of the influence of fiber–matrix interphase failure on carbon/epoxy material properties. Compos Struct 159:625–635
69. Hassanzadeh-Aghdam MK, Ansari R, Darvizeh A (2019) Micromechanics-based thermoelastic analysis of polyimide nanocomposites containing 3D randomly oriented carbon nanotubes. Iran J Sci Technol, Trans Mech Eng 43(1):27–39
70. Chiew C, Malakooti MH (2022) Micromechanics modeling of multifunctional EGaIn-polymer composites. In: Nondestructive characterization and monitoring of advanced materials, aerospace, civil infrastructure, and transportation XVI 12047:157–163

Nanocomposites: Homogenization and Kinematic Relations

Desalegn Atalie, Rotich Gideon, Kilole Tesfaye, and Peng-Cheng Ma

Abstract Nanocomposites have attracted great attention in industrial and academic fields for a variety of applications, and their manufacturing methods are progressing rapidly. The ability to design nanocomposites with the tailored mechanical, biological, magnetic, electrical, optical, thermal, transport, and other properties was superior to some standard filler-reinforced materials, thus allowing the rapid development of this novel material. As a result, unique fabrication and dispersion methods are continually being generated to produce nanocomposites with the suitable properties for the intended end use. In addition to the characteristics of the individual parts of nanocomposite, interface is crucial in increasing or restricting the overall performance of the nanocomposite. Nanocomposites have a substantial number of interactions between the constituent intermixed phases because of the large surface area of the nanostructures. Therefore, the homogenization is one of the most crucial techniques in the nanocomposite manufacturing and this affects the performance of the finished product by introducing unintended fractures and faults in the material. The main drawback of using nanocomposite materials is the toughness and performance impact caused by the addition of nanoparticles to the bulk matrix. This is mostly caused by an inadequate understanding on the formulation, structural relationships, and the necessity for exfoliation and dispersion of nano-particles. This chapter discusses in detail about the types of nanocomposites, their common matrices, reinforcements, homogenization techniques, governing equations, characterization procedures, specific applications, and product performances. The chapter

D. Atalie · P.-C. Ma (✉)
Laboratory of Environmental Science and Technology, Key Laboratory of Functional Materials and Devices for Special Environments, The Xinjiang Technical Institute of Physics and Chemistry, Chinese Academy of Sciences, Urumqi 830011, China
e-mail: mapc@ms.xjb.ac.cn

Center of Material Science and Optoelectronics Engineering, University of Chinese Academy of Sciences, Beijing 100049, China

D. Atalie · K. Tesfaye
Ethiopian Institute of Textile and Fashion Technology, Bahir Dar University, Bahir Dar 6000, Ethiopia

R. Gideon
Clothing and Textiles, South Eastern Kenya University, Kitui, Kenya

© The Author(s), under exclusive license to Springer Nature Singapore Pte Ltd. 2023
H. Abdellaoui et al. (eds.), *Mechanics of Nanomaterials and Polymer Nanocomposites*, Smart Nanomaterials Technology,
https://doi.org/10.1007/978-981-99-2352-6_12

also discusses the relationship between the homogeneity of nanocomposite materials and kinematic features as well as the future outlook of nanocomposite materials.

Keywords Nanocomposites · Nano fillers · Homogenization · Characterization · Kinematic

1 Introduction

Nanocomposites are heterogeneous (multiphase) materials composed of at least one nanoscale phase (known as nanofiller) dispersed in a second phase (known as matrix) to obtain a combination of the individual properties of its constituents. Many distinct nanoparticles that are entrapped inside nanocomposites can be presented in a bulk material, which may consist of a combination of soft and hard nanomaterials, two soft nanomaterials, or two hard nanomaterials. Each component of the composite contributes to the bulk properties of the end material [1]. In other terms, composites refer to mixtures of materials that are dispersed across a continuous matrix and whose constituent parts and interfaces between different parts can be physically differentiated [2]. Materials science and engineering have evolved significantly as a result of the discovery of nanocomposites with enhanced physical, mechanical, and chemical properties [3]. Nanocomposite manufacturing systems and products are growing rapidly due to their remarkable properties for a wide range of end-use. Researchers are also giving attention to nanoscale investigation rather than macro-scale composites. Nanocomposites can provide opportunities to produce new and suitable products for solving problems ranging from the pharmaceutical, filtration industry, biomedical, food packaging, electronic, to the automobiles and energy industries [4].

In recent years, there has been an increase in multi-functional nanocomposites, which have strong mechanical properties, moisture resistance, high electrical conductivity, thermal insulation, redox reactivity, and bridgeable and suitable property for the required application. These nanocomposites also have high surface-to-volume ratios for loading biomolecules like cellulases. It is well documented that the characteristics of a composite may be greatly influenced by a variety of parameters, including surface characteristics, nanoparticle size distribution, geometric shape, and dispersion state. Nanocomposites may be divided into three classes based on the type of matrix, i.e., polymer, ceramic, and metal matrix. Polymer nanocomposites are gaining popularity as a result of nanoparticles' present commercial availability. These composites show greatly enhanced mechanical qualities including strength, hydrocarbon permeability, dimensional and thermal stability, optical features, flame retardant, chemical resistance, as well as electrostatic, magnetic, and electrical capabilities. The term "ceramic nanocomposites" refers to materials that contain various nanophase, including ceramic, a carbonaceous nanophase, or a metal nanophase in a ceramic matrix. Ceramic nanocomposites are being employed more often in the creation of novel processing methods that make it possible to produce goods for use in both commercial and scientific settings. With the use of a high-energy ball milling technique, nanoparticles may be created. Metallic nanocomposites are made up of reinforcement and materials having at least one nanoscale dimension. In the automotive industry, metallic nanocomposites are typically constructed from low-density metals like magnesium (Mg), aluminum (Al), and titanium (Ti). In polymeric, ceramic and metallic nanocomposites, the formulation of nanocomposites their structural interactions, and the requirement of exfoliation and particle dispersion all have a significant impact on the efficiency of the production process as well as the performance of final products.

2 Historical Background

The importance of composite materials for the structural integrity of living and nonliving systems cannot be emphasized. The bones that sustain a bird's anatomy and the lofty skyscrapers that dot our cities are both examples of nanocomposites that may be found in many forms. Therefore, breakthroughs in composite material science and technology are vital for the growth of human civilisation. Before we define nanocomposite, there is a need to be brief on the most common terms, which are nanoscience and nanotechnology. Nanoscience is a deviation of materials science, physics, and biology, which deal with the operation of materials at atomic and molecular levels. On the other hand, nanotechnology can be defined as the ability to manipulate, measure, control, assemble and manufacture matter at less than 100 nm [5]. Therefore, nanoscience is the study of nanomaterials and their characteristics, while nanotechnology is using those materials to create unique materials with new or modified properties. Nanocomposites, nanoparticles, nanofibers and nanoclays are included in nanomaterials. Nanocomposite is a multiphase solid material where one of the phases has one, two or three dimensions of nanoscale structures with reappearance distances between the different phases that make up the material [4].

Many technologies are derived from nature and the history of nanomaterial is also the history of nature because many nanostructured materials are generated from volcanic eruptions, seashells, early meteorites and skeletons [6]. Gold, Silver and Iron oxides are some of the examples of naturally nanostructured elements [7]. United States of America physicist and Nobel Prize awardee Richard Feynman introduce the concept of nanotechnology in 1959 during his presentation on "There's Plenty of Room at the Bottom" at the California Institute of Technology during the American Physical Society's annual meeting [5]. Since Feynman first identified this new field of study, it has attracted the attention of many scientists. There are several alternatives available today for manufacturing, processing, and creating of nanostructures.

3 Nanocomposite Classification

Based on their matrix nanocomposite materials are classified into three categories:

i. Polymer matrix nanocomposites
ii. Ceramic matrix nanocomposites
iii. Metal matrix nanocomposites

a. **Polymeric nanocomposites**

The research of polymer nanocomposites (PNCs) can be dated back to Toyota's investigate in the early 1990s [8]. Advantageously, that nanofillers often offer greater property improvement on polymers for a given concentration than more conventional micrometer-sized fillers. The existence of polymer between layers, the strength of bonding at the filler layer/polymer interaction, and the dispersion of filler layers within the polymer matrix are crucial considerations in PNCs. The goal of creating stronger nanocomposites may be realized if adequate answers to these problems could be found. The rapidly increasing surface area of the filler could not be handled since there wouldn't be enough polymer molecules remaining. The production of nanocomposites with a high filler loading may be possible by in-situ polymerization and solid intercalation. Currently, it is well-accepted that an exfoliated structure with very wide filler-to-filler interlayer spacing improves the mechanical characteristics of nanocomposites. This is accurate when there is little filler loading [9]. PNCs have enormouly promise, particularly when considered in light of the explosion of readily accessible functional nanoparticles, which make it possible to create polymers with hitherto unrealized qualities [10, 11]. Table 1 shows the most common matrix and reinforcement of polymer, ceramic and metal nanocomposites.

b. **Ceramic nanocomposites**

The most prevalent ceramic nanocomposite architectures are made up of a micron-sized matrix that contains nanoparticles. The nano dispersions can fill both inter and intra-granular locations as well as be embedded inside the matrix grains, positioned at grain borders, or any combination of them [12]. Ceramic nanocomposites are used in several fields of applications due to their outstanding performance including corrosion resistance, high abrasion resistance, and temperature oxidation than that of metals in high-temperature conditions. The coatings made of nanocomposites have a metastable nanostructure. Ceramic matrix composites are intended to increase the toughness of ordinary ceramics, whose big limitation is brittleness.

c. **Metallic nanocomposites**

Metal nanocomposites (MNCs) are materials that combine at least two physically and chemically distinct phases, one of them being metal, in a way that gives them qualities that none of the individual phases alone can. In a metallic matrix, typically there are two phases present: a fibrous phase and a particle phase. A few examples include the cutting tools and oil drilling inserts made of cobalt (Co) particulate and tungsten

Table 1 Raw materials of nanocomposites

Composite type	Matrix/reinforcement	References
Polymer matrix nanocomposites	Polypropylene/montmorillonite	Manias et al. [18], Ma et al. [19], Morlat-Therias et al. [20], Ben Ammar and Fakhfakh [21]
	Polypropylene/montmorillonite/CaCO3	Zare et al. [22]
	Polyester/TiO_2	Koltsakidou et al. [23], Camargo et al. [24]
	Polymer/layered double hydroxides	Mallakpour et al. [25]
	Polymer/CNT	Iqbal et al. [26], Lankone et al. [27], Zare and Rhee [28]
	Mechanoresponsive fluorescence TPE-4N, polydimethylsiloxane/basalt fiber	Meng et al. [29]
	N-Isopropylacrylamide and N, N-Methylenebisacrylamide/cotton	Chen et al. [30]
	Epoxy/CNT	Ma et al. [31]
	Polydimethylsiloxane/tetraphenylethene (TPE) derivatives with 3-butenloxy	Yang et al. [32]
	Polyethylene/jute fiber	Hossen et al. [33, 34]
	HDPE/aramide fiber	Sahu et al. [35, 36], Dias et al. [37], Kord and Roohani [38]
	Polystyrene/Al_2O_3	Bahgat Radwan et al. [39], Bhavsar et al. [40]
	Polystyrene/cellulose fiber	Neves et al. [41], Mohamed et al. [42]
	Polycarbonate/ZrO_2 and PC/palladium	Rostamiyan and Ferasat [43], Nouh et al. [44]
	PVC/NiO, PVC/TiO2 and PVC/GO	Moustafa et al. [45], Abdel-Gawad et al. [46], Taha et al. [47]
	PU/Cellulose and PU/aramide fibers	Ke et al. [48], Vaithylingam et al. [49], Urbina et al. [50], Yazdi et al. [51], Shokraei et al. [52]
Ceramic matrix nanocomposites	B4C/TiB_2	Khajehzadeh et al. [53], Mikeladze et al. [54], Zhang et al. [55], Guo et al. [56], Saeedi Heydari and Baharvandi [57], Chkhartishvili et al. [58]
	B4C/TiB_2 and SiC	Zhang et al. [55], Song and Zhang [59], He et al. [60]

(continued)

Table 1 (continued)

Composite type	Matrix/reinforcement	References
	B4C/SiC	Roghani et al. [61], Zhang et al. [62], Matović et al. [63]
	PVA-MgO/SiC	Ahmed et al. [64, 65]
	MgO/SiC–C	Qi et al. [66], Wei et al. [67]]
	$MoSi_2/ZrO_2$	Zhu et al. [68]
	$Al_2O_3/NdAlO_3$	Mathur et al. [69], Jin and Gao [70]
	Al_2O_3/Mo	Riyas et al. [71], Ni et al. [72], Song et al. [73]
	Mullite/SiC	Chen et al. [74], Zhou et al. [75], Xie et al. [76]
	Si_3N_4/SiC	Wang et al. [77], Zhang et al. [78], Yan et al. [79], Yang et al. [80]
	Al_2O_3/SiC	Dehgahi et al. [81], Momohjimoh et al. [82], Klement et al. [83], Gevorkyan et al. [84], Suresh et al. [85], Jiang et al. [86]
	Al_2O_3/ZrO_2	Yu et al. [87, 88], Duntu et al. [89], Koltsov [90], Park and Choi [91], Santos et al. [92], Koltsov et al. [93], Yaghoubi et al. [94]
	Al_2O_3/W	Ni et al. [95], Shi et al. [96], Yuan et al. [97]
Metal matrix nanocomposites	CNT/Sb	Rikhtegar et al. [98]
	α-Fe/$Fe_{23}C_6$/Fe_3B	Din et al. [99]
	CNT/SnSb	Chen et al. [100]
	Cu/Nb	Xiang et al. [101], Thyagatur and Mushongera [102]
	CNT/Fe_3O_4	Hosseini et al. [103], Kundu et al. [104]
	Ni-Al_2O_3	Raghavendra et al. [105], Ma et al. [106], Dehestani et al. [107]
	Al/SiC	Khdair and Fathy [108], Yaghobizadehet al. [109], Mei et al. [110]
	Al/AlN	Zhe et al. [111], Riquelme et al. [112], Gostariani et al. [113]

(continued)

Table 1 (continued)

Composite type	Matrix/reinforcement	References
	Cu/Al_2O_3	Sadoun and Fathy [114], Sadoun et al. [115], Najjaret al. [116]
	Ag/Au	Sunwoo et al. [117], Fazleeva et al. [118], Park et al. [119], Zhang et al. [120], Mostafa et al. [121], Choi et al. [122], Emam et al. [123], Wang et al. [124], Newase and Bankar [125]
	Al/CNT	Singh et al. [126], Jargalsaikhan et al. [127], Sharma et al. [128]
	CNT/Si–SnSb	Nithyadharseni et al. [129]

carbide (WC), the composites made by reinforcing A1 matrix with SiC particles used in automotive, thermal management, and aerospace applications, and the composites made from Al matrix reinforced with a continuous Al_2O_3 fiber used in power transmission lines. These MNCs are produced from nanoscale reinforcing materials and a ductile alloy or metal matrix. These materials have high modulus and strength as well as ceramic and metal qualities including toughness and ductility. As a result, the development of materials with high service temperatures and shear/compression strengths is well-suited for these nanocomposites. They have a great deal of potential for use in a wide range of fields, such as the production of structural materials and the automotive and aerospace sectors [10, 13, 14, 15].

4 Raw Materials

Al_2O_3, SiN, SiC, B4C, MoSi2 and MgO are the most common raw materials and they are used as a matrix in CMNC. In all types of nanocomposite, reinforcements can be iron and metal powders on Nanoscale. Silica, clays, and TiO_2 are preferable for crystalline reinforcement and abundantly available in nanoparticle size with interaction properties [16]. Magnesium, Aluminum, Titanium, Nickel, and Cobalt are commonly used as the matrix in MNCs, whereas Sb, SnSb, SiC, Au, and AlN can be used as reinforcing elements. When selecting the material, one must consider that the reinforcing phase must be firmly bonded to the matrix material. Polymeric matrices are classified into two: thermoplastic and thermoset. In polymeric nanocomposite manufacturing, a thermoplastic matrix such as polypropylene, polyethylene, HDPE and Polystyrene are commonly used. Epoxy, Polyester, Polyurethane, Phenolic and Vinylester thermosets matrix also extensively used in PNCs Polymer matrix nanocomposites [17].

5 Nanofillers

An essential factor that significantly affects the interfacial characteristics of nanocomposites is nanoparticle size and type. The major benefit of nanoparticles in nanocomposites is diminished by the huge interfacial area of tiny nanoparticles. According to [14, 15], nanoparticles may be divided into four categories: organic-based nanomaterials, inorganic-based nanomaterials, carbon-based nanomaterials, and composite-based nanomaterials.

a. Organic-based nanomaterials: The carbon-free organic materials, such as polymers, cyclodextrin, micelles, dendrimers, protein-based, exosomes, and liposomes, are used to create the organic-based nanomaterials.
b. Inorganic-based nanomaterials: inorganic nanoparticles (INPs) play a significant role in our daily lives as medications, antiseptics, and imaging tools. Metal-based inorganic nanomaterials are Au, Cd, Ag, Zn, Cu, Al, Fe and Pb nanomaterials, whereas examples of metal oxide-based inorganic nanomaterials are CuO, ZnO, TiO_2, Fe_2O_3, $MgAl_2O_4$, CeO_2, Fe_3O_4 and SiO_2 etc. Metal, inorganic dendrimers, silicon, bioinorganic hybrids, and organic–inorganic hybrids are some of the most promising INPs currently being used. An inorganic nanomaterial can be made up of either a metal or a non-metal element, or it can be a compound made of an oxide, chalcogenide, hydroxide or phosphate. These materials have a wide range of applications in the area of photonics, electronics, biological sensors, medicinal equipment and chemicals.
c. Carbon-based nanomaterials: Carbon-based nanomaterials include graphene, nanodiamonds, fullerene, carbon nanotube, graphene oxide, an activated carbon, and carbon black. Among these, graphene and carbon nanotubes are likely the ones utilized mostly in chemical analysis. The extraordinary technical instruments known as carbon-based nanomaterials have exceptional qualities such as chemical adaptability, high mechanical strength and thermal conductivity, appealing optical features, etc.
d. Composite-based nanomaterials: Nanomaterials having composite structures made up of two or more components with unique chemical and physical characteristics are known as composite nanoparticles. These nanomaterials feature complex structures, such as a metal–organic framework, and can be made by the combination of metal, metal oxide, organic and carbon nanomaterials [130].

If nanoscale fillers of various shapes and sizes were combined with a polymeric matrix, very large interfacial areas would be resulted [131]. Nanofillers can be categorized into three types based on their nano dimension: 1D (nanoplatelet), 2D (nanofiber), and 3D (nanoparticulate), as schematically shown in Fig. 1 [132].

a. One-dimensional nanofillers

One-dimensional (1D) nanofillers are fillers with a dimension of less than 100 nm. They often take the shape of sheets that range in size from hundreds to thousands of nanometers long and range in thickness from one to a few nanometers [133].

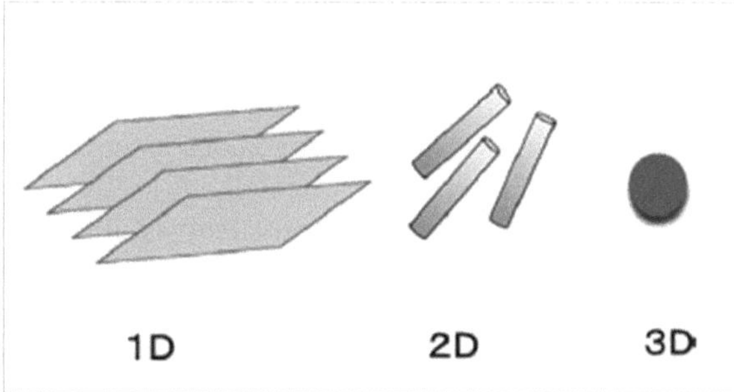

Fig. 1 Type of nanofillers: based on Nano-dimension. Reproduced from reference [132]

Materials like nano graphene platelets and montmorillonite clay are examples of one-dimensional (1D) nanofillers. They may also be found as branching structures, nanowalls, nanoplates, nanodisks, nanosheets, and nano prisms. There are several significant examples, including ZnO nanoplatelets, graphite nanoplatelets (GNP), ZnO nanodiscs, amphiphilic graphene platelets, Fe_3O_4nanodiscs, carbon nanowall, and ZnO nanosheets. 1D nanofillers are widely used in biosensors, microelectronics, sensors, coatings and biomedical applications [134]. Due to their dimensionality, 1D nanofillers are very useful in electrical and thermal applications because they exhibit remarkable magnetic, electrical, and optical capabilities.

b. Two-dimensional nanofillers

Nanofillers that have two dimensions with less than 100 nm are called two-dimensional (2D) nanofillers. The attainment of a homogenous dispersion of 2D nanofillers in the polymer matrix plays a crucial role in the creation of high-performance and multipurpose nanocomposites because of their exceptional physical characteristics. A novel approach to achieve homogenous dispersion of 2D nanofillers in the polymer matrix is the synthesis of 2D nanosheets into hybrid nanofillers [135, 136]. 2D nanofillers are mostly in the shape of tubes, fibers, or filaments. Cellulose whiskers, carbon nanotubes (CNTs), boron carbon nitride tubes, boron nitride (BN) tubes, gold or silver nanotubes, black phosphorus, 2D graphene, and clay nanotubes are the most common examples of 2D nanofillers. In the fabrication of polymer nanocomposites, several 2D fillers are available in the form of natural sepiolite clay fibers [137, 138], gold nanowires [139, 140], nanotubes [141, 142], cerium dioxide [143], carbon fibers [144, 145], cellulose fibers [146, 147], titanium dioxide [148], zinc oxide [149, 150], copper oxide [151, 152] and silica [153, 154]. 2D nanofillers are very useful in electronics [155], sensors [156], catalysis [157], energy [158], photocatalysts [159], and nanocontainers [160].

c. Three-dimensional nanofillers

Three-dimensional (3D) nanofillers have five domains at the nanoscale level and are substantially tetragonal particles. They are typically spherical and cubical in form and are known as nanoparticles or zero dimensional nanoparticles. Nanogranules, nanospheres, and nanocrystals are some other terms for 3D nanofillers that are often used. Quantum dots, carbon black, silicon carbide, nanosilica, nanoalumina, nanotitanium oxide, Polyhedral Oligomeric Silsesquioxane, nanomagnesium hydroxide, and semiconductor nanoclusters are a few examples of 3D nanofillers. Due to their unique features, 3D nanofillers are crucial in the manufacturing of polymer nanocomposites. Some of these nanoparticles, such as TiO_2, Ag, SiO_2, Fe_3O_4, and ZnO, have great opacity to visible region, excellent stability, high refractive index, wettability, UV resistance, nontoxicity, strong catalytic properties, and low cost. They provide features that enable them to be employed in applications including separation and purification, coatings, and biomedical materials when paired with an appropriate polymer matrix [161, 162].

6 Manufacturing Methods

In this chapter, the many methods used to create polymer nanocomposites are compiled and explained. All methods strive to create nanocomposite materials that are evenly dispersed and free of aggregations. The most popular methods for creating polymer nanocomposite include mixing, in-situ polymerization, melting, selective laser sintering, and electrospinning. The problem of nanoparticle aggregation in melt-mixing approach has been known to be eliminated by the use of water, plasma-assisted mechanochemistry and atomic layer deposition. High-frequency sonication also plays a part in the mixing processes. The manufacturing of thermodynamically stable nanocomposites is made possible by in-situ polymerization. An efficient technique that works well for creating porous materials is electrospinning. Additionally, the creation of nanocomposites using selective laser sintering offers apparent advantages to solving the aggregation issue [163].

6.1 *In-situ Polymerization*

In recent years, the use of in-situ polymerization techniques have significantly increased to produce nanocomposite materials with enhanced nanoparticle dispersion and distribution. A neat monomer, many neat monomers, or a solution of monomer are often mixed with nanomaterials, and then polymerization occurs in the presence of the dispersed nanomaterials [164]. Additionally, in-situ, polymerization may be utilized to create nanocomposites since polymers are insoluble in all solvents and are neither ideal for melt processing due to low temperature stability nor for solution casting due to their suitability for melt processing. In a typical in-situ polymerization procedure, nanoparticles are distributed in a monomer or monomer solution, and when the monomer is polymerized using conventional polymerization methods, nanocomposite materials are produced. Because deposition mechanisms in monomers may go along extremely quickly, high dispersion of the nanoparticles in the monomer is crucial for well-dispersed molecules in the subsequently produced polymer.

6.2 *Melt Intercalation*

The common and established method for creating thermoplastic polymer nanocomposites is melt intercalation. To achieve uniform dispersion, entails annealing the polymer matrix at high temperatures, adding the filler, and then kneading the composite, as shown in Fig. 2. If the stacked surfaces are sufficiently suitable, the selected organic polymer will penetrate the interlayer space and produce an intercalated nanocomposite.

6.3 *Exfoliation-Adsorption*

The layered complex is exfoliated using a solvent that the polymer is soluble in. Some layered compounds interact with weak forces between the layers, which makes exfoliating them with the appropriate solvents straightforward. Once the solvent has gone, the exfoliated layers can be piled once again once the polymer has adhered to them. The intercalation of the polymer results in the formation of a well-ordered multilayer structure.

6.4 *Melt-Spinning*

As shown in Fig. 3, the basic operation of melt-spinning involves extruding molten polymer into a large number of continuous filaments, which are subsequently quenched by a crossflowing air stream. The orifices of a spinneret, which determine the output material's fineness, are utilized in this process. In order to produce metallic glasses, materials with nanocrystals, or materials with exceedingly small grains, the melt-spinning method is widely utilized. The rate of cooling, which may be represented by continuous cooling transformation curves, affects how the products of melt-final spinning are structured [166].

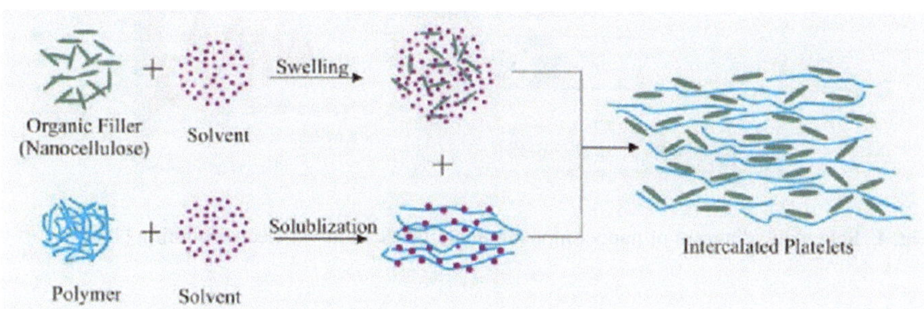

Fig. 2 Melt intercalation process for preparation of nanocomposite [165]

Fig. 3 Schematic of the melt-spinning process [167]

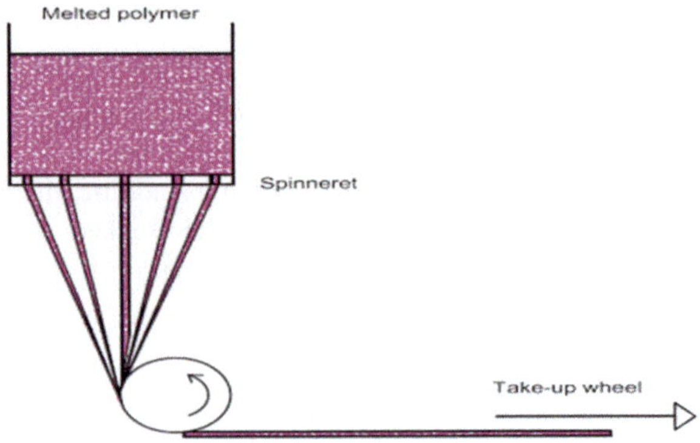

6.5 Electrospinning

In electrospinning, a polymer solution or melt is transformed into fibers by the application of an electric force. The diameters of electrospun fibers range from nanometers to microns, and they have large surface areas. In an electrospinning device, a spinneret, collecting plate, and high voltage source are frequently used to discharge a polymer solution or melt jet toward an opposing polarity collector. For this procedure, a few kV of direct current (DC) is needed to generate fibers (See Fig. 4). This approach relies on stronger electrical repulsive forces than the jet surface tension does. The conical form of the polymer solution droplet that emerges from a syringe is due to surface tension that is stronger than electrostatic attraction. The fiber may release through the needle tip as a result of this. The evaporation of the solvent causes the nanofiber to solidify. Both horizontal and vertical electrospinning methods have been used. The electrospun nanofibers feature a large specific surface area, a smooth surface, a high aspect ratio, and negligible porosity [168].

Fig. 4 Schematic diagram of nanocomposite fibers fabrication via electrospinning [169]

6.6 Sol–Gel

A three-dimensional network structure is produced using the wet chemical process known as "sol–gel" by first producing an inorganic colloidal suspension (sol) and then gelating the sol in a continuous gel. For instance, metal oxide nanoparticles may be made using the sol–gel process, which is a more chemical approach [170]. In the sol–gel process, the precursor of the filler is combined with the polymer in a suitable solvent in place of the filler or nanomaterial, followed by hydrolysis and the precipitation of the nanomaterial. Either pH changes or hydrothermal processes are used to produce precipitation. This technique can only be used to create nanocomposites using nanoparticles that already exist as chlorides. Additionally, compared to the blending approach, the cost of the sol–gel technique comprises many phases, which raises the expense of the procedure. The sol–gel procedure offers greater linkages between the nanomaterial and the matrix than the blending method does. Figure 5 shows a schematic of the sol–gel method.

7 Nanocomposite Homogenization Techniques

Nanocomposite materials are made up of at least two components with different properties, i.e., matrix and fillers. The differences in the functional groups between these components cause weak interaction and inadequate dispersion of fillers in a polymeric matrix. Mostly, the surface treatment of fillers by functionalization increases their dispersion in the matrix. The incorporation of higher filler concentration induces agglomerates formation and endangers the homogeneity further and diminishes the properties of the produced nanocomposites [171]. Efficient load transfer, thermal, and electrical conductivity of nanocomposite depend upon the extent of homogeneity of the mixture. Nanocomposite materials with insufficient homogeneity could not have potential applications and have a negative impact if implemented for certain functions [172].

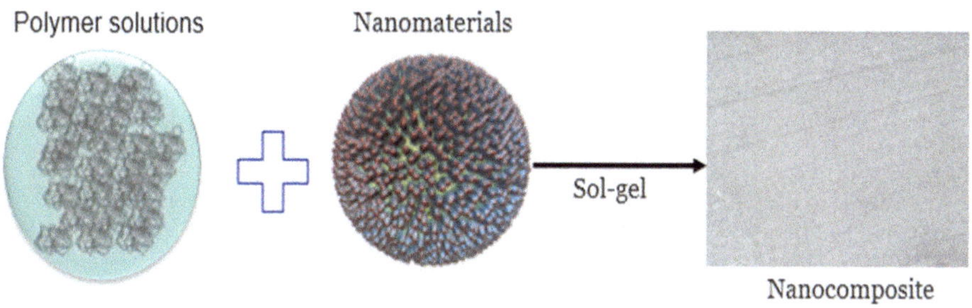

Fig. 5 Sol–gel nanocomposite preparation

Based on the nature of the phase components, several nanocomposite preparation techniques have been developed to encourage better filler distribution and obtain improved homogenization. The in-situ polymerization, melt intercalation, solution mixing, melt-mixing, electrospinning, and sol–gel are among the techniques used to disperse nanofillers in nanocomposites. In in-situ polymerization, nanofillers are initially dispersed in a suitable monomer solution and then the nanocomposites are polymerized. The in-situ polymerization process can be initiated through the diffusion of an initiator or catalyst in the solution or by the application of heat [173]. In melt intercalation, compatible nanofillers are introduced into the molten polymer in the absence of solvent which makes it a safe and environment-friendly process [174]. This method is widely used in large-scale production such as extrusion and injection molding techniques. Solution mixing is a process where nanofillers are introduced into a suitable solvent and gently blended with the aid of high-shear homogenization or sonication or magnetic/mechanical stirring to form homogeneous nanocomposites. Finally, the dissolution of the polymer within the solution is removed through evaporation or coagulation to obtain pure nanocomposites [175]. Melt-mixing is a method where nanofillers are directly mixed with the thermoplastic matrix in the required weight percentage and subjected to a heated twin-screw extruder with a high-shear mixer [176]. The nanofillers are uniformly dispersed within the viscous liquid of thermoplastic matrix due to the high-shear mixing process. This processing of nanocomposites is widely used in industrial mass production due to its economic viability, environment-friendly, and the possibility to be integrated into the molding machine to obtain the desired geometrical shape. Figure 6 shows the dispersion of spherical fillers in a matrix.

However, there is a high probability of agglomerates formation in all methods due to the strong force of attraction between the nanofillers typically when it is incorporated in large quantities. Hence, the surface characteristics of the nanofillers and matrix must be emphasized and proper surface treatment of the nanofillers needs to be carried out to ensure uniform dispersion of nanofillers within the polymeric matrix [175]. As most of polymeric matrices are hydrophobic while inorganic

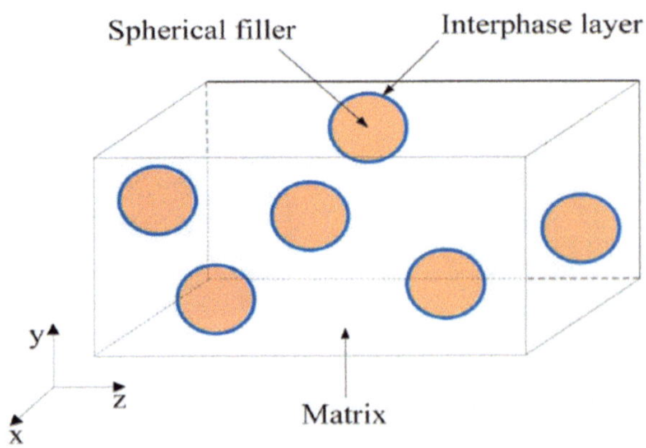

Fig. 6 Spherical fillers dispersed in the matrix [131]

nanofillers are hydrophilic, surface modifications are aimed to increase hydrophobicity of the nanofillers to make them compatible with the polymeric matrices. Nanofillers' surfaces can be modified through physical methods or chemical reactions between the nanofillers and surface modifiers.

i. ***Physical surface modifications***

Low molecular weight surfactants are used to coat the nanofillers to form electrostatic, hydrogen bonding, and van der Waals forces that allow its adherence to the polymeric matrix. The ionic bonds formed between nanofillers and surfactant leads to the reduction of nanofillers surface energy which results in a low tendency of attraction between nanofillers and reduced chances of agglomerates formation [177]. Encapsulation is another physical surface modification where the nanofillers are encapsulated in a polymerized polymer or polymeric component grown over the nanofillers and forms uniform coating through the in-situ method. This functional group enables the formation of appropriate bonding between the nanofillers and the polymeric matrix. The physical surface modification has a drawback as polymeric components coated over the nanofillers can be desorbed due to the low bond strengths [175].

ii. ***Chemical surface modifications***

This method is more viable than that of the physical method as it enables the formation of strong bonds such as the covalent bonds between nanofillers and matrix through chemical reactions. Coupling agent treatment is among the easy and economical methods used for the surface modification of nanofillers chemically. Coupling agents such as silane, zirconate, and titanate are employed to improve the wettability and dispersion of the nanofillers within matrices. However, coupling agent treatment methods have shown limitations in terms of achieving uniformly dispersed nanofillers in nanocomposites [178]. Another method called the grafting of macromolecules on the nanofillers' surface is used to overcome such limitations. The intrinsic weak interface in nanocomposites can be efficiently tailored through covalent grafting of the nanofillers and polymeric matrix [179]. There are two types of grafting: "grafting to" or "grafting-from". A thin polymer brush layer which determines the surface properties is formed on the nanofillers in both techniques. However, highly reactive end groups on pre-formed polymer chemically react with functional groups of nanofillers in "grafting to" whereas functional monomers adhere on the nanofillers followed by polymerization from the surface in the "grafting-from" technique. The latter technique is a more easy and reliable method. Ultrasonication is another method used to disrupt the physical and/or chemical interactions between the nanofillers by break-up aggregates formation and contributes to the homogeneous dispersion of nanofillers in nanocomposites. Some of the nanocomposites produced from various types of nanofillers dispersed within different polymeric matrices, dispersion techniques used, and resulting properties are summarized in Table 2.

Table 2 Summary of the homogenization of various types of nanocomposites

Matrix/reinforcement	Dispersion techniques	Properties	References
Epoxy resin/amine-functionalized multi-graphene layer (AMLG) and amine-functionalized multi-walled carbon nanotube (AMWCNT)	Mechanical stirring	Uniform dispersion and homogeneity of graphene/MWCNT, minimum poly dispersity index, and maximum tensile properties at 0.5 wt% filler content	Shukla and Sharma [180]
Polydimethylsiloxane/graphene nanoplatelets	Microfluidization	Highest filler dispersion homogeneity, enhanced thermal stability	Łapińska [171]
Polylactic acid/zirconium phosphate	Melt-mixing	~25 times impact strength than pure PLA, and excellent toughness	Zhu et al. [181]
Ultra-high molecular weight polyethylene (UHMWPE)/cellulose Nanofiber	Melt-mixing	Improved yield strength, elongation at break, Young's modulus, toughness and crystallinity	Sharip et al. [182]
Polypropylene/graphene oxide nanoplatelets	Melt-mixing	Enhanced crystallization kinetics (α-to mesophase), 44–93% nucleation efficiency	Carmeli et al. [183]
Polycaprolactone/graphene oxide	Melt-mixing	Increase ~10% Young's modulus and decrease in ductility	Minář et al. [184]
	In-situ polymerization	higher crystallinity, nanofiller dispersion and exfoliation, ~45% increase in Young's modulus, and decrease in ductility	
Polystyrene/organoclay	In-situ polymerization	Improved thermal stability, impact resilience, and stiffness	Mrah and Meghabar [185]

(continued)

Table 2 (continued)

Matrix/reinforcement	Dispersion techniques	Properties	References
Nylon 66/reduced graphene oxide	In-situ polymerization	Enhanced interfacial energy and improved ultimate tensile strength	Duan et al. [186]
Epoxy/graphene nanosheets	Chemical functionalization with 4-nitrobenzenediazonium salt	Improved tensile strength, elongation at break, and thermal conductivity	Yao et al. [174]
Epoxy/zinc oxide (ZnO) nanoparticles and nanotubes	In-situ curing	Improved tensile strength, toughness by 105%, and flexibility	Salahuddin and El-Kemary [187]

8 Homogenization Measurement Methods

The degree of dispersion of nanofillers in the matrix can be investigated through the analysis of microstructural images obtained from high resolution microscopy such as scanning electron microscope (SEM). This method is mainly used to perform the qualitative analysis of homogeneity even though it can be explored to calculate the quantitative values of homogeneity [188]. The SEM and dynamic light scattering (DLS) are mostly implemented to analyze particle size and its distribution in the nanocomposites through advanced testing methods [189, 190]. Degree of dispersion and nanocomposite materials properties are highly dependent on the particle size. As the particle size reduces, the theoretical surface coverage of the nanofillers increases [191]. However, it is difficult to use this value in practical usage as the reduction of the particle size beyond a certain limit cannot be quantified due to the limited resolution of SEM imaging. Although there is a wide acknowledgement of nanoparticles' significance, their characterization is still mostly qualitative and dependent on the subjective perception of typical TEM images. Numerous quantitative techniques have been put suggested, however, none have received widespread adoption due to their lack of generality and convenience. Although the uniform dispersion seems implausible, it is popular in research and is valuable for describing reasons. The ideal dispersion is frequently described using the homogeneous dispersion. Every filler particle in this model is equally spaced from its four nearest neighbours, as seen in Fig. 7a. This dispersion reduces the size of the unreinforced polymer domains and efficiently compartmentalizes damage; the reinforcing was likely necessary due to the polymer matrix's subpar characteristics.

The volume of the reinforced polymer is reduced by variations in homogeneity, while the size of the non–reinforced polymer domains is increased. In addition to the uniform distribution as a dispersion characterize tool, inter-particle distance as a dispersion characterization is also applicable as shown in Fig. 4. The spacing between particles serves as a spectrum indicator of the dispersion state and indicates the size

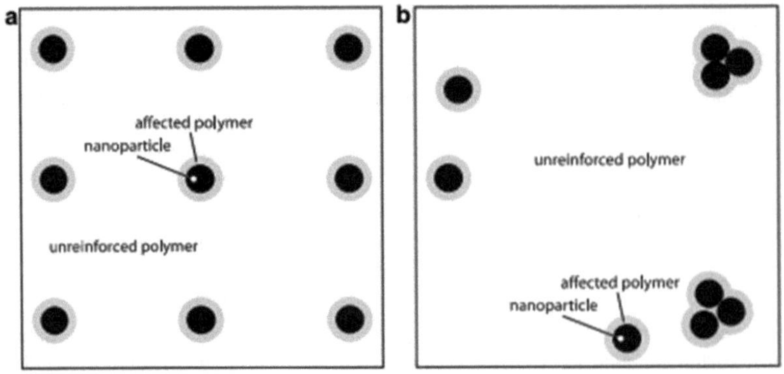

Fig. 7 Illustrations of two possible dispersion states for nanoparticles within a polymer matrix: **a** a uniform dispersion, **b** a more realistic dispersion with two small clusters of particles [192]

of the non–reinforced polymer. A homogeneous dispersion of 30 nm particles at 6% loading is shown in Fig. 8. The distance between the centers of the particles is 105 nm, and the field of vision is 2 mm. There are 75 nm between particle tips. The free-space length obtained via statistical simulation is 73 nm. Larger squares will fit between particles even when the probability is less than the bare minimum required to generate a zero mode. It is more probable that randomly positioned boxes will fit inside when the box size is decreased.

(i) ***Measurement of homogeneity by DLS***

The DLS test method is used to characterize the affinities of filler particles with matrices and the homogeneity in nanocomposites. To reduce the rate of nanofillers re-agglomeration, it is required to form a well-stabilized and smallest average particle size at low solids concentration and low viscosity during the DLS test. The heterogeneity of particle sizes in a nanocomposite mixture is measured by its dispersity. The poly dispersity index (PDI) can be estimated from a cumulants analysis of the DLS measured intensity by taking three numbers code from each sample as shown in Eq. 1 [180]. The measured widths of particles play an important role in determining particle size distributions. Particles with small widths have favourable particle size distributions, where as big particles act as a center for an error like crack initiation point. The values $\chi(16, \text{i})$ and $\chi(84, \text{i})$ represents the positive and negative particle size standard deviation, respectively, were chosen by statistical means where $\chi(50, \text{i})$ is median particle size [193].

$$PDI = \frac{\chi(16, i)}{\chi(84, i)} \tag{1}$$

where $\chi(\text{n}, \text{i})$: n represent percentage particles to be considered among total sample and i is particular parameter of considered particle. Value of i can be 0, 1, 2 and 3 corresponding to the number, length, area and volume, respectively, of the particle.

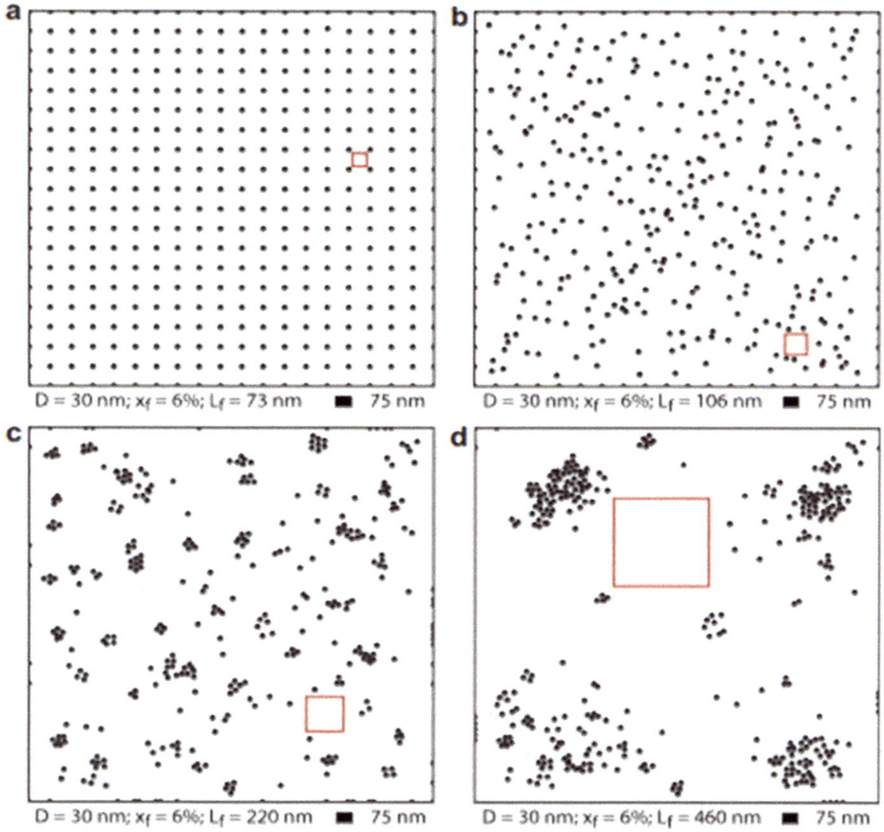

Fig. 8 Varying dispersions of 30 nm particles at 6% loading **a** uniform dispersion—Lf ¼ 73 nm; **b** random dispersion—Lf ¼ 106 nm; **c** clustered dispersion—Lf ¼ 220 nm; **d** agglomerated dispersion—Lf ¼ 460 nm. Open squares have a length equal to the free-space length which increases as the dispersion worsens [192]

(ii) ***Measurement of homogeneity by image analysis***

The degree of distribution or homogeneity in nanocomposite materials can be quantified by analysis of SEM microstructures to verify the results obtained from the DLS test. The microstructural image analysis technique to quantify the homogeneity of nanofillers in nanocomposite materials is based upon the free-path spacing which is the distance between two neighbor particles. Multiple straight lines in horizontal, h and vertical, v, total 2N, are drawn on a cross-sectional microstructure of a nanocomposite and processed by using the Image J software [194]. All the free-path spacing arithmetic mean ($\overline{X}$) and the standard deviation (s) values are calculated according to Eqs. 2 and 3, respectively.

$$\overline{X} = \frac{(\overline{X}_{h1} + \overline{X}_{h2} + \ldots + \overline{X}_{hN}) + (\overline{X}_{v1} + \overline{X}_{v2} + \ldots + \overline{X}_{vN})}{2N}. \quad (2)$$

$$s = \left(\frac{[\sum_{i=1}^{n} (\overline{X}_{hi} - \overline{X})^2] + [\sum_{i=1}^{n} (\overline{X}_{vi} - \overline{X})^2]}{2N - 1} \right)^{-1/2} \tag{3}$$

The degree of homogeneity is maximum when all the arithmetic mean values of the free-path spacing are similar on each line as well as the mean $(\overline{X})$ value. However, the mean values differ from each other and from the mean $(\overline{X})$ of all free-path spacing values in the case of an inhomogeneous distribution of particles. The degree of variation of one variable having different mean values among multiple free-path spacing can be assessed by the coefficient of variation, C_v. TheC_v also known as the distribution index (D index) is an accepted proposed homogeneity analysis method and can be determined using Eq. 4.

$$D\,index = C_v = \left(\frac{s}{\overline{\overline{X}}} \right) \tag{4}$$

The D index is zero for a perfectly homogeneous nanocomposites (maximum homogeneity) as all free-path spacing values between the nanofillers are the same. If there are certain levels of particle agglomeration in a nanocomposite material, the D index diverges from zero and possesses up to 1.0 or above value. The D index value of above 1.0 in certain analyzed regions indicates that the standard deviation is higher than the mean value and this shows that there is extremely high agglomeration formation of nanofillers in that area. Thus, the free-path spacing, D index, method is used to efficiently quantify the homogeneity of nanofillers in nanocomposites and provides the possibility of analyzing the dispersion level of nanofillers in different nanocomposite samples.

9 Nanocomposite Kinematics

In addition to the uniform dispersion of nanofillers in the polymer matrix, the interfacial interaction between the nanofillers and polymeric matrices plays a significant role in enhancing the mechanical properties of nanocomposites [188, 189]. A strong interfacial interaction between these components results in improved mechanical properties of nanocomposites [195]. The improvements in mechanical properties can be predicted through appropriate theoretical modeling and validated through experimental results. Measurements of certain parameters during the experimental test are desired to design a model. A designed model is used to develop a necessary theory to compare the predicted behavior with experimental results through simulation [196]. The interdependence between the experimental test results and predicted modeling are summarized in Fig. 9. Mechanical properties of nanocomposites can be predicted through computational modeling methods by considering their analytical micromechanics methods.

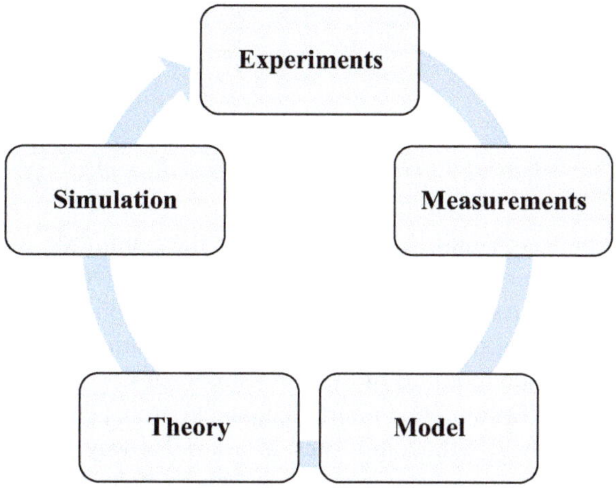

Fig. 9 A schematic diagram of correlation between modeling and experimental data [196]

To predict the mechanical properties of the nanofillers reinforced nanocomposites, several micromechanical models have been proposed. Among these, common models used to predict elastic modulus and yield stress of nanocomposites are described in this chapter.

a. **Halpin–Tsai Model**

The Halpin–Tsai model is a semi-empirical model that has been developed by Halpin and Kardos [197] to estimate the elastic modulus and yield stress of nanocomposite materials. The general Halpin–Tsai Equation can be written as shown in Eq. 5:

$$\frac{E}{E_m} = \frac{1 + \xi\eta\varphi_f}{1 - \eta\varphi_f} \text{where } \xi = 2x\frac{L}{D} \tag{5}$$

where E is the longitudinal or transverse modulus, E_m is the elastic modulus of the matrix, φ_f is the volume fraction of the nanofillers. The parameter ξ depends on the geometry of the nanofillers., The value of $\xi = 2$ for the transverse modulus while ξ can be determined from the longitudinal modulus where D and L are the diameters and the length of the nanofiller, respectively. η is calculated using Eq. 6:

$$\eta = \frac{\left(E_f/E_m\right) - 1}{\left(E_f/E_m\right) + \xi} \tag{6}$$

where E_f and E_m are the elastic modulus of the nanofiller and matrix, respectively.

To make a more precise prediction of nanocomposite elastic modulus, a modified form of the Halpin–Tsai model was proposed. In this model, three important factors are incorporated into the Eq. 6: an orientation factor, f_R, for the random dispersion of the nanofillers, waviness efficiency factor, f_ω, for nanofillers waviness, and agglomeration efficiency factor, f_A, for the nanofillers agglomerated state [198]:

$$\eta = \frac{\left(f_R f_\omega f_A E_f / E_m\right) - 1}{\left(f_R f_\omega f_A E_f / E_m\right) + \xi} \tag{7}$$

$$f_\omega \, and \, f_A \, can \, be \, defined \, as: \; f_\omega = 1 - \frac{A}{W}, \; and \; f_A = \exp\left(-\alpha \varphi_f{}^{\beta}\right)$$

whereas A and W are defined as the amplitude and half-wavelength of a wavy nanofiller, respectively. The α and β parameters are related to the degree of agglomeration. As the new form of Halpin–Tsai model takes into account of the nanofillers agglomeration state, waviness, and orientation it gives a more realistic prediction of the elastic modulus of the nanocomposites.

The Halpin–Tsai micromechanical model can be employed to predict the tensile strength of nanocomposites in a similar way to the elastic modulus. The tensile strength of nanocomposites, σ, can be determined by using the new form of the Halpin Tsai method as shown in Eq. 8:

$$\frac{\sigma}{\sigma_m} = \frac{1 + \xi \eta \varphi_f}{1 - \eta \varphi_f}, \; \eta = \frac{\left(f_R f_\omega f_A \sigma_f / E_m\right) - 1}{\left(f_R f_\omega f_A \sigma_f / E_m\right) + \xi} \tag{8}$$

where σ_m and σ_f represents the tensile strength of the matrix and nanofillers, respectively.

b. **Richeton-Ji Model for the Elastic Modulus**

The Richeton-Ji Model was developed by incorporating the Richeton model into the Ji model, which takes into account the interface bonding between the matrix and nanofillers [199]. The RJ model for the reference instantaneous moduli of the nanocomposite E^{ref}_{ic} as a function of the one of the matrices E^{ref}_{im} is expressed as [200]:

$$\frac{E^{ref}_{ic}}{E^{ref}_{im}} = \left[(1-\alpha) + \frac{\alpha - \beta}{(1-\alpha) + \alpha(h-1)/\ln(h)} + \frac{\beta}{\left(1-\alpha\right) + (\alpha - \beta)(h+1)/2 + (E_f / E^{ref}_{im})} \right]^{-1}. \tag{9}$$

The subscript i = 1,2,3 represents beta, glass, and flow transition, respectively, where α and β are correlated to the nanofillers volume fraction, φ_f. The E^{ref}_{ic} and transition temperatures have a dependence on the frequency (f) or strain rate ($\acute{\varepsilon}$) and can be determined from the following Equation:

$$E_i = E_i{}^{ref}\left[1 + s.\log_{10}(f/f^{ref})\right]; \alpha = \sqrt[2]{[2(\tau/t_c) + 1]\varphi_f}; \beta = \sqrt[2]{\varphi_f}$$

where, the superscript “ref” indicates a reference value, and s is the sensitivity constant of the modulus to frequency for a specified matrix. Whereas, the three

E_{im}^{ref} are the instantaneous stiffness of the pure polymeric matrix at a reference strain rate ($\acute{\varepsilon}$); h is the stiffness ratio; E_f is the Young's modulus of the nanofillers; τ and t_c are thicknesses of the interphase and particles, respectively; nanocomposites instantaneous moduli, E_{ic}^{ref} is calculated by using Eq. 9.

Finally, the Young's modulus of the nanocomposites as a function of temperature and strain rate E(T, $\acute{\varepsilon}$) can be determined by using the Richeton model shown in Eq. 10.

$$E(T, \varepsilon') = E_1(\varepsilon') - E_2(\varepsilon').\exp\left(-\left(\frac{T}{T_\beta(\varepsilon')}\right)^{m1}\right) + E_2(\varepsilon') - E_3(\varepsilon').\exp\left(-\left(\frac{T}{T_g(\varepsilon')}\right)^{m2}\right) + E_3(\varepsilon').\exp\left(-\left(\frac{T}{T_f(\varepsilon')}\right)^{m3}\right). \tag{10}$$

where, T_β—beta transition, T_g—glass transition, T_f—flow transition, and mi is Weibull moduli (mi) used to represent the statistics of the bond breakage. The material behavior is defined by Weibull parameters: m1 affects the slope of the secondary relaxation, m2 affects the slope of the glass transition region, and m3 affects the slope of the flow region.

10 Nanocomposite Homogenization and Kinematic Relation

To validate the results obtained from the predicted model for nanocomposites, a comparison with the results obtained from the experimental analysis is done. Accordingly, the correlation between the two results has been described in several studies. Modified Halpin–Tsai micromechanical model showed best fit to the experimental Young's modulus of the melt-mixed nanocomposites of short-chain branched-polyethylene reinforced with graphene nanoplatelets [201]. Another study by Hassanzadeh-Aghdam et al. also implemented a modified Halpin–Tsai model to predict the mechanical properties of the carbon nanotube reinforced polymeric nanocomposites and verified the model predictions with certain experimental works [198]. They found that the proposed approach predictions best fit the experimental data as it takes into consideration the random orientation, waviness and agglomeration of the nanofillers.

The Richeton-Ji model applied to polypropylene filled with nanoclay matches the experimental data very well [202]. Effects of porosity formation in the nanocomposites mechanical properties is also described in this study. The Young's modulus of polypropylene nanoclay highly decreased as the porosity content increased within the nanocomposites. The Young's modulus dependence on the temperature, frequency/strain rate, and volume fraction of organoclay is analyzed by incorporating 2-phase and 3-phase nanocomposites statistical model approaches in micromechanical model [200]. The nanocomposite elastic modulus predicted by both phases showed a good agreement with experimental results.

11 Nanocomposite Materials Characterization

To comprehend and assess the many properties of nanocomposites as well as to determine whether the material is appropriate for the intended end product, characterization of the nanocomposite materials is required. Characterization techniques will vary based on their intended end product behavior. Generally, the following techniques are used.

11.1 Surface Morphology of Nanocomposite

Analytical imaging, which uses high-end microscopes to produce images of commodities, materials, and objects that are undetectable to the naked eye, includes surface morphology as one of its components. A sophisticated kind of high resolution imaging is used for imaging analysis. These images show the exposed surface of the sample or item. These kinds of analyzes can be carried out for comparison purposes, to identify the surface morphology of a produced sample, and to determine the fracture level following specific mechanical testing. There are several morphology analysis methods including optical microscopy, X-ray diffraction, interferometry, dynamic light scattering, and porosimetry. Surface morphology can be analyzed using transmission electron microscopy (TEM), scanning electron microscope (SEM), atomic force microscopy (AFM), Energy-Dispersive X-ray Spectroscopy (EDX) and X-ray Diffraction (XRD) used for crystal structural analyzes of nanocomposites.

In-situ fractographic analysis

In-situ composites are multiphase materials in which the reinforcing phase develops during the composite's production inside the matrix. SEM can be used to analyze the material's in-situ fracture (see Figs. 10 and 11) development and propagation.

Morphological analysis of nanocomposite

The morphology of the CNT sponge at different scales was visualized using SEM. CNTs were twisted and connected to one another, as shown in (Fig. 8c). Figure 8d shows that the epoxy matrix's surface was clear, smooth, and only seldom marked. SEM images also showed that the surface was smooth (E and F in Fig. 12). However, focusing on the sponge's structure showed the broken surface all around CNT sponge (Fig. 12i).

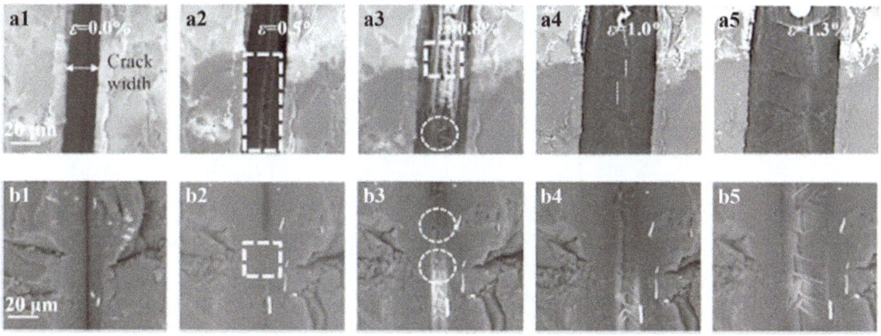

Fig. 10 In-situ observation of crack generation and propagation in an epoxy matrix (**a**) and corresponding nanocomposites (**b**), the applied strain ε is 0%, 0.5%, 0.8%, 1.0% and 1.3%, corresponding to the sample number 1–5, all images were in same scale bar). Reproduced from reference [31]

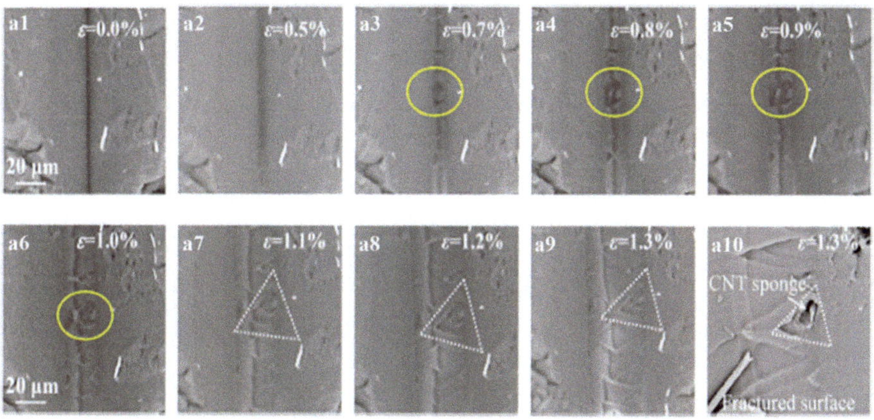

Fig. 11 In-situ observation of void generation and propagation in CNT/epoxy nanocomposites (The applied strain ε is 0%–1.3%, corresponding to the sample number 1–9, all images were in the same scale bar). Reproduced from reference [31]

11.2 Mechanical Properties Nanocomposites

a. Tensile strength

Tensile strength, commonly referred to as fracture strength, refers to the highest stress that a material can endure before breaking. The tensile strength of the composites decreases as they become brittle, as shown in Fig. 13, however, it was discovered that it improved when nanofillers present were increased from 0.2 to 0.6 weight percent, which is 19% less than composites without nanofillers.

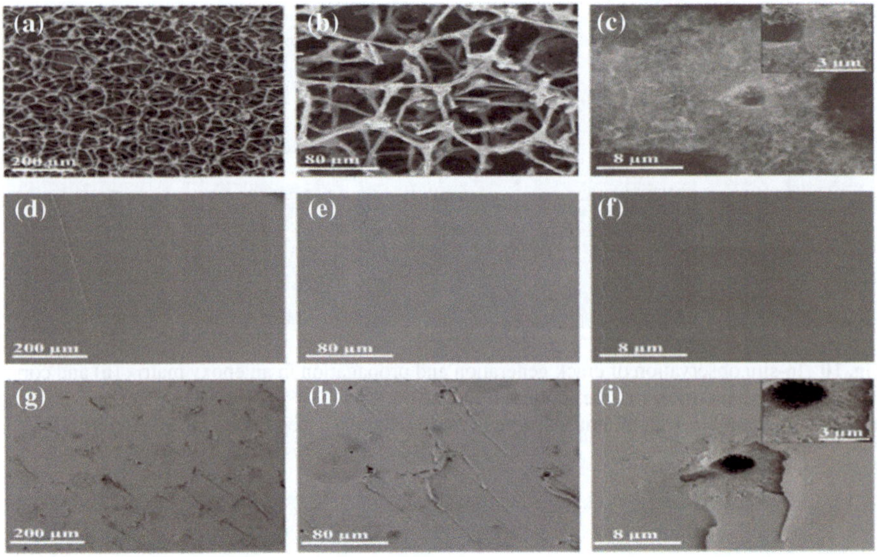

Fig. 12 Fracture morphologies of different samples (**a–c**: CNT sponge; **d–f**: Neat epoxy; **g–i**: CNT/epoxy nanocomposites). Reproduced from reference [31]

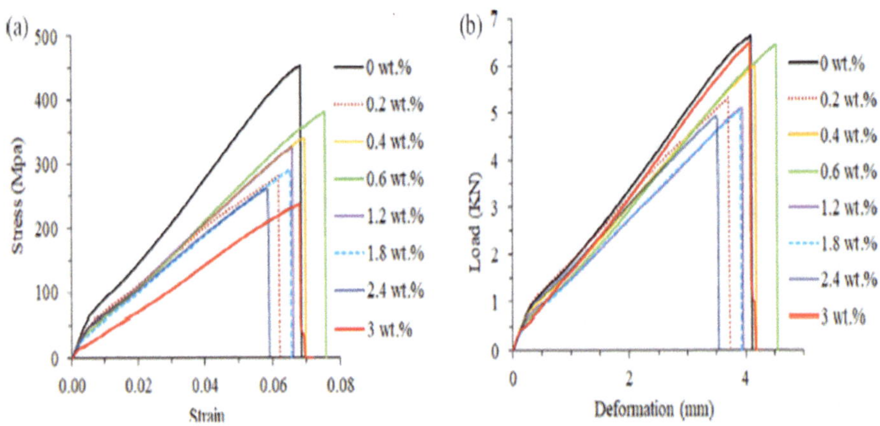

Fig. 13 **a** Stress–strain curve and **b** load-deformation curve obtained during tensile loading for composites with different percentages of nano-Al2O3 particles.). Reproduced from reference [203]

b. Fracture toughness

The ability of a structural material to resist cracking and be controlled by the amount of effort required to destroy it is known as fracture toughness. The term "fracture toughness" refers to fragile materials' resistance to the spread of faults under an applied stress, and it makes the assumption that the longer the flaw, the lower the force required to produce fracture. The material's fracture toughness affects whether a

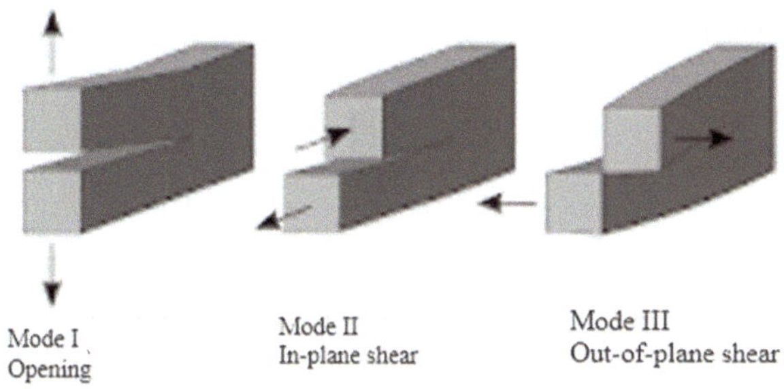

Fig. 14 Three types of fracture toughness

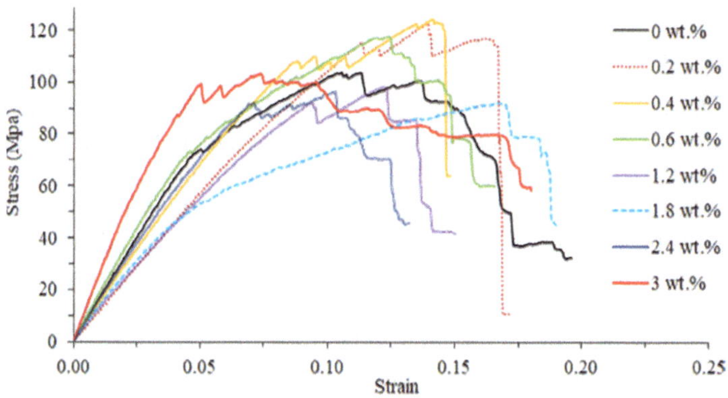

Fig. 15 **a** Stress–strain curve and **b** load-deformation curve obtained during flexural test for composites with different percentages of nano-Al2O3 particles Reproduced from reference [203] with the kind permission of the authors]

fault may lead to fracture. As seen in Fig. 14, the fracture toughness may be divided into three categories: Mode I, II, and III. A shear force operating perpendicular to the fracture front and parallel to the crack plane characterizes mode II, sliding mode, mode I, and opening mode, respectively. Mode III—Tearing mode-shear stress parallel to fracture front and parallel to crack plane.

c. Flexural strength

The stress in a material right before it bends in a flexure test is known as flexural strength, often referred to as bend strength, modulus of rupture or transverse rupture strength. Figure 15 shows that flexural strength of the composite sample increased up to 0.4 weight percent addition as the filler content increased, which may be the result of the filler and epoxy working well together. However, as filler concentration increased, flexural strength decreased due to inadequate nanofiller dispersion with in matrix.

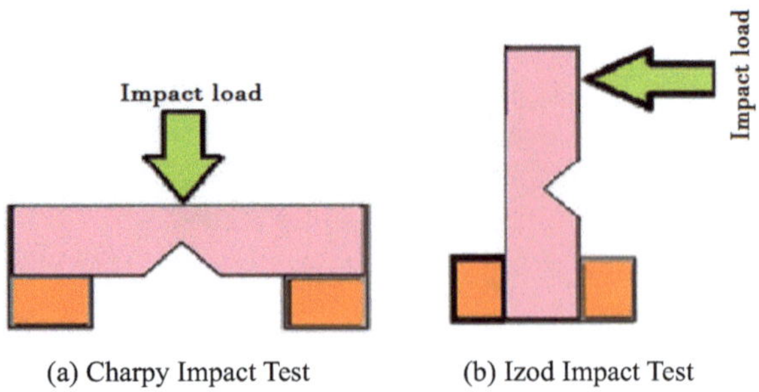

Fig. 16 Impact strength test methods

d. Impact strength

The impact strength of the material is determined by the amount of energy needed to break or fracture the specimen. To conduct the experiment, the pendulum arm is set at a particular position correspondent to an energy setting. The arm is released and its hammer end is allowed to hit the center of the specimen. For a single impact test, the three most popular types of test are the Charpy V-notch test, the Izod test and the Tensile Impact test. A tensile impact test measures the toughness of polymers using an impact tester for plastic according to the ISO 8256 standard. Polymers that are too flexible or thin for a Charpy or Izod impact test generally require an uniaxial tensile impact test with a relatively high strain rate. The Izod impact test is a standard test that measures the impact energy needed to fracture a material. This test helps engineers and scientists assess the fracture properties of a given part or component. In the Charpy impact test, the sample lies flat on the test bed as a simple beam. Both ends of the specimen get secured before the moment of impact. As shown in Fig. 16, during the Izod impact test, the sample is in a vertical cantilevered position. Only the bottom end of the specimen gets locked in place.

e. Inter-laminar shear strength

The inter-laminar shear strength (ILSS) of nanocomposites is the shear strength between lamination planes of composites and is assessed using the short beam shear test [203]. As well as ASTM D2344, standards EN ISO 14130, EN 2563, and ISO 14130 offer details on the performance of the test. As shown in Fig. 17. ILSS can perform at I-beam test, double notched, v-notched and short beam shear three-point bending positions.

12 Nanocomposite Applications

Nanotechnology is considerably progressing, if not completely transforming, in a number of businesses and technical disciplines, including information technology, homeland security, healthcare, energy, transportation, food safety, and environmental research [205]. Recently, nanocomposite materials became a constitutive

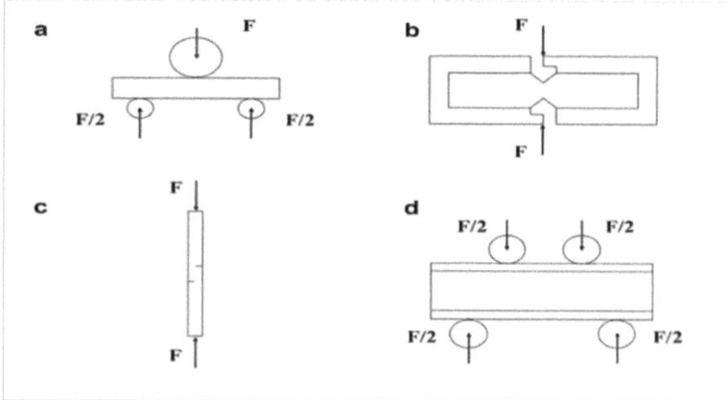

Fig. 17 Inter-laminar shear test methods: **a** short beam shear three-point bending, **b** v-notched shear, **c** double notched shear, **d** sandwich core shear test and I-beam four-point bending [204]

part of human life due to their highly enhanced properties including mechanical, thermal, electrical, etc., compared to the corresponding pure polymeric matrix. Thus, nanocomposites are widely used in applications such as food packaging [206, 207], medical [208,209] electronics [210, 211], aerospace [212, 213], energy harvestings devices and sensors [214, 215], and microwave absorption and shielding devices [216, 217] and many more [218–231].

A. **Food packaging**

Nanomaterials are employed in food packaging to increase barrier qualities, extend the shelf life, impart antibacterial action, and freshness of packaged food. In addition to being lighter than traditional composites, polymer nanocomposites have a number of advantages over neat polymers, including the ability to achieve high degrees of stiffness and strength with a significant reduction in the amount of high-density material required, superior barrier properties, and perhaps mechanical and thermal properties.

B. **Biomedical**

Nanocomposite hydrogels are sophisticated biomaterials with several pharmacological and biological applications. Versatility, flexibility, soft structure, and stimuli-responsive are the advantages of hydrogels. In biomedical devices such as stem cell engineering, actuators, sensors, drug delivery systems, and regenerative medicine, these hydrogel networks are used.

C. **Electro-analysis**

The nanocomposites-based sensors have superior sensitivity, electrical conductivity, selectivity, response, and performance relative to neat polymers and other nanomaterials. Specifically, nanocomposites' applications have been verified for biosensing, chemical sensing, strain sensing, and gas sensing.

D. **Wastewater treatment**

Nanocomposites are adept at removing viruses, bacteria, and inorganic and organic pollutants from wastewater because of their precise binding activity, which includes absorption, chelation, and ion exchange.

E. **Automotive**

Nanocomposite can be used in a variety of auto parts, including fuel cells, tires, paint, glass, and batteries. Many nanomaterials have been employed in automobiles, including tungsten nanospheres, nanostructured boric acid, graphene, and copper nanoparticles. Fluid lubricants can have their mechanical characteristics improved by adding nanoparticles, which has a number of positive economic effects.

13 Nanocomposites in Economic and Environmental Perspective

Recently, nanocomposites have proven to hold promise for environmental remediation. Nanomaterials may spur economic growth and enhance the capacity and quality in industrial sectors, and it has made a substantial contribution to society's welfare and molding the nature of modern living. It has the potential to profoundly alter societal dynamics, economic conditions, and human life. One of the promises made about the potential effects of nanotechnology is that it will encourage more sustainable and environmentally friendly economic growth [232]. The future of the global packaging sector is polymer nanocomposites. Companies will use this technology to boost their product's stability and survival along the supply chain once manufacturing and material costs are lower in order to offer superior quality to their consumers while saving money [233].

Polymeric nanocomposites might play an ever-more-important role in improving environmental parameters related to the removal of heavy metal ions and organics from waterways and soil [234]. In addition to this, environmental problems might also be mitigated by the use of these natural materials. All of the biopolymer prerequisites will enable the production of nanometer-sized fibers with a regulated fiber diameter, without bead flaws, and with a particular surface area. It has been demonstrated recently that substituting at least one natural component with a synthetic one can improve environmental protection without sacrificing the original features of these materials. In this sense, the utilization of these natural polymers as a workable substitute for a clean and safe environment is supported by nanocomposite methodology. The next generation of materials to offer a mix of performance and environmental friendliness is "green" nanocomposites based on renewable resources. These natural or biodegradable nanocomposites are referred regarded as "sustainable" materials since they are recyclable, and stable in use but may be "triggered" to decay when composted, ecologically friendly, and economically feasible. Future nanocomposite manufactures and the user will take into account all necessary factors, including the economics, environment, energy and life cycle analysis, in order to attain "sustainable." [235].

14 Challenges and Future Prospective of Nanocomposites

The main drawbacks of using nanocomposite materials are the toughness and performance impact caused by the addition of nanoparticles to the bulk matrix of the material. This is mostly caused by an inadequate understanding of the formulation, structural relationships, and the necessity for exfoliation and dispersion of particles. In various research projects, solution mixing procedures are used to swiftly screen for the production of polymer nanocomposites. In solution-mixed polymers, ultra-sonification is widely employed to enhance the dispersion of nanofiller [236]. The present issue with the development of polymer nanocomposites reinforced with layered inorganic fillers is the challenge of producing nanostructured composites at high filler loading. Increasing the filler content frequently results in the production of micro-structured filler/polymer composites. The solvent assisted approach and melt intercalation may have issues manufacturing nanocomposites with a high filler content among the four synthetic procedures designed to manufacture clay/polymer nanocomposites. In nanocomposites, a high filler loading might lead to a small filler interlayer gap. An inorganic filler particle's surface area can expand by hundreds or thousands of times when the stacked layers are separated [237].

Nanocomposites, which have a 25% yearly growth rate due to their multifunctionality as well as distinct design prospects and features, are the materials of the twenty-first century. According to the product, the nanocomposites category accounted for a major share of more than 62% and led the worldwide nanotechnology market in 2021. According to Precedence Research, the global nanotechnology market size reached US$ 85.39 billion in 2021 and is anticipated to rake up US$ 288.71 billion by 2030 [238] and the market for nanocomposites is forecasted to grow at a Compound Annual Growth Rate (CAGR) of 15.2 percent from 2022 to 2027, reaching US$11.3 billion [239]. This is because the nanocomposite system facilitates the fabrication of high-quality, high-performing products at lower costs by generating complex structures at the atomic or molecular level, nanocomposite has swiftly gained interest since its beginnings. All industrial sectors, including the agricultural, transportation, pharmaceutical, communication, manufacturing, and others, use this innovative and it is a highly useful technology. The rising use of nanomaterials in several end use verticals, such as the healthcare, aerospace, electronics, and textiles industries, is expected to drive growth in the global nanotechnology market during the course of the projected period.

Natural nanocomposites combine great adaptability, flexibility, and multifunctionality with high robustness and tolerance for failure [240], and have good sustainability [241]. In recent years, natural polymers like starch, cellulose acetate, polylactic acid (PLA), etc. have seen widespread application. These materials are preferable from an environmental point of view [2, 242–245]. The worldwide packaging industry's future is in polymer nanocomposites and they will be able to provide clients with products of greater quality while economizing. Due to environmental concerns and landfills for waste, the next generation of nanomaterials will be biodegradable bio-based nanocomposites.

15 Conclusion

Nanoscale particles are integrated into a matrix of conventional material to form nanocomposites. The use of nanoparticles dramatically improves a variety of properties, including toughness, mechanical strength, thermal conductivity and electrical conductivity. Materials with nanostructures exhibit great qualities that are uncommon in bulk materials with similar compositions. In recent years, both the commercial and academic communities have been interested in nanocomposite materials. This is due to the rapid rise in interest in nanocomposite systems since their inception, which makes it easier to fabricate high-quality, high-performing goods at cheaper prices by creating complex structures at the atomic or molecular level. This cutting-edge and very practical technology is used across all industrial sectors, including agriculture, transportation, manufacturing, and the pharmaceutical industry. Nanocomposite homogeneity is one of the most crucial steps in the creation of nanomaterials. If the dispersion process is not carried out correctly, it might create undesirable flaws and fractures into the material, which can affect how well the finished product functions. The main drawbacks of using nanocomposite materials are the toughness and performance effect caused by the addition of nanoparticles to the bulk matrix of the material. This is mostly caused by an ignorance of formulation, structural correlations, and exfoliation and dispersion needs for particles. Several nanocomposite preparation strategies have been created based on the nature of the phase components in order to impact the filler distribution and achieve superior homogeneity. Nanofillers are dispersed in nanocomposites using a variety of processes, including in-situ polymerization, melt intercalation, solution and melt-mixing, electrospinning, and sol–gel. Nanofillers are first disseminated in an appropriate monomer solution during in-situ polymerization, after which the nanocomposites are polymerized. The two most popular models for predicting the elastic modulus and yield stress of nanocomposites are the Halpin–Tsai Model and the Richeton-Ji Model. Natural polymer nanocomposite materials are superior in the long run from an economic and environmental perspective. Additionally, biodegradable, bio-based nanocomposites are anticipated to make up the next generation of nanomaterials.

Acknowledgments This project was supported by the Key Research and Development Program of Xinjiang (Project No.2022B01023-1), the Natural Science Fund for Distinguished Young Scientist in Xinjiang (Project No. 2022D01E88), and the West Light Foundation of the Chinese Academy of Sciences (Project No. 2019-XBQNXZ-B-010).

References

1. Ray SS, Iroegbu AOC, Bordado JC (2020) Polymer-based membranes and composites for safe, potable, and usable water: a survey of recent advances. Chem Afr 3(3):593–608
2. Iroegbu AOC, Ray SS (2022) Recent developments and future perspectives of biorenewable nanocomposites for advanced applications. Nanotechnol Rev 11(1):1696–1721

3. Viswanathan V, Laha T, Balani K, Agarwal A, Seal S (2006) Challenges and advances in nanocomposite processing techniques. Mater Sci Eng R Rep 54(5–6):121–285. https://doi.org/10.1016/j.mser.2006.11.002
4. Omanović-Mikličanin E, Badnjević A, Kazlagić A, Hajlovac M (2020) Nanocomposites: a brief review. Health Technol (Berl) 10(1):51–59. https://doi.org/10.1007/s12553-019-00380-x
5. Bayda S, Adeel M, Tuccinardi T, Cordani M, Rizzolio F (2019) The history of nanoscience and nanotechnology: from chemical–physical applications to nanomedicine. Molecules 25(1):112
6. Karak N (2018) Fundamentals of nanomaterials and polymer nanocomposites. Elsevier Inc.
7. Sharma VK, Filip J, Zboril R, Varma RS (2015) Natural inorganic nanoparticles–formation, fate, and toxicity in the environment. Chem Soc Rev 44(23):8410–8423
8. Safdari M, Al-Haik MS (2018) A review on polymeric nanocomposites: effect of hybridization and synergy on electrical properties. Carbon-Based Polym Nanocompos Environ Energy Appl 113–146
9. Do TL, Montanheiro A et al (2020) A brief review concerning the latest advances in the influence of nanoparticle reinforcement into polymeric-matrix biomaterials. J Biomater Sci Polym Ed 31(14):1869–1893
10. Vaia RA, Wagner HD (2004) Framework for nanocomposites. Mater Today 7(11):32–37. https://doi.org/10.1016/S1369-7021(04)00506-1
11. Zhao J, Wu L, Zhan C, Shao Q, Guo Z, Zhang L (2017) Overview of polymer nanocomposites: computer simulation understanding of physical properties. Polymer (Guildf) 133:272–287
12. Palmero P (2015) Structural ceramic nanocomposites: a review of properties and powders' synthesis methods. Nanomaterials 5(2):656–696
13. Thostenson ET, Li C, Chou TW (2005) Nanocomposites in context. Compos Sci Technol 65(3–4):491–516. https://doi.org/10.1016/j.compscitech.2004.11.003
14. Camargo PHC, Satyanarayana KG, Wypych F (2009) Nanocomposites: synthesis, structure, properties and new application opportunities. Mater Res 12:1–39
15. Chawla KK (2012) Metal matrix composites. In: Composite materials. Springer, pp 197–248
16. Babu Valapa R, Loganathan S, Pugazhenthi G, Thomas S, Varghese TO (2017) An overview of polymer-clay nanocomposites. Elsevier Inc.
17. Yashas Gowda TG, Sanjay MR, Subrahmanya Bhat K, Madhu P, Senthamaraikannan P, Yogesha B (2018) Polymer matrix-natural fiber composites: an overview. Cogent Eng 5(1):1446667
18. Manias E, Touny A, Wu L, Strawhecker K, Lu B, Chung TC (2001) Polypropylene/montmorillonite nanocomposites. Review of the synthetic routes and materials properties. Chem Mater 13(10):3516–3523
19. Ma J, Zhang S, Qi Z, Li G, Hu Y (2002) Crystallization behaviors of polypropylene/montmorillonite nanocomposites. J Appl Polym Sci 83(9):1978–1985
20. Morlat-Therias S, Mailhot B, Gonzalez D, Gardette J-L (2005) Photooxidation of polypropylene/montmorillonite nanocomposites. 2. Interactions with antioxidants. Chem Mater 17(5):1072–1078
21. Ben Ammar L, Fakhfakh S (2020) Optical and dielectric properties of polypropylene/montmorillonite nanocomposites. Funct Compos Struct 2(4):45003
22. Zare Y, Rhee KY, Park S-J (2020) Simple models for interphase characteristics in polypropylene/montmorillonite/CaCO3 nanocomposites. Phys Mesomech 23(2):182–188
23. Koltsakidou A, Terzopoulou Z, Kyzas GZ, Bikiaris DN, Lambropoulou DA (2019) Biobased poly (ethylene furanoate) polyester/TiO2 supported nanocomposites as effective photocatalysts for anti-inflammatory/analgesic drugs. Molecules 24(3):564
24. Camargo PHC, Satyanarayana KG, Wypych F (2009) Nanocomposites: Synthesis, structure, properties and new application opportunities. Mater Res 12(1):1–39. https://doi.org/10.1590/S1516-14392009000100002
25. Mallakpour S, Radfar Z, Hussain CM (2021) Current advances on polymer-layered double hydroxides/metal oxides nanocomposites and bionanocomposites: fabrications and applications in the textile industry and nanofibers. Appl Clay Sci 206:106054

26. Iqbal A, Saeed A, Ul-Hamid A (2021) A review featuring the fundamentals and advancements of polymer/CNT nanocomposite application in aerospace industry. Polym Bull 78(1):539–557
27. Lankone RS, Wang J, Ranville JF, Fairbrother DH (2017) Photodegradation of polymer-CNT nanocomposites: effect of CNT loading and CNT release characteristics. Environ Sci Nano 4(4):967–982
28. Zare Y, Rhee KY (2017) Accounting the reinforcing efficiency and percolating role of interphase regions in tensile modulus of polymer/CNT nanocomposites. Eur Polym J 87:389–397
29. Meng BW, Yue X, Zhang YR, Yang FH, Xing D, Ma PC (2021) Direct visualization of interfacial debonding in FRP structure using an AIE molecule. Compos Commun 27(March). https://doi.org/10.1016/j.coco.2021.100816
30. Chen J, Shen C, Yang S, Rana M, Ma PC (2017) Acid and temperature dual-responsive cotton fabrics with polymer coating. Compos Commun 4(March):10–15. https://doi.org/10.1016/j.coco.2017.03.003
31. Ma Q, Hao B, Ma PC (2021) In-situ characterization on the fracture behavior of three dimensional polymer nanocomposites reinforced by CNT sponge. Compos Sci Technol 217(August):2022. https://doi.org/10.1016/j.compscitech.2021.109132
32. Yang F, Hao B, Yue X, Ma P (2022) Polymer chemistry. Polym Chem 13:3126–3135. https://doi.org/10.1039/d2py00396a
33. Hossen MF et al (2017) Improved thermal properties of jute fiber-reinforced polyethylene nanocomposites. Polym Compos 38(7):1266–1272
34. Hossen MF, Asraf M, Zakaria CM, and others (2022) Review on the development of jute polyethylene nanocomposites as a function of fiber chemical treatments. Micro Nanosyst 14(2):166–186
35. Sahu SK, Badgayan ND, Sreekanth PSR (2022) Rheological properties of HDPE based thermoplastic polymeric nanocomposite reinforced with multidimensional carbon-based nanofillers. Biointerface Res Appl Chem 12(4):5709–5715
36. Sahu SK, Badgayan ND, Sreekanth PSR (2019) Understanding the influence of contact pressure on the wear performance of HDPE/multi-dimensional carbon filler based hybrid polymer nanocomposites. Wear 438
37. Dias RR, Zattera AJ, Pereira IM, Soares BG (2021) Ballistic impact performance of composite laminates based on high-density polyethylene/montmorillonite nanocomposite and Aramide fiber. Polym Compos 42(10):5586–5597
38. Kord B, Roohani M (2017) Water transport kinetics and thickness swelling behavior of natural fiber-reinforced HDPE/CNT nanocomposites. Compos Part B Eng 126:94–99
39. Bahgat Radwan A, Abdullah AM, Mohamed AMA, Al-Maadeed MA (2018) New electrospun polystyrene/Al2O3 nanocomposite superhydrophobic coatings; synthesis, characterization, and application. Coatings 8(2):65
40. Bhavsar S, Singh NL, Suryanarayana SV, Murthy KVR (2019) Proton beam induced modification of luminescence properties of polystyrene/Al2O3 polymer nanocomposites. J Fluoresc 29(4):1007–1012
41. Neves RM, Lopes KS, Zimmermann MVG, Poletto M, Zattera AJ (2019) Characterization of polystyrene nanocomposites and expanded nanocomposites reinforced with cellulose nanofibers and nanocrystals. Cellulose 26(7):4417–4429
42. Mohamed GR, Mahmoud RK, Fahim IS, Shaban M, Abd El-Salam HM, Mahmoud HM (2021) Bio-composite thermal insulation materials based on banana leaves fibers and polystyrene: physical and thermal performance. J Nat Fibers 1–16
43. Rostamiyan Y, Ferasat A (2017) High-speed impact and mechanical strength of ZrO2/polycarbonate nanocomposite. Int J Damage Mech 26(7):989–1002
44. Nouh SA, Benthami K, Barakat MME (2020) Gamma-induced changes in the optical and structural properties of palladium/polycarbonate nanocomposite membrane. Radiat Eff Defects Solids 175(1–2):41–53
45. Moustafa H, Morsy M, Ateia MA, Abdel-Haleem FM (2021) Ultrafast response humidity sensors based on polyvinyl chloride/graphene oxide nanocomposites for intelligent food packaging. Sensors Actuators A Phys 331:112918

46. Abdel-Gawad NMK, El Dein AZ, Mansour DEA, Ahmed HM, Darwish MMF, Lehtonen M (2017) Enhancement of dielectric and mechanical properties of polyvinyl chloride nanocomposites using functionalized TiO 2 nanoparticles. IEEE Trans Dielectr Electr Insul 24(6):3490–3499
47. Taha TA, Hendawy N, El-Rabaie S, Esmat A, El-Mansy MK (2020) Fluorescence and dielectric spectroscopy identification of polyvinyl chloride/NiO nanocomposites. J Mol Struct 1212:128162
48. Ke K, Bonab VS, Yuan D, Manas-Zloczower I (2018) Piezoresistive thermoplastic polyurethane nanocomposites with carbon nanostructures. Carbon NY 139:52–58
49. Vaithylingam R, Ansari MNM, Shanks RA (2017) Recent advances in polyurethane-based nanocomposites: a review. Polym Plast Technol Eng 56(14):1528–1541
50. Urbina L et al (2019) Hybrid and biocompatible cellulose/polyurethane nanocomposites with water-activated shape memory properties. Carbohydr Polym 216:86–96
51. Yazdi M, Asl VH, Pourmohammadi M, Roghani-Mamaqani H (2019) Mechanical properties, crystallinity, and self-nucleation of carbon nanotube-polyurethane nanocomposites. Polym Test 79:106011
52. Shokraei N, Asadpour S, Shokraei S, Nasrollahzadeh Sabet M, Faridi-Majidi R, Ghanbari H (2019) Development of electrically conductive hybrid nanofibers based on CNT-polyurethane nanocomposite for cardiac tissue engineering. Microsc Res Tech 82(8):1316–1325
53. Khajehzadeh M, Ehsani N, Baharvandi HR, Abdollahi A, Bahaaddini M, Tamadon A (2020) Thermodynamical evaluation, microstructural characterization and mechanical properties of B4C–TiB2 nanocomposite produced by in-situ reaction of nano-TiO2. Ceram Int 46(17):26970–26984
54. Mikeladze A, Chkhartishvili L, Tsagareishvili O, Chedia R, Darchiashvili M (2018) Ceramic composites B4C--TiB2 in nanocrystalline state: Production, structure, and mechanical properties. In: Book of abstracts of the 3rd international conference on inorganic materials science modern technologies and methods. Georgian National Academy of Sciences Press, Tbilisi, pp 62–65
55. Zhang X et al (2019) High-performance B4C–TiB2–SiC composites with tuneable properties fabricated by reactive hot pressing. J Eur Ceram Soc 39(10):2995–3002
56. Guo W et al (2021) Microstructure and mechanical properties of B4C–TiB2 ceramic composites prepared via a two-step method. J Eur Ceram Soc 41(14):6952–6961
57. Saeedi Heydari M, Baharvandi H (2019) Comparison the effect of sintering temperature on the synthesis and properties of B4C-TiB2 nanocomposites by sol gel and insitu methods. Adv Process Mater Eng 13(4):1–15
58. Chkhartishvili L, Tsagareishvili O, Mikeladze A, Chedia R, Kvatchadze V, Ugrekhelidze V (2021) Highly stable boron carbide--based nanocomposites. Handb Nanomater Nanocomposites Energy Environ Appl 327–351
59. Song Q, Zhang Z-H (2022) Microstructure and self-healing mechanism of B4C--TiB2--SiC composite ceramic after pre-oxidation behavior. Ceram Int
60. He Q, Wang A, Liu C, Wang W, Wang H, Fu Z (2018) Microstructures and mechanical properties of B4C-TiB2-SiC composites fabricated by ball milling and hot pressing. J Eur Ceram Soc 38(7):2832–2840
61. Roghani H, Tayebifard SA, Kasraee K, Asl MS (2020) Volume combustion synthesis of B4C–SiC nanocomposites in tubular and spark plasma furnaces. Ceram Int 46(18):28922–28932
62. Zhang X et al (2017) Densification behaviour and mechanical properties of B4C–SiC intergranular/intragranular nanocomposites fabricated through spark plasma sintering assisted by mechanochemistry. Ceram Int 43(2):1904–1910
63. Matović B et al (2021) Characterization of B4C-SiC ceramic composites prepared by ultra-high pressure sintering. J Eur Ceram Soc 41(9):4755–4760
64. Ahmed H, Hashim A, Abduljalil H (2019) Analysis of structural, electrical and electronic properties of (polymer nanocomposites/silicon carbide) for antibacterial application. Egypt J Chem 62(4):767–776

65. Ahmed H, Abduljalil HM, Hashim A (2019) Structural, optical and electronic properties of novel (PVA–MgO)/SiC nanocomposites films for humidity sensors. Trans Electr Electron Mater 20(3):218–232
66. Qi X, Luo X, Zhang L, Wang S, Zhao J (2022) In situ synthesis and interfacial bonding mechanism of SiC in MgO--SiC--C refractories. Int J Appl Ceram Technol
67. Wei Y, Xu H, Li X, Jiang Y (2015) Mechanism of SiC formation in the synthesis of MgO-SiC-C powder from magnesite or magnesia. Interceram-Int Ceram Rev. 64(3):122–124
68. Zhu L, Ren X, Wang X, Kang X, Zheng R, Feng P (2021) Microstructure and high-temperature oxidation resistance of MoSi2-ZrO2 composite coatings for Niobium substrate. J Eur Ceram Soc 41(2):1197–1210
69. Mathur S, Veith M, Shen H, Hüfner S, Jilavi MH (2002) Structural and optical properties of NdAlO3 nanocrystals embedded in an Al2O3 matrix. Chem Mater 14(2):568–582
70. Jin XH, Gao L (2003) Microstructure and mechanical properties of NdAlO3/Al2O3 nanocomposite. Mater Sci Eng A 354(1–2):326–330
71. Riyas AH, Geethanjali CV, Arathy S, Anil A, Shibli SMA (2022) Exploration and tuning of Al2O3/Mo composite for enhancement of anti-corrosion and tribological characteristics in zinc phosphate conversion coatings. Appl Surf Sci 593:153370
72. Ni X, Zhao J, Sun J, Gong F, Li Z (2017) Effects of metal binder on the microstructure and mechanical properties of Al2O3-based micro-nanocomposite ceramic tool material. Int J Miner Metall Mater 24(7):826–832
73. Song J, Su Y, Fan H, Zhang Y, Hu L (2017) A novel design to produce high-strength and high-toughness alumina self-lubricated composites with enhanced thermal-shock resistance—Part I: mechanical properties and thermal shock behavior of Al2O3/Mo-Al2O3 laminated composites. J Eur Ceram Soc 37(1):213–221
74. Chen P, Xiao P, Li Z, Tang X, Liao J, Li Y (2022) Wet-oxygen corrosion resistance and mechanism of bi-layer Mullite/SiC coating for Cf/SiC composites. Ceram Int 48(21):31428–31437
75. Zhou W et al (2020) Fabrication, characterization and thermal-insulation modeling of foamed mullite-SiC ceramics. J Alloys Compd 829:154523
76. Xie M, Wu X, Liu J, Zhang K (2017) In-situ synthesis and textural evolution of the novel carbonaceous SiC/mullite aerogel via polymer-derived ceramics route. Ceram Int 43(13):9896–9905
77. Wang Y et al (2021) In situ formation of Si3N4–SiC nanocomposites through polymer-derived SiAlCN ceramics and spark plasma sintering. Ceram Int 47(15):22049–22054
78. Zhang W, Wang S, Ding H, Zhai X, Yang D (2021) Improved crystallisation of polysilazane-derived Si3N4/SiC nanocomposites with Fe2O3 catalyst. Micro Nano Lett 16(7):399–404
79. Yan S et al (2021) Fabrication and tribological characterization of laser textured engineering ceramics: Si3N4, SiC and ZrO2. Ceram Int 47(10):13789–13805
80. Yang H, Li Q, Wang Z, Wu H, Wu Y, Cheng X (2021) Effect of different sintering additives on the microstructure, phase compositions and mechanical properties of Si3N4/SiC ceramics. ES Mater Manuf 15:65–71
81. Dehghahi S, Amini R, Alizadeh M (2017) Microstructure and corrosion resistance of Ni-Al2O3-SiC nanocomposite coatings produced by electrodeposition technique. J Alloys Compd 692:622–628
82. Momohjimoh I, Saheb N, Hussein MA, Laoui T, Al-Aqeeli N (2020) Electrical conductivity of spark plasma sintered Al_2O_3--SiC and Al_2O_3-carbon nanotube nanocomposites. Ceram Int 46(10):16008–16019
83. Klement R, Švančárek P, Parchoviansk\'y M, Sedláček J, Galusek D (2018) Al2O3-SiC nanocomposites: preparation, microstructure, and properties. In: Advances in ceramic matrix composites, Elsevier, pp 49–92
84. Gevorkyan E et al (2021) Special features of manufacturing cutting inserts from nanocomposite material Al2O3-SiC. J Instrum 16(10):P10015
85. Suresh S, Gowd GH, Kumar MLS (2019) Mechanical and wear behavior of Al 7075/Al2O3/SiC/mg metal matrix nanocomposite by liquid state process. Adv Compos Hybrid Mater 2(3):530–539

86. Jiang D, Yu J (2018) Fabrication of Al2O3/SiC/Al hybrid nanocomposites through solidification process for improved mechanical properties. Metals (Basel) 8(8):572
87. Yu W et al (2022) Effects of CeO2 on the phase, microstructure and mechanical properties of Al2O3-ZrO2 (CeO2) nanocomposite ceramics (AZC-NCs) by solid solution precipitation. Ceram Int
88. Yu Y et al (2021) High-density nanoprecipitation mechanism and microstructure evolution of high-performance Al2O3/ZrO2 nanocomposite ceramics. J Eur Ceram Soc 41(10):5269–5279
89. Duntu SH, Agyapong J, Ahmad I, Islam M, Boakye-Yiadom S (2022) Combined effect of low ZrO2 content and carbon nanostructures on mechanical and wear properties of hot-pressed Al2O3/ZrO2 hybrid nanocomposites. Ceram Int
90. Koltsov I (2021) Thermal stability of polymeric carbon nitride (PCN)-Al2O3--ZrO2 nanocomposites used in photocatalysis. J Therm Anal Calorim 1–8
91. Park SJ, Choi DJ (2017) Synthesis of porous Al2O3/ZrO2 nanocomposites by chemical vapour deposition. Adv Appl Ceram 116(5):236–241
92. Santos KH et al (2020) Influence of the cold plasma treatment on the Al2O3/ZrO2 nanocomposites surfaces. Appl Surf Sci 531:147206
93. Koltsov I et al (2018) Mechanism of reduced sintering temperature of Al2O3–ZrO2 nanocomposites obtained by microwave hydrothermal synthesis. Mater (Basel) 11(5):829
94. Yaghoubi A, Ramazani A, Taghavi Fardood S (2020) Synthesis of Al2O3/ZrO2 nanocomposite and the study of its effects on photocatalytic degradation of reactive blue 222 and reactive yellow 145 dyes. ChemistrySelect 5(32):9966–9973
95. Ni X, Zhao J, Gong F, Li G (2021) Thermal shock and thermal fatigue resistance of Al2O3/(W, Ti) C/TiN/Mo/Ni multidimensional graded ceramics. Int J Appl Ceram Technol 18(3):972–980
96. Shi S, Cho S, Goto T, Sekino T (2022) The effects of microstructure on mechanical and electrical properties of W dispersed Al2O3 ceramics. Int J Appl Ceram Technol 19(3):1746–1755
97. Yuan L-J et al (2021) Comparison of sintering behavior and reinforcing mechanisms between 3Y-TZP/Al2O3 (w) and 12Ce-TZP/Al2O3 (w) composites: combined effects of lanthanide stabilizer and Al2O3 whisker length. J Eur Ceram Soc 41(1):706–718
98. Rikhtegar F, Shabestari SG, Saghafian H (2017) Microstructural evaluation and mechanical properties of Al-CNT nanocomposites produced by different processing methods. J Alloys Compd 723:633–641
99. Din SH, Shah MA, Sheikh NA, Butt MM (2020) Nano-composites and their applications: a review. Charact Appl Nanomater 3(1):40–48
100. Chen WX, Lee JY, Liu Z (2003) The nanocomposites of carbon nanotube with Sb and SnSb0. 5 as Li-ion battery anodes. Carbon N. Y. 41(5):959–966
101. Xiang S, Yang X, Qiu Y, Li J, Liang Y, Xu J (2022) Effects of interface density on mechanical and electrical properties of annealed Cu–Nb nanocomposite wires. Mater Sci Eng A 856:144001
102. Thyagatur A, Mushongera LT (2022) Effect of interface orientation and loading direction on the mechanical response of Cu-Nb multilayered nanocomposites. J Mater Eng Perform 1–7
103. Hosseini MG, Shahryari E, Yardani Sefidi P (2021) Polyaniline grafted chitosan/GO-CNT/Fe3O4 nanocomposite as a superior electrode material for supercapacitor application. J Appl Polym Sci 138(38):50976
104. Kundu M, Bhardwaj H, Pandey MK, Krishnan P, Kotnala RK, Sumana G (2019) Development of electrochemical biosensor based on CNT–Fe3O4 nanocomposite to determine formaldehyde adulteration in orange juice. J Food Sci Technol 56(4):1829–1840
105. Raghavendra CR, Basavarajappa S, Sogalad I (2018) Study on influence of surface roughness of Ni-Al2O3 nano composite coating and evaluation of wear characteristics. IOP Conf Ser: Mater Sci Eng 310(1):12112
106. Ma CY, Zhao DQ, Xia FF, Xia H, Williams T, Xing HY (2020) Ultrasonic-assisted electrodeposition of Ni-Al2O3 nanocomposites at various ultrasonic powers. Ceram Int 46(5):6115–6123

107. Dehestani M, Khayati GR, Sharafi S (2021) An improved optimization model to predict the microhardness of Ni/Al2O3 nanocomposite coatings prepared by electrodeposition: a hybrid artificial neural network-modified particle swarm optimization approach. Measurement 179:109423
108. Khdair AI, Fathy A (2020) Enhanced strength and ductility of Al-SiC nanocomposites synthesized by accumulative roll bonding. J Mater Res Technol 9(1):478–489
109. Yaghobizadeh O, Baharvandi HR, Ahmadi AR, Aghaei E (2019) Development of the properties of Al/SiC nano-composite fabricated by stir cast method by means of coating SiC particles with Al. SILICON 11(2):643–649
110. Mei XM et al (2021) Enhanced strengthening of Al–SiC nanocomposites containing a uniform dispersion of dense nanoparticles fabricated by a hybrid accumulative roll-bonding process. Mater Sci Eng A 799:140217
111. Zhe J-F, Wu Y-L, Xie G-W, Liu X-F (2022) Microstructure and thermal stability of heterostructured Al-AlN nanocomposite. Acta Met Sin 0
112. Riquelme A, Rodrigo P, Escalera-Rodriguez MD, Rams J (2019) Effect of the process parameters in the additive manufacturing of in situ Al/AlN samples. J Manuf Process 46:271–278
113. Gostariani R, Bagherpour E, Rifai M, Ebrahimi R, Miyamoto H (2018) Fabrication of Al/AlN in-situ nanocomposite through planetary ball milling and hot extrusion of Al/BN: microstructural evaluation and mechanical behavior. J Alloys Compd 768:329–339
114. Sadoun AM, Fathy A (2019) Experimental study on tribological properties of Cu–Al2O3 nanocomposite hybridized by graphene nanoplatelets. Ceram Int 45(18):24784–24792
115. Sadoun AM, Mohammed MM, Elsayed EM, Meselhy AF, El-Kady OA (2020) Effect of nano Al2O3 coated Ag addition on the corrosion resistance and electrochemical behavior of Cu-Al2O3 nanocomposites. J Mater Res Technol 9(3):4485–4493
116. Najjar IMR, Sadoun AM, Alsoruji GS, Abd Elaziz M, Wagih A (2022) Predicting the mechanical properties of Cu--Al2O3 nanocomposites using machine learning and finite element simulation of indentation experiments. Ceram Int 48(6):7748–7758
117. Sunwoo S-H et al (2020) Stretchable low-impedance nanocomposite comprised of Ag–Au core–shell nanowires and Pt black for epicardial recording and stimulation. Adv Mater Technol 5(3):1900768
118. Fazleeva RR, Nasretdinova GR, Gubaidullin AT, Evtyugin VG, Yanilkin VV (2022) The two-step electrosynthesis of nanocomposites of Ag, Au, and Pd nanoparticles with iron (ii) oxide-hydroxide. New J Chem 46(5):2380–2392
119. Park T-H, Lee H, Lee J, Jang D-J (2017) Morphology evolution of Ag/Au nanocomposites via temperature-controlled galvanic exchange to enhance catalytic activity. RSC Adv 7(13):7718–7724
120. Zhang H, Xing Z, Pan M, Wang H-B, Liu Y-M (2020) Highly sensitive and selective electrochemical determination of 4-aminophenol based on flower-like Ag-Au nanocomposites modified glassy carbon electrode. J Electrochem Soc 167(12):126504
121. Mostafa AM, Mwafy EA, Awwad NS, Ibrahium HA (2021) Catalytic activity of Ag nanoparticles and Au/Ag nanocomposite prepared by pulsed laser ablation technique against 4-nitrophenol for environmental applications. J Mater Sci Mater Electron 32(9):11978–11988
122. Choi S et al (2018) Highly conductive, stretchable and biocompatible Ag–Au core–sheath nanowire composite for wearable and implantable bioelectronics. Nat Nanotechnol 13(11):1048–1056
123. Emam HE (2019) Arabic gum as bio-synthesizer for Ag–Au bimetallic nanocomposite using seed-mediated growth technique and its biological efficacy. J Polym Environ 27(1):210–223
124. Wang R et al (2017) A microfluidic chip based on an ITO support modified with Ag-Au nanocomposites for SERS based determination of melamine. Microchim Acta 184(1):279–287
125. Newase S, Bankar AV (2017) Synthesis of bio-inspired Ag–Au nanocomposite and its anti-biofilm efficacy. Bull Mater Sci 40(1):157–162

126. Singh LK, Bhadauria A, Laha T (2021) Understanding the effect of bimodal microstructure on the strength–ductility synergy of Al–CNT nanocomposites. J Mater Sci 56(2):1730–1748
127. Jargalsaikhan B, Bor A, Lee J, Choi H (2020) Al/CNT nanocomposite fabrication on the different property of raw material using a planetary ball mill. Adv Powder Technol 31(5):1957–1962
128. Sharma A, Gupta G, Paul J (2021) A comprehensive review on the dispersion and survivability issues of carbon nanotubes in Al/CNT nanocomposites fabricated via friction stir processing. Carbon Lett. 31(3):339–370
129. Nithyadharseni P et al (2015) Electrochemical studies of CNT/Si–SnSb nanoparticles for lithium ion batteries. Mater Res Bull 70:478–485
130. Luo G, Du L, Wang Y, Wang K (2013) Composite Nanoparticles. In: Li (ed) Encyclopedia of microfluidics and nanofluidics. Springer US, Boston, MA, pp 1–9
131. Al-Mutairi NH, Mehdi AH, Kadhim BJ (2022) Nanocomposites materials definitions, types and some of their applications: a review. Eur J Res Dev Sustain 3(2)
132. Jacoby M (2004) Composite materials. Chem Eng News 82(35):34–39. https://doi.org/10.1021/cen-v082n035.p034
133. Akpan EI, Shen X, Wetzel B, Friedrich K (2019) Design and synthesis of polymer nanocomposites. In: Polymer composites with functionalized nanoparticles, Elsevier, pp 47–83
134. Vatanpour V et al (2022) Cellulose acetate in fabrication of polymeric membranes: a review. Chemosphere 133914
135. Liu W, Ullah B, Kuo C-C, Cai X (2019) Two-dimensional nanomaterials-based polymer composites: fabrication and energy storage applications. Adv Polym Technol 2019
136. Liu Y, Wang Y, Zhang C, Liu T (2019) 2D nanosheet-constructed hybrid nanofillers for polymer nanocomposites with synergistic dispersion and function. APL Mater 7(8):80904
137. Mohd Zaini NA, Ismail H, Rusli A (2017) Short review on sepiolite-filled polymer nanocomposites. Polym Plast Technol Eng 56(15):1665–1679
138. Charradi K et al (2019) Amelioration of PEMFC performance at high temperature by incorporation of nanofiller (sepiolite/layered double hydroxide) in Nafion membrane. Int J Hydrogen Energy 44(21):10666–10676
139. Danlée Y, Mederos-Henry F, Hermans S, Bailly C, Huynen I (2020) Ranking broadband microwave absorption performance of multilayered polymer nanocomposites containing carbon and metallic nanofillers. Front Mater 7:214
140. Lu J, Han T, Dai J (2020) Preparation of gold-decorated silver nanowires for improving conductivity of electrically conductive adhesives. J Mater Sci Mater Electron 31(17):14601–14607
141. Sanusi OM, Benelfellah A, Hocine NA (2020) Clays and carbon nanotubes as hybrid nanofillers in thermoplastic-based nanocomposites–a review. Appl Clay Sci 185:105408
142. Sanusi OM et al (2021) Influence of montmorillonite/carbon nanotube hybrid nanofillers on the properties of poly (lactic acid). Appl Clay Sci 201:105925
143. Gofman I, Nikolaeva A, Yakimansky A, Ivanova O, Baranchikov A, Ivanov V (2019) Unexpected selective enhancement of the thermal stability of aromatic polyimide materials by cerium dioxide nanoparticles. Polym Adv Technol 30(6):1518–1524
144. Cho B-G, Lee J-E, Hwang S-H, Han JH, Chae HG, Park Y-B (2020) Enhancement in mechanical properties of polyamide 66-carbon fiber composites containing graphene oxide-carbon nanotube hybrid nanofillers synthesized through in situ interfacial polymerization. Compos Part A Appl Sci Manuf 135:105938
145. Bisht A, Dasgupta K, Lahiri D (2019) Investigating the role of 3D network of carbon nanofillers in improving the mechanical properties of carbon fiber epoxy laminated composite. Compos Part A Appl Sci Manuf 126:105601
146. Onyszko M et al (2022) The cellulose fibers functionalized with star-like zinc oxide nanoparticles with boosted antibacterial performance for hygienic products. Sci Rep 12(1):1–13
147. Norrahim MNF et al (2021) Performance evaluation of cellulose nanofiber with residual hemicellulose as a nanofiller in polypropylene-based nanocomposite. Polym (Basel) 13(7):1064

148. Al-Taweel SS, Saud HR, Kadhum AAH, Takriff MS (2019) The influence of titanium dioxide nanofiller ratio on morphology and surface properties of TiO2/chitosan nanocomposite. Results Phys 13:102296
149. Tripathi M, Kumar A (2018) Zinc oxide nanofiller-based composite polymer gel electrolyte for application in EDLCs. Ionics (Kiel) 24(10):3155–3165
150. Ramesan MT, Siji C, Kalaprasad G, Bahuleyan BK, Al-Maghrabi MA (2018) Effect of silver doped zinc oxide as nanofiller for the development of biopolymer nanocomposites from chitin and cashew gum. J Polym Environ 26(7):2983–2991
151. Rani P, Ahamed MB, Deshmukh K (2021) Dielectric and electromagnetic interference shielding performance of graphene nanoplatelets and copper oxide nanoparticles reinforced polyvinylidenefluoride/poly (3, 4-ethylenedioxythiophene)-block-poly (ethylene glycol) blend nanocomposites. Synth Met 282:116923
152. Al-Gunaid MQA, Saeed AMN (2018) Effects of the electrolyte content on the electrical permittivity, thermal stability, and optical dispersion of poly (vinyl alcohol)--cesium copper oxide--lithium perchlorate nanocomposite solid-polymer electrolytes. J Appl Polym Sci 135(8):45852
153. Watanabe R, Sugahara A, Hagihara H, Sakamoto K, Nakajima Y, Naganawa Y (2020) Polypropylene-based nanocomposite with enhanced aging stability by surface grafting of Silica Nanofillers with a silane coupling agent containing an antioxidant. ACS Omega 5(21):12431–12439
154. Maharana SM, Pandit MK, Pradhan AK (2021) Influence of fumed silica nanofiller and stacking sequence on interlaminar fracture behaviour of bidirectional jute-kevlar hybrid nanocomposite. Polym Test 93:106898
155. Shen X, Zheng Q, Kim J-K (2021) Rational design of two-dimensional nanofillers for polymer nanocomposites toward multifunctional applications. Prog Mater Sci 115:100708
156. Karimzadeh Z, Mahmoudpour M, Rahimpour E, Jouyban A (2022) Nanomaterial based PVA nanocomposite hydrogels for biomedical sensing: Advances toward designing the ideal flexible/wearable nanoprobes. Adv Colloid Interface Sci 102705
157. Zhang H, Shin B-G, Lee D-E, Yoon K-B (2020) Preparation of PP/2D-nanosheet composites using MoS2/MgCl2-and BN/MgCl2-bisupported ziegler-natta catalysts. Catalysts 10(6):596
158. Wen F et al (2022) Enhanced energy storage performance of polymer nanocomposites using hybrid 2D ZnO@ MoS2 semiconductive nano-fillers. Chem Eng J 430:132676
159. Mutharasi Y, Kaleekkal NJ, Arumugham T, Banat F, Kapavarapu MSRS (2020) Antifouling and photocatalytic properties of 2-D Zn/Al layered double hydroxide tailored low-pressure membranes. Chem Eng Process Intensif 158:108191
160. Liu T, Li W, Zhang C, Wang W, Dou W, Chen S (2021) Preparation of highly efficient self-healing anticorrosion epoxy coating by integration of benzotriazole corrosion inhibitor loaded 2D-COF. J Ind Eng Chem 97:560–573
161. Liu H et al (2021) Synergistic effects of cellulose nanocrystals-organic montmorillonite as hybrid nanofillers for enhancing mechanical, crystallization, and heat-resistant properties of three-dimensional printed poly (lactic acid) nanocomposites. Polym Eng and Sci 61(12):2985–3000
162. Li X et al (2022) Combined effect of volume fractions of nanofillers and filler-polymer interactions on 3D multiscale dispersion of nanofiller and Payne effect. Compos Part A Appl Sci Manuf 152:106722
163. Kamal A, Ashmawy M, Algazzar AM, Elsheikh AH (2022) Fabrication techniques of polymeric nanocomposites: a comprehensive review. Proc Inst Mech Eng Part C J Mech Eng Sci 236(9):4843–4861
164. Sandra Paszkiewicz AS (2019) Graphene-based nanomaterials and their polymer nanocomposites. In: Nanomaterials and polymer nanocomposites
165. Bharimalla AK, Deshmukh SP, Vigneshwaran N, Patil PG, Prasad V (2017) Nanocellulose-polymer composites for applications in food packaging: Current status, future prospects and challenges. Polym Plast Technol Eng 56(8):805–823

166. Rong C, Shen B (2018) Nanocrystalline and nanocomposite permanent magnets by melt spinning technique. Chinese Phys B 27(11):117502
167. Asmatulu WSK (2019) Introduction to electrospun nanofibers Ramazan. In: Synthesis and applications of electrospun nanofibers
168. Kausar A (2021) Polymeric nanocomposite via electrospinning: Assessment of morphology, physical properties and applications. J Plast Film Sheeting 37(1):70–92
169. Wang W, Li H, Yang Y, Zhao F, Li H, Xu K (2021) Enhanced thermal decomposition, laser ignition and combustion properties of NC/Al/RDX composite fibers fabricated by electrospinning. Cellulose 28(10):6089–6105
170. Guglielmi M, Kickelbick G, Martucci A (2014) Sol-gel nanocomposites. Springer
171. Łapińska A, Grochowska N, Antonowicz J, Michalski P (2022) Influence of the filler distribution on PDMS—graphene based nanocomposites selected properties. Sci Rep 12:1–10. https://doi.org/10.1038/s41598-022-23735-3
172. Sanjuan-Navarro L, Moliner-Martínez Y, Campíns-Falcó P (2022) The state of art of nanocarbon black as analyte in a variety of matrices: a review. TrAC - Trends Anal Chem 157:116769. https://doi.org/10.1016/j.trac.2022.116769
173. Xu T, Liao MT, Han W (2016) Polyacrylamide/metakaolinite nanocomposites by in-situ intercalative polymerization. Part Sci Technol 34(5):602–607. https://doi.org/10.1080/02726351.2015.1096873
174. Yao H, Hawkins SA, Sue HJ (2017) Preparation of epoxy nanocomposites containing well-dispersed graphene nanosheets. Compos Sci Technol 146:161–168. https://doi.org/10.1016/j.compscitech.2017.04.026
175. Hiremath A, Murthy AA, Thipperudrappa S, Bharath KN (2021) Nanoparticles filled polymer nanocomposites: a technological review. Cogent Eng 8(1):1991229. https://doi.org/10.1080/23311916.2021.1991229
176. Chaka KT (2022) Fused deposition modeling of polypropylene-aluminium silicate dihydrate microcomposites. E-Polymers 22(1):87–98. https://doi.org/10.1515/epoly-2022-0014
177. Hussain T, Ahmad MN, Nawaz A, Mujahid A, Bashir F, Mustafa G (2017) Surfactant incorporated co nanoparticles polymer composites with uniform dispersion and double percolation. J Chem 2017:7191590. https://doi.org/10.1155/2017/7191590
178. Tong Y, Zhang L, Bass P, Rolin TD, Cheng ZY (2018) Influence of silane coupling agent on microstructure and properties of CCTO-P(VDF-CTFE) composites. J Adv Dielectr 8(2):1–9. https://doi.org/10.1142/S2010135X1850008X
179. Yang S (2022) Interface and interphase of nanocomposites tailored by covalent grafting of carbon nanotube: hierarchical multiscale modeling. Int J Mech Sci 220:107160. https://doi.org/10.1016/j.ijmecsci.2022.107160
180. Shukla MK, Sharma K (2019) Measurement of dispersion and homogeneity of functionalized graphene/CNT hybrid epoxy composite. SN Appl Sci 1(9):1–7. https://doi.org/10.1007/s42452-019-1031-1
181. Zhu C, Lu X, Li Y, Deng Y, Lu J, Liu Z, Wu H, Tong Y (2022) Enhancing toughness of PLA/ZrP nanocomposite through methacrylate copolymer. Polym (Basel) 14:3748
182. Sharip NS et al (2021) Melt-vs. Non-melt blending of complexly processable ultra-high molecular weight polyethylene/cellulose nanofiber bionanocomposite. Polym (Basel) 13(3):1–15. https://doi.org/10.3390/polym13030404
183. Carmeli E et al (2022) Crystallization kinetics of melt-mixed 3D hierarchical graphene/polypropylene nanocomposites at processing-relevant cooling rates. Compos Part B Eng 247:110287. https://doi.org/10.1016/j.compositesb.2022.110287
184. Minář J, Doležal J, Brožek J (2022) The effect of nanocomposite synthesis and the drying procedure of graphene oxide dispersion on the polycaprolactone/graphene oxide nanocomposite properties. Polym Polym Compos 30:1–12. https://doi.org/10.1177/09673911211068923
185. Mrah L, Meghabar R (2021) In situ polymerization of styrene–clay nanocomposites and their properties. Polym Bull 78(7):3509–3526. https://doi.org/10.1007/s00289-020-03274-5

186. Duan X, Yu B, Yang T, Wu Y, Yu H, Huang T (2018) In situ polymerization of nylon 66/reduced graphene oxide nanocomposites. J Nanomater 2018:1047985. https://doi.org/10.1155/2018/1047985
187. Salahuddin NA, El-Kemary M (2017) High-performance flexible epoxy/ZnO nanocomposites with enhanced mechanical and thermal properties. Polym Eng Sci 57(9):1–10. https://doi.org/10.1002/pen.24520
188. Yakaboylu GA, Sabolsky EM (2017) Determination of a homogeneity factor for composite materials by a microstructural image analysis method. J Microscopy 266(3):263–272. https://doi.org/10.1111/jmi.12536
189. Lu PJ et al (2018) Methodology for sample preparation and size measurement of commercial ZnO nanoparticles. J Food Drug Anal 26(2):628–636. https://doi.org/10.1016/j.jfda.2017.07.004
190. De Kanter M et al (2016) Enabling the measurement of particle sizes in stirred colloidal suspensions by embedding dynamic light scattering into an automated probe head. Meas J Int Meas Confed 80:92–98. https://doi.org/10.1016/j.measurement.2015.11.024
191. Hupfeld T et al (2020) Analysis of the nanoparticle dispersion and its effect on the crystalline microstructure in carbon-additivated PA12 feedstock material for laser powder bed fusion. Mater (Basel) 13(15):1–18. https://doi.org/10.3390/ma13153312
192. Khare HS, Burris DL (2010) A quantitative method for measuring nanocomposite dispersion. Polym (Guildf) 51(3):719–729
193. Nolte H, Schilde C, Kwade A (2012) Determination of particle size distributions and the degree of dispersion in nanocomposites. Compos Sci Technol 72(9):948–958. https://doi.org/10.1016/j.compscitech.2012.03.010
194. Schneider CA, Rasband WS, Eliceiri KW, Instrumentation C (2017) NIH Image to Image J: 25 years of Image Analysis. Nat Methods 9(7):671–675
195. Alasfar RH, Ahzi S, Barth N, Kochkodan V, Khraisheh M, Koç M (2022) A review on the modeling of the elastic modulus and yield stress of polymers and polymer nanocomposites: effect of temperature, loading rate and porosity. Polym (Basel) 14(3):360. https://doi.org/10.3390/polym14030360
196. Valavala PK, Odegard GM (2005) Modeling techniques for determination of mechanical properties of polymer nanocomposites. Rev Adv Mater Sci 9(1):34–44
197. Hablot E, Matadi R, Ahzi S, Avérous L (2010) Renewable biocomposites of dimer fatty acid-based polyamides with cellulose fibres: thermal, physical and mechanical properties. Compos Sci Technol 70(3):504–509. https://doi.org/10.1016/j.compscitech.2009.12.001
198. Hassanzadeh-Aghdam MK, Jamali J (2019) A new form of a Halpin-Tsai micromechanical model for characterizing the mechanical properties of carbon nanotube-reinforced polymer nanocomposites. Bull Mater Sci 42(3):1–8.https://doi.org/10.1007/s12034-019-1784-6
199. Ji XL, Jing JK, Jiang W, Jiang BZ (2002) Tensile modulus of polymer nanocomposites. Polym Eng Sci 42(5):983–993. https://doi.org/10.1002/pen.11007
200. Wang K, Ahzi S, Matadi Boumbimba R, Bahlouli N, Addiego F, Rémond Y (2013) Micromechanical modeling of the elastic behavior of polypropylene based organoclay nanocomposites under a wide range of temperatures and strain rates/frequencies. Mech Mater 64:56–68. https://doi.org/10.1016/j.mechmat.2013.04.009
201. Kourtidou D, Tsongas K, Grigora M, Tzetzis D, Bikiaris DN, Chrissafi K (2021) Applied sciences on the improved mechanical properties of ball-milled GNPs reinforced short chain branched-polyethylene nanocomposite : micromechanical modeling and fractography study. Appl Sci 11:9420
202. Alasfar R, Said A, Wang N (2020) KuiBarth, modeling the mechanical response of polymers and nano- filled polymers: Effects of porosity and fillers content. Appl Polym 2020:1–16. https://doi.org/10.1002/app.49545
203. Vinay SS, Sanjay MR, Siengchin S, Venkatesh CV (2021) Effect of Al2O3 nanofillers in basalt/epoxy composites: mechanical and tribological properties. Polym Compos 42(4):1727–1740

204. Gagani AI, Krauklis AE, Sæter E, Vedvik NP, Echtermeyer AT (2019) A novel method for testing and determining ILSS for marine and offshore composites. Compos Struct 220:431–440
205. Smith DM, Simon JK, Baker JR Jr (2013) Applications of nanotechnology for immunology. Nat Rev Immunol 13(8):592–605
206. Sarfraz J, Gulin-Sarfraz T, Nilsen-Nygaard J, Pettersen MK (2021) Nanocomposites for food packaging applications: an overview. Nanomaterials 11(1):1–27.https://doi.org/10.3390/nano11010010
207. Li J et al (2022) Emerging food packaging applications of cellulose nanocomposites: a review. Polym (Basel) 14(19):4025. https://doi.org/10.3390/polym14194025
208. Shokrani H et al (2022) Polysaccharide-based nanocomposites for biomedical applications: a critical review. Nanoscale Horizons 7(l):1136–1160. https://doi.org/10.1039/d2nh00214k
209. Ashrafizadeh M et al (2022) Doxorubicin-loaded graphene oxide nanocomposites in cancer medicine: stimuli-responsive carriers, co-delivery and suppressing resistance. Expert Opin Drug Deliv 19(4):355–382. https://doi.org/10.1080/17425247.2022.2041598
210. Díez-Pascual AM, Rahdar A (2022) Graphene-based polymer composites for flexible electronic applications. Micromachines 13(7):1–18. https://doi.org/10.3390/mi13071123
211. Namkoong M et al (2022) "Moldable and transferrable conductive nanocomposites for epidermal electronics", *npj Flex.* Electron 6(1):1–9. https://doi.org/10.1038/s41528-022-00170-y
212. Bhat A et al (2021) Review on nanocomposites based on aerospace applications. Nanotechnol Rev 10(1):237–253. https://doi.org/10.1515/ntrev-2021-0018
213. Romero-Fierro D, Bustamante-Torres M, Bravo-Plascencia F, Esquivel-Lozano A, Ruiz J-C, Bucio E (2022) Recent trends in magnetic polymer nanocomposites for aerospace applications: a review. Polym (Basel) 14(19):4084. https://doi.org/10.3390/polym14194084
214. Delgado-Alvarado E et al (2022) Recent progress of nanogenerators for green energy harvesting: performance, applications, and challenges. Nanomaterials 12(15):2549. https://doi.org/10.3390/nano12152549
215. Mishra K et al (2022) Ionic liquid-based polymer nanocomposites for sensors, energy, biomedicine, and environmental applications: roadmap to the future. Adv Sci 9(26):1–50. https://doi.org/10.1002/advs.202202187
216. Afzali SS, Hekmatara SH, Seyed-Yazdi J, Hosseini SMBM (2022) Tuned MWCNT/CuO/Fe3O4/Polyaniline nanocomposites with exceptional microwave attenuation and a broad frequency band. Sci Rep 12(1):1–12. https://doi.org/10.1038/s41598-022-13210-4
217. Alamri S, Rajhi AA, Tran N, Senthil TS, Kumar Narukullapati B, Mohanavel V (2022) Tunable microwave absorption and shielding effectiveness in the nanocomposite of 3D hierarchical flower-like Co3O4 and rod-like polyindole. J Magn Magn Mater 555:169363. https://doi.org/10.1016/j.jmmm.2022.169363
218. Liu Y, Wu X, Liu M, Huang Y, Huang Z (2020) Microstructure and mechanical properties of B4C–TiB2–SiC composites fabricated by spark plasma sintering. Ceram Int 46(3):3793–3800
219. Wen Q, Tan Y, Zhong Z, Zhang H, Zhou X (2017) High toughness and electrical discharge machinable B4C-TiB2-SiC composites fabricated at low sintering temperature. Mater Sci Eng A 701:338–343
220. Ahmed H, Hashim A, Abduljalil HM (2020) Determination of optical parameters of films of PVA/TiO2/SiC and PVA/MgO/SiC nanocomposites for optoelectronics and UV-detectors. Ukr J Phys 65(6):533
221. Guo Y et al (2020) TPU/PLA nanocomposites with improved mechanical and shape memory properties fabricated via phase morphology control and incorporation of multi-walled carbon nanotubes nanofillers. Polym Eng Sci 60(6):1118–1128
222. Onyszko M et al (2020) Few layered oxidized h-BN as nanofiller of cellulose-based paper with superior antibacterial response and enhanced mechanical/thermal performance. Int J Mol Sci 21(15):5396

223. Alves Z, Nunes C, Ferreira P (2021) Unravelling the role of synthesis conditions on the structure of zinc oxide-reduced graphene oxide nanofillers. Nanomaterials 11(8):2149
224. Vahidi G, Bajwa DS, Shojaeiarani J, Stark NM (2022) Experimental investigation into the direct feeding of coupling agent, cellulose nanocrystals, and nano zinc oxide in high-density polyethylene. Compos Part C Open Access 100287
225. Gibril ME, Ahmed K, Lekha P, Sithole B, Khosla A, Furukawa H (2022) Effect of nanocrystalline cellulose and zinc oxide hybrid organic–inorganic nanofiller on the physical properties of polycaprolactone nanocomposite films. Microsyst Technol 28(1):143–152
226. Aslam M, Kalyar MA, Raza ZA (2019) Effect of separate zinc, copper and graphene oxides nanofillers on electrical properties of PVA based composite strips. J Electron Mater 48(2):1116–1121
227. Yao F et al (2022) Superb electromagnetic shielding polymer nanocomposites filled with 3-dimensional p-phenylenediamine/aniline copolymer nanofibers@ copper foam hybrid nanofillers. Compos Part B Eng 245:110236
228. Said AR, Nawar AG, Elsayed AE, Abd-Allah MA, Kamel S (2021) Enhancing electrical, thermal, and mechanical properties of HV cross-linked polyethylene insulation using silica nanofillers. J Mater Eng Perform 30(3):1796–1807
229. Ahmadi M, Janakiram S, Dai Z, Ansaloni L, Deng L (2018) Performance of mixed matrix membranes containing porous two-dimensional (2D) and three-dimensional (3D) fillers for CO2 separation: a review. Membranes (Basel) 8(3):50
230. Fu S, Sun Z, Huang P, Li Y, Hu N (2019) Some basic aspects of polymer nanocomposites: a critical review. Nano Mater Sci 1(1):2–30. https://doi.org/10.1016/j.nanoms.2019.02.006
231. Fatimah I et al (2022) Magnetic-silica nanocomposites and the functionalized forms for environment and medical applications: a review. Inorg Chem Commun 137(December 2021):109213. https://doi.org/10.1016/j.inoche.2022.109213
232. Kumar A, Gupta K, Dixit S, Mishra K, Srivastava S (2019) A review on positive and negative impacts of nanotechnology in agriculture. Int J Environ Sci Technol 16(4):2175–2184
233. Downing-Perrault A (2005) Polymer nanocomposites are the future. Univ. Wisconsin-Stout, Menomonie, Wis, USA
234. Matei E et al (2022) Natural polymers and their nanocomposites used for environmental applications. Nanomaterials 12(10):1707
235. Drzal LT, Misra M, Mohanty A (2005) Sustainable biodegradable green nanocomposites from bacterial bioplastic for automotive applications. In: Proceedings of the US EPA 2004 STAR progress review workshop--nanotechnology and the environment II. Washington, DC (USA): Published by US Environmental Protection Agency. Report number EPA/600/R-05/089, pp 23–25
236. Chow WS, Mohd Ishak ZA (2020) Smart polymer nanocomposites: a review. Express Polym Lett 14(5)
237. Gao F (2002) The future prospect of polymer nanocomposites in reinforcement application. E-Polymers 4:1–7
238. Precedence Research (2022) Nanotechnology Market Size to Hit US$ 288.71 Billion by 2030
239. Future® MR (2022) Nanocomposites Market. https://www.marketresearchfuture.com/sample_request/8280
240. Abdellaoui H, Bouhfid R, and others (2020) Review of nanocellulose and nanohydrogel matrices for the development of sustainable future materials. In: Sustainable nanocellulose and nanohydrogels from natural sources. Elsevier. pp 155–176
241. Rangappa SM, Siengchin S, Dhakal HN (2020) Green-composites: ecofriendly and sustainability. Appl Sci Eng Prog 13(3):183–184
242. Rajeshkumar G, Seshadri SA, Ramakrishnan S, Sanjay MR, Siengchin S, Nagaraja KC (2021) A comprehensive review on natural fiber/nano-clay reinforced hybrid polymeric composites: materials and technologies. Polym Compos 42(8):3687–3701
243. Green M, Chen X (2019) Recent progress of nanomaterials for microwave absorption. J Mater 5(4):503–541. https://doi.org/10.1016/j.jmat.2019.07.003

244. Sethulekshmi AS, Jayan JS, Saritha A, Joseph K (2022) Recent developments in natural rubber nanocomposites containing graphene derivatives and its hybrids. Ind Crops Prod 177:114529
245. Ilyas RA et al (2022) Natural-fiber-reinforced chitosan, chitosan blends and their nanocomposites for various advanced applications. Polym (Basel) 14(5):874

MIX
Papier aus verantwortungsvollen Quellen
Paper from responsible sources
FSC® C105338

If you have any concerns about our products,
you can contact us on
ProductSafety@springernature.com

In case Publisher is established outside the EU,
the EU authorized representative is:
Springer Nature Customer Service Center GmbH
Europaplatz 3, 69115 Heidelberg, Germany

Printed by Libri Plureos GmbH
in Hamburg, Germany